The Interior of the Earth

*In his hand are the deep places of the earth:*
*the strength of the hills is his also*

Ps. 95. 4

The Earth from space, taken during the Apollo 8 mission, December, 1968. (By courtesy of NASA)

# The Interior of the Earth

Martin H. P. Bott

*Professor of Geophysics, University of Durham*

Edward Arnold

© Martin H. P. Bott 1971
First published 1971

Reprinted 1972

First published in Great Britain by
Edward Arnold (Publishers) Ltd.
25 Hill Street
London W1X 8LL

ISBN: 0 7131 2274 9

# Preface

During the last decade there has been a revolution in the Earth sciences, mainly stemming from new discoveries in geophysics. The study of the Earth's crust and mantle has received a great impetus from the Upper Mantle Project and from national nuclear test detection programmes. For the first time the broad pattern of lateral variations in the properties of the upper mantle is being systematically investigated. Palaeomagnetic and oceanographical research has shown that large horizontal displacements have affected both continental and oceanic crust. An important outcome of these new discoveries is that a unified theory of the Earth's primary surface features is emerging at last. It is this aspect that is of fundamental importance to geology, because it provides a new framework for the understanding of well-known but previously unrelated observations in tectonic geology, petrology and stratigraphy.

The book aims to present the modern picture of the Earth's interior, including the structure and the processes which occur in it. This must be based on a synthesis of the well-established older knowledge with the newer discoveries. The physical and chemical structure of crust, mantle and core are described in the first five chapters. The last four chapters deal with processes within the Earth. The interior supply of heat is considered to be the main source of energy as described in Chapter 6. This is followed by a geometrical discussion of the large horizontal displacements which affect the ocean floor and move the continents about, and the relation of these movements to the origin of the Earth's primary surface features. The mechanism of fracture and flow which allows large horizontal and vertical movement to occur, and the primary mechanism which converts thermal energy into strain energy form the subjects of the last two chapters.

The book is mainly concerned with deep structure and mechanism and consequently details of surface geology have been kept to a minimum. The origin of the primary surface features such as ocean ridges, young fold mountains, trenches and rift valleys is described in relation to their deep structure and the ocean-floor spreading hypothesis, but detailed description of them has not been given. The origin of secondary tectonic features such as sedimentary basin subsidence has been omitted.

The book has been written for advanced students of geology and geophysics and for Earth scientists requiring a general account of solid-Earth geophysics. With this in view, the mathematical treatment has been omitted although indispensable formulae are included.

In a rapidly progressing branch of science, it is certain that further important advances will have been made before the time of publication. Every effort has been made to keep the book broadly up to date at the time of going to press, but in a book of this scope it is quite impossible to give full coverage of the increasingly voluminous literature. Selection has been necessary but this has been done in an attempt to give a balanced picture of the present state of knowledge.

M. H. P. BOTT

Durham 1970

# Acknowledgments

I wish to express my grateful thanks to Dr. D. Davies and to Dr. G. A. L. Johnson for critically reading the manuscript and for making many helpful suggestions which have been incorporated. I am also most grateful to my colleagues in the University of Durham for specialist advice on a number of topics during the preparation of the book. My wife has assisted with the proofs and indexing, and Dr. P. B. A. Attewell, Professor G. M. Brown, Dr. C. H. Emeleus, Dr. J. H. Holland and Dr. R. E. Long read sections of the manuscript and helped with the proofs. Mr. G. Dresser assisted with photography and Mrs. Pamela Howe typed part of the manuscript. I am also indebted to the publishers for their courtesy and helpful advice during production of the book. I would like to thank all of these and many others who have given me practical assistance, advice and stimulating discussion over many years on the subjects covered in this book.

I am grateful to the following authors and publishers for permission to use material in Figures and Tables:

H. E. Agger and E. W. Carpenter (Figure 2.12(d)); S. Akimoto and H. Fujisawa (Figure 4.24); D. L. Anderson (Figure 4.31); D. L. Anderson and C. B. Archambeau (Figure 8.14); D. L. Anderson and R. O'Connell (Figure 8.15); M. Barazangi and J. Dorman (Figure 8.18); M. J. Berry and G. F. West (Figure 2.5); F. Birch (Figure 5.7); B. A. Bolt (Figures 5.2(b) and 5.3); Sir Edward Bullard (Figures 4.20 and 4.21); Sir Edward Bullard and H. Gellman (Figures 5.15, 5.16 and 5.17); Sir Edward Bullard, J. E. Everett and A. G. Smith (Figure 7.1); B. Caner and W. H. Cannon (Figure 4.23); J. Capon, R. J. Greenfield and R. T. Lacoss (Figures 4.3(b) and 4.4); S. P. Clark, Jr. (Table 2.2); S. P. Clark, Jr. and A. E. Ringwood (Figures 4.25 and 4.30); S. Coron (Figures 2.19 and 2.21); A. Cox and R. R. Doell (Figure 5.13); K. M. Creer (Figures 5.11(c), 7.3 and 7.5); M. D. Crittenden, Jr. (Figure 8.12); D. Davies (Figure 1.3); R. Dearnley (Figure 9.11); C. L. Drake and R. W. Girdler (Figure 2.25); J. P. Eaton (Figure 2.4); L. Egyed (Figure 9.1); J. W. Elder (Figures 6.8 and 9.4); J. Ewing and M. Ewing (Figure 7.17); M. Ewing (Figures 3.3 and 3.5); K. Fuchs and others (Figures 2.20 and 2.21); R. W. Girdler (Figures 2.22 and 2.23); A. L. Hales and I. S. Sacks (Figure 2.11); W. B. Harland, A. G. Smith and B. Wilcock (Table 1.3); D. E. Hayes (Figure 3.16); B. C. Heezan (Figure 3.1); J. R. Heirtzler, G. O. Dickson, E. M. Herron, W. C. Pitman, III and X. Le Pichon (Figures 7.13, 7.14 and 7.15); J. R. Heirtzler, X. Le Pichon and J. G. Baron (Figure 3.8); W. A. Heiskanen (Figure 2.24); D. P. Hill and L. C. Pakiser (Figure 2.12(b)); A. P. Holder (Figure 2.6); E. Irving (Figures 5.11 and 9.2); B. Isacks and P. Molnar (Figure 8.28); B. Isacks, J. Oliver and L. R. Sykes (Figures 7.9 and 8.26); J. C. Jaeger (Figure 8.3); D. E. James and J. S. Steinhart (Figure 2.8); B. R. Julian and D. L. Anderson (Figure 4.13 (a) and (b)); D. G. King-Hele (Figure 1.1); M. G. Langseth (Figure 6.2); M. G. Langseth, P. J. Grim and M. Ewing (Figure 6.12); G. V. Latham and others (Figure 1.12); W. H. K. Lee and S. Uyeda (Figures 6.3, 6.4, 6.5, 6.6 and 6.13; Table 6.2); Miss I. Lehmann (Figure 4.11); X. Le Pichon (Figures 7.16 and 7.20; Table 7.2); A. E. Maxwell and others (Figures 7.18 and 7.19; Table 7.1); G. J. F. MacDonald (Figures 1.11, 6.7(a), 6.10 and 6.11); D. P. McKenzie (Figure 8.9); P. J. Melchior (Figures 5.4 and 5.5); H. W. Menard (Figure 3.6); S. Moorbath (Figure 4.29); S. Mueller and M. Landisman (Figure 2.10); W. H. Munk and G. J. F. MacDonald (Figures 1.8 and 1.9); J. E. Nafe and C. L. Drake (Figure 2.27); M. Niazi and D. L. Anderson (Figure 4.12); S. R. Nockolds (Table 2.1); J. Oliver and B. Isacks (Figure 8.27); N. D. Opdyke, B. Glass, J. D. Hays and J. Foster (Figure 7.10); E. Orowan (Figure 8.2); L. C. Pakiser (Figure 2.9); W. C. Pitman, III and J. R. Heirtzler (Figure 7.12); F. Press (Figure 2.26; Table 2.3); R. H. Rapp (Figure 1.2); A. E. Ringwood and D. H. Green (Figure 2.28; Table 2.5); S. K. Runcorn (Figures 9.12 and 9.13); C. T. Scrutton (Figure 1.10); S. W. Smith (Figure 4.9); T. J. Smith, J. S. Steinhart and L. T. Aldrich (Figure 2.12(c)); J. S. Steinhart and R. P. Meyer (Figure 2.3); L. R. Sykes (Figures 8.22, 8.24 and 8.25); M. Talwani (Figure 3.2); M. Talwani, X. Le Pichon and M. Ewing (Figures 3.11, 3.12 and 3.13); M. Talwani, G. H. Sutton and J. L. Worzel (Figure 3.15); M. N. Töksoz, M. A. Chinnery and D. L. Anderson (Figures 4.10, 4.14, 4.15, 4.16 and 4.19); J. R. Truscott (Figure 4.3(a)); W. Tucker, E. Herrin and H. W. Freedman (Figure 4.17); F. J. Vine (Figure 7.11); J. Tuzo Wilson (Figures 7.6, 7.7 and 7.8); J. L. Worzel (Figure 3.14); J. S. V. van Zijl, K. W. T. Graham and A. L. Hales (Figure 5.12); Academic Press (Figures 1.4, 1.5, 3.1, 4.11, 4.28, 5.1, 5.2(a), 8.11, 9.12 and 9.13); American Association for the Advancement of Science (Figures 1.12, 4.31, 7.10, 7.11, 7.12, 7.17, 7.18, 7.19, 8.21 and 8.23; Table 7.1); American Geophysical Union (Figures 1.2, 2.4, 2.5, 2.8, 2.9, 2.12(b), 2.12(c), 2.28, 3.11, 3.12, 3.13, 3.15, 4.9, 4.12, 4.24, 4.25, 4.30, 6.2, 6.3, 6.4, 6.5, 6.6, 6.7(a), 6.8, 6.10, 6.11, 6.12, 6.13, 7.9, 7.13, 7.14, 7.15, 7.16, 7.20, 8.14, 8.22, 8.24, 8.25, 8.26, 8.27, and 9.4; Tables 2.5, 6.2 and 7.2); British Astronomical Association (Table 1.4); Butterworths (Figures 3.14 and 8.6); Cambridge University Press (Figures 1.8 and 1.9); The Director, Carnegie Institution of Washington (Figures 2.3 and 5.9); Colston Research Society (Figures 3.14 and 8.6); Elsevier Publishing Company (Figures 3.2 and 3.16); Geological Society of America (Figures 2.26, 5.13, 8.19 and 8.20; Tables 2.1, 2.2 and 2.3); Geological Society of London (Table 1.3); The Editor, Geologische Rundschau (Figure 9.1); The Director, Institute of Geological Sciences (Figure 4.20(a)); Interscience Publishers (Figure 2.27); The Director, Isostatic Institute I.A.G. (Figure 2.18); Liverpool Geological Society (Figures 2.6 and 2.7); McGraw-Hill Book Company (Figures 2.24 and 3.6); Methuen (Figure 8.3); The Editor, Nature (Figures 4.23, 5.2(b), 7.6, 7.7, 7.8, 8.28 and 9.13); Thomas Nelson and Sons (Figures 7.2 and 7.4); Oliver and Boyd (Figures 4.29, 8.4 and 8.5); Plenum Press (Figure 1.11); Pergamon Press (Figures 1.10, 2.22, 2.23, 3.8, 5.4, 5.5 and 9.11); Princeton University Press (Figures 8.16 and 8.17); The Director, Royal Aircraft Establishment (Figure 1.1(b)); Royal Astronomical Society (Figures 2.10, 2.11, 2.12(d), 2.25, 3.3, 3.5, 3.7, 4.3(a), 4.10, 4.14, 4.15, 4.16, 4.19, 4.20(b), 4.21, 5.7, 5.12, 8.9, 8.15, 9.7, 9.8 and 9.9); Royal Geological Society of Cornwall (Figures 2.15 and 2.16); Royal Society (Figures 5.11(c), 5.15, 5.16, 5.17, 7.1, 7.3, 7.5 and 8.2); Editor, Science Journal (Figure 1.3); Editor, Scientific American (Figure 1.1(a)); Seismological Society of America (Figures 4.13(a) and (b), 4.17, 5.3 and 8.18); Society of Exploration Geophysicists (Figures 4.3(b) and 4.4); U.S. Coast and Geodetic Survey (Figure 4.7); Director, U.S. Geological Survey (Figure 8.12); University of Pittsburgh Press (Figure 5.18); Washington Academy of Sciences (Figure 9.3); John Wiley and Sons (Figures 5.11 and 9.2); Yorkshire Geological Society (Figure 2.17).

M. H. P. B.

# Contents

# 1 The broad structure and origin of the Earth

## 1.1 Introduction

The rapid acceleration in scientific advance during the second half of the twentieth century has placed man on the threshold of the exploration of space. But the deep interior of the Earth remains as inaccessible as ever. This is the realm of solid-Earth geophysics, which must still depend on observations made at or near the Earth's surface. Despite this limitation, there has been a major revolution in our knowledge of the Earth's interior quietly taking place over the last ten years or so. At last, we think we are beginning to have some real understanding of the processes which occur within the Earth to produce surface conditions outstandingly different from those of the other inner planets and the Moon. How has this come about? It is mainly the result of the introduction of new experimental and theoretical techniques into geophysics, together with opportunities to make observations on a much wider scale than before. It is to be expected that this new impetus will continue for some years to come and that further knowledge of the Earth's interior will be a by-product of planetary exploration.

At the outset, let us summarize some of these recent advances in geophysics. Palaeomagnetic studies have given a firm basis to the theory of continental drift, whereas before it was a subject of inconclusive debate. Ocean-floor surveys have led to the discovery of the Earth's largest uplifted feature—the ocean ridge system—previously largely unknown. Oceanic investigations have also demonstrated the surprising youth of the ocean-floor and have led to the hypothesis that the ocean crust is forming and spreading laterally from the crest regions of the ridges, thereby providing a mechanism for continental drift and for the origin of the Earth's major surface features. Seismology has been revitalized by new techniques and vast improvement in the world-wide network of stations, multiplying our knowledge of the Earth's interior especially above 1000 km depth but also below. No longer can the Earth be treated as a rigid body possessing radial symmetry, for large lateral variations have been shown to occur in crust and mantle. One of the most encouraging features of the present state of knowledge is that the old controversies within geology and geophysics are melting away as an overall pattern of processes in the crust and mantle is beginning to emerge.

Before turning to the study of the structure and processes of the Earth's interior, we shall examine in this introductory chapter some of the features of the Earth as a planet, its shape, layering, age and origin.

## 1.2 The shape of the Earth

It was Isaac Newton who first showed that the Earth, because of its rotation, ought to be an ellipsoid of revolution slightly flattened at the poles (i.e. an oblate spheroid). As a result of early observations, the French geodesist Cassini inclined to the opposite view that the equator is flattened and the poles bulge like an egg. To settle this controversy, French geodetic expeditions were sent to Peru and Lapland in the 1730's to measure the radius of curvature of the Earth's surface near the equator and near a pole. They did this by measuring the length of an arc along the Earth's surface in a north-south direction by triangulation, and by measuring the difference between the directions of the vertical at each end of the arc by astronomical surveying. This enabled the radius of curvature to be calculated near pole and equator. The result conclusively proved that the Earth is flattened at the poles as predicted by Newton.

The theoretical argument for flattening of the poles is as follows. The surface of a rotating mass of fluid of uniform density acted on by its own gravitation and centrifugal force alone is an ellipsoid of revolution. An increase of density towards the centre such as is known to occur within the Earth would only cause a departure from a true ellipsoid of 3 m maximum deviation. Thus the Earth's shape ought to be very nearly an ellipsoid of revolution flattened at the poles provided it is in hydrostatic equilibrium.

It will be shown later that the Earth is not completely in hydrostatic equilibrium (p. 245). Nevertheless, the hydrostatic hypothesis is sufficiently close to reality to make it reasonable to treat the Earth's surface as an ellipsoid of revolution. Consequently it is convenient to divide the theory of the shape of the Earth into two parts: (1) determining the shape and dimension of the ellipsoid which gives the best fit to the sea-level surface (this ellipsoid is called the spheroid); and (2) determining deviations of the sea-level surface (which is called the geoid) from the spheroid. The shape of the spheroid is defined by its flattening according to the following equation

$$f = (a - c)/a,$$
where $f$ = degree of flattening,
$a$ = mean equatorial radius
and   $c$ = polar radius.

The old method of determining the Earth's flattening was to measure the radius of curvature by geodetic surveying as described above. A much better method nowadays is to use the variation of the Earth's gravity field with latitude. This method assumes that the surface is an ellipsoid of revolution and that the vertical direction is everywhere perpendicular to the surface (i.e. it is an equipotential surface). Then the variation of gravity at sea-level with latitude is given by (JEFFREYS, 1959, Chapter 4)

$$g_\phi = g_e\{1 + (\tfrac{5}{2}m - f + \tfrac{15}{4}m^2 - \tfrac{17}{14}fm)\sin^2\phi + (\tfrac{1}{8}f^2 - \tfrac{5}{8}fm)\sin^2 2\phi\}$$
where $g_\phi$ = gravity at latitude $\phi$,
$g_e$ = gravity at the equator
and   $m$ = ratio of centripetal acceleration to gravity at sea-level on the equator.

This formula neglects third and higher powers of $f$ and $m$ and corresponding product terms.

Studies on the orbits of artificial satellites have greatly improved our knowledge of the global features of the Earth's gravity field, thereby improving the accuracy of our estimate of the flattening by a factor of thirty. The equatorial bulge produces an observable influence on the orbit. If the satellite moves towards the north-east as it crosses the equator in a northward direction, then the point at which the orbit crosses the equator will move progressively westwards (Figure 1.1(a)). The rate at which the orbit moves towards the west enables the flattening to be calculated.

The value of $f$ calculated by Newton for an Earth of uniform density was 1/230. The eighteenth-century French geodetic expeditions obtained observational values ranging from 1/310 to 1/178. Before the satellite observations, a typical estimate of $f$ obtained by JEFFREYS (1959) based on surface gravity measurements was 1/279·3, with an accuracy of about 1 part in 300. Observations on satellite orbits yield an estimate of 1/298·25 with an accuracy of 1 part in 30 000 (KOZAI, 1964). The Earth's mean equatorial radius is 6378·160 km and the polar radius is 6356·775 km.

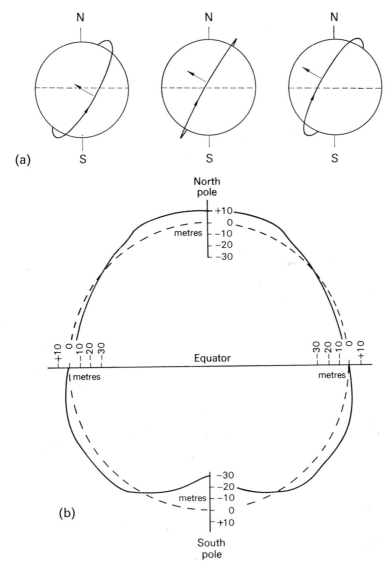

**Fig. 1.1** (a) The Earth's equatorial bulge makes the orbit of an eastward moving artificial satellite regress towards the west. The rate of regression enables the flattening of the Earth to be calculated. Redrawn from KING-HELE (1967), *Scientific American*, **217**, No. 4, p. 70. (b) Height of the geoid (solid line) relative to a spheroid of flattening 1/298.25 (broken line) assuming the Earth to be axially symmetrical about the polar axis. Note exaggeration. Redrawn from KING-HELE (1969), *Royal Aircraft Establishment Technical Memorandum Space 130*, Fig. 4.

Turning to the deviation of the geoid from the spheroid, this also can be best estimated from variations of gravity over the Earth's surface, using a theorem of Stokes (JEFFREYS, 1959, chapter 4). Here, too, studies on the orbits of artificial satellites have produced a great improvement. They yield a much more accurate picture of the broader variations of gravity than surface observations do at present. However, surface measurements of gravity are still the best method for relatively local variations of gravity and the geoid, say over distances less than about 2000 km. One of the more widely publicized aspects of 'satellite geodesy' is that the Earth is 'pear-shaped' (KING-HELE, 1967).

In fact the pear-shaped deviation of the geoid from the spheroid is less than 20 m, whereas the equatorial bulge is over 20 km. The average deviation of the geoid from the spheroid around an arbitrary line of longitude is shown in Figure 1.1(b). A map showing the deviations of the geoid over the Earth's surface using satellites and surface gravity observations has been constructed by RAPP (1968) and is shown in Figure 1.2.

The study of the shape of the Earth is one branch of geophysics which has greatly benefited from the introduction of a new technique, in this case the technique being the use of artificial satellites. The increased accuracy has important repercussions on our knowledge of the state of stress within the Earth's interior, showing that the flattening is slightly greater than would be expected for a 'hydrostatic Earth' (p. 245) and that there are significant lateral variations of density within the mantle (p. 147).

## 1.3  The Earth's mass and moments of inertia

The mass of the Earth can be estimated from the value of gravity at the surface, after correction for the small contribution to gravity caused by rotation. The method in simplified form is as follows. The Earth is treated as a radially symmetrical sphere, not rotating, when it can be shown that the gravitational attraction at an external point is given by $GM/r^2$, where $r$ is the distance from the centre, $M$ is the mass and $G$ is the gravitational constant. The gravitational attraction (gravity) is accurately known and so is the radius. $G$ has been determined by experiment to an accuracy of about $0.03\%$. By allowing for the shape of the Earth, or using the whole of the gravity field over the surface, the assumption of radial symmetry can be dropped. Thus $M$ can be calculated to an accuracy of about $0.03\%$, the limitation being the experimental accuracy of $G$. For this reason the determination of $G$ is sometimes spoken of as 'weighing the Earth'. From the mass and volume the mean density can also be obtained. The results are

$$\text{mass} = 5{\cdot}977 \times 10^{27}\text{ g},$$
$$\text{mean density} = 5{\cdot}517\text{ g/cm}^3.$$

If the Earth is assumed to be symmetrical about the polar axis, then there are two principal moments of inertia, defined as $A$ about an equatorial axis and $C$ about the polar axis. These can be estimated from astronomical observations and the observed flattening through two steps as follows (o'KEEFE, 1965):

(1) The Earth's axis does not remain in a fixed direction in space, but it precesses about a fixed direction describing a cone of angle 23° 27'. The period of precession is 25 735 y which is accurately known from astronomical observations. From the theory of the motion of a symmetrical body, the period of procession enables the precession constant $H = (C - A)/C$ to be obtained, yielding a value of $0.003275$ for the Earth (COOK, 1967).

(2) The quantity $(C - A)$ can be estimated from a theoretical relation connecting the principal moments of inertia of a body and the second harmonic of the external gravity potential which is known as MacCullagh's formula (RAMSEY, 1940, p. 87). In turn, the second harmonic of gravity potential can be expressed in terms of the flattening, rotation, mass and dimension of the Earth. The relation is (COOK, 1967).

$$(C - A)/Ma^2 = \tfrac{1}{3}(f - \tfrac{1}{2}m) = 0{\cdot}00108265.$$

Combination of the two above equations enables $A$ and $C$ to be estimated. The flattening $f$ is the least accurately known term in the equations and it therefore defines the accuracy of the estimates of the principal moments of inertia. The improvement in the estimate of the flattening stemming

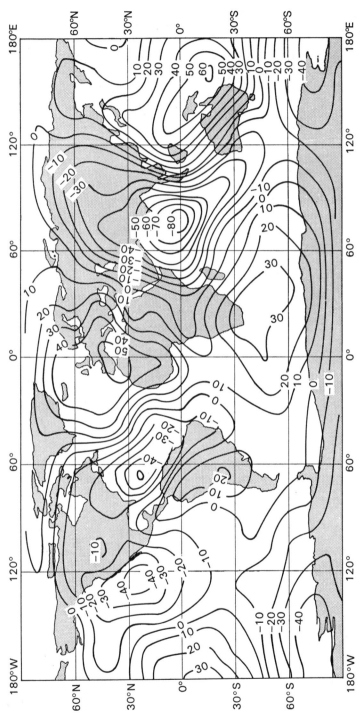

**Fig. 1.2** Map of the geoid referred to a flattening of 1/298·25 based on satellite and surface gravity observations, aided by use of isostatic model anomalies where surface observations are absent, and incorporating up to and including the 14th degree and order spherical harmonic terms. The contours are given in metres above the spheroid. A high geoid corresponds to relatively high density rocks beneath, and a low geoid to low density rocks (probably within the mantle, p. 149). Redrawn from RAPP (1968), *J. geophys. Res.*, **73**, 6560.

from observations of satellite orbits also improves the accuracy of estimates of the moments of inertia. Thus $C/Ma^2$ is found to be 0·3306. If the Earth were a uniform sphere it would be 0·4.

Both the mean density and the moment of inertia show that there must be a strong increase in density with depth within the Earth. Knowledge of the mean density and moment of inertia is important in studying the internal density distribution within the Earth, because any acceptable model of density must satisfy these observations. The moments of inertia are also important because they are needed to calculate the shape the Earth would have if it were in perfect hydrostatic equilibrium (p. 245).

## 1.4 Internal layering of the Earth

The branch of geophysics dealing with the origin and propagation of elastic waves within the Earth is called seismology. It is seismology which has revealed the internal layering of the Earth. It has therefore provided the framework for most other types of investigation of the interior, and is of fundamental importance to geophysics. Instrumental seismology was begun near the beginning of the present century and a few years later the three main subdivisions of the Earth had been discovered. A second stage of rapid advance in seismology was stimulated by the need to detect underground nuclear explosions and by the Upper Mantle Project. This has vastly increased our knowledge of the Earth's interior since about 1960, by revealing much greater detail in the layering and in showing up lateral variations. It is, however, with the early discoveries of seismology that this section is concerned.

When an earthquake or an explosion occurs within the Earth, part of the energy released takes the form of elastic waves which are transmitted through rocks with a definite velocity depending on density and elastic moduli. There are two main types, body waves and surface waves (Figure 1.3). Body waves conform to the laws of geometrical optics, being reflected and refracted at interfaces where the velocity changes. The two types of body waves correspond to transmission of (1) compressions and rarefactions ($P$ waves), and (2) shear displacement ($S$ waves). The velocities are given by

$$V_P = \sqrt{\left(\frac{k + \frac{4}{3}\mu}{\rho}\right)} \text{ and } V_S = \sqrt{\frac{\mu}{\rho}},$$

where $V_P$ = velocity of $P$ waves,
$V_S$ = velocity of $S$ waves,
$k$  = bulk modulus,
$\mu$  = rigidity modulus
and  $\rho$  = density.

It follows that $P$ waves always travel faster than $S$ waves, and that $S$ waves cannot be propagated through liquid. In general, denser rocks have higher body wave velocities since the elastic moduli increase with increasing density more rapidly than the density itself does.

Surface waves are restricted to the vicinity of a free surface, or in exceptional circumstances, an internal interface. The two main types are Rayleigh waves, with the particle motion confined to the vertical plane containing the direction of propagation, and Love waves, with motion in a horizontal direction perpendicular to the direction of propagation. These are described in more detail in Chapter 4.

Elastic waves are detected by seismographs, which respond to ground displacement or velocity depending on design. Short-period instruments (about 1s period) are used to detect body waves, and long-period instruments (15s or longer period) are used for surface waves. A normal seismograph station has three short-period and three long-period seismographs, to detect the three components of ground motion.

The elastic waves emanating from an earthquake are usually taken to originate from a single point within the Earth, although more realistically the source is often extended over a distance of 10–100 km. This point is called the *focus* (or *hypocentre*). The point on the surface vertically above the focus is the *epicentre*. It will be shown later (Chapter 8) that most earthquake foci are above a depth of 100 km but that deep events occur down to about 700 km depth. The location of the focus and epicentre of a given earthquake is determined from the arrival times of seismic waves at a selection of seismological observatories—the more the better. Recent improvements to the world-wide network of stations, supplemented by the use of computers, have considerably increased the accuracy of locating earthquake foci.

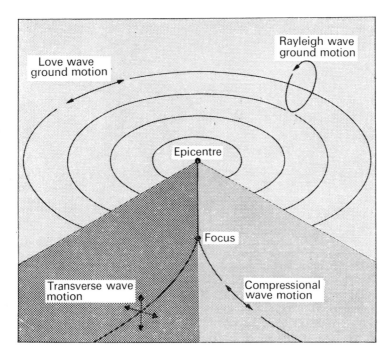

Love wave ground motion

Rayleigh wave ground motion

Epicentre

Focus

Transverse wave motion

Compressional wave motion

**Fig. 1.3**   Diagram illustrating the focus and epicentre of an earthquake, and showing the fundamental types of seismic wave originating from the earthquake. Redrawn from DAVIES (1968), *Science Journal*, Nov. 1968, p. 79.

Before about 1950, most of the major discoveries of seismology came from studies of the time of travel of body waves from earthquakes. The Earth was assumed to possess radial symmetry and thus many different earthquakes could be used to build up a table of travel-times of $P$ and $S$ waves over the range of possible angular distances from the epicentre (0° to 180°).

The existence of a central core within the Earth was deduced by OLDHAM (1906) as the result of his observation that $P$ waves recorded near the angular distance of 180° from the earthquake epicentre arrived much later than expected. This was attributed to delay introduced by passage through a low velocity core. The discovery was later substantiated by the following more detailed evidence (Figure 1.4). Up to an angular distance of about 103° from the epicentre, $P$ and $S$ waves are observed and indicate a progressive increase in velocity with depth through the major shell of the Earth which is known as the *mantle*. However, between 103° and 142°, *P* and *S* waves are both substantially absent, causing a 'shadow zone'. From 142° to 180° delayed *P* arrivals occur but *S* is absent. This shows that about half way to the centre of the Earth there is a major discontinuity beneath which the *P* velocity drops abruptly and *S* is absent, suggesting a fluid *core* in contrast to the solid, overlying *mantle*. The depth of the discontinuity between can be obtained by a variety of methods, including

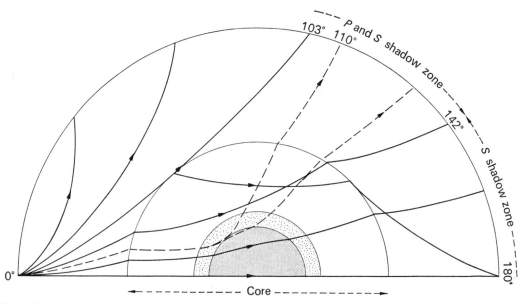

**Fig. 1.4**   Selected ray paths for *P* waves passing through the Earth. The *P* and *S* shadow zones are shown, and the dashed ray paths represent the weak *P* arrivals within the shadow zone which provide the main evidence for an inner high *P* velocity division of the core. Redrawn from GUTENBERG (1959), *Physics of the Earth's interior*, p. 104, Academic Press.

use of reflections from it. In 1914, Gutenberg obtained a depth of 2900 km for it (although recent revisions reduce this estimate by 10–20 km). The corresponding value for the mean radius of the core is 3470 km. The core-mantle boundary is known as the *Gutenberg discontinuity*.

A second major discontinuity at shallow depth was discovered by MOHOROVIČIĆ (1909) through study of seismograms of the Yugoslavia earthquake of October 8, 1909, out to distances of a few hundred kilometres. He observed two *P* and two *S* pulses, and he interpreted these as direct and refracted *P* and *S* arrivals caused by a low velocity *crust*, about 50 km thick, overlying a higher velocity substratum now called the *mantle*. The intervening boundary is called the *Mohorovičić discontinuity* (or just the *Moho*). Seismic refraction studies using artificial explosions have greatly amplified and extended this discovery (p. 31). Early studies on the dispersion of surface waves suggested that the crust is thinner beneath oceans than continents, and this also has been substantiated by later refraction surveys.

The three major subdivisions of the 'solid' Earth were known by 1910. Between the world wars, the main effort of seismology was devoted to obtaining and refining the velocity-depth distribution throughout the Earth for *P* and *S* waves. This depended on building up detailed knowledge of the travel-times of the waves covering all angular distances from the epicentre. If radial symmetry is assumed for the Earth, then, provided velocity increases continuously with depth, the velocity-depth curves can be computed from the $P$ and $S$ travel-time curves by a mathematical process described in 4.2. This method was used by JEFFREYS (1939a) to construct the velocity-depth curves for $P$ and $S$ through the mantle. Jeffreys obtained the velocity distribution in the core by trial and error.

The velocity-depth curves of JEFFREYS (1939a) and GUTENBERG (1959) are shown in Figure 1.5.

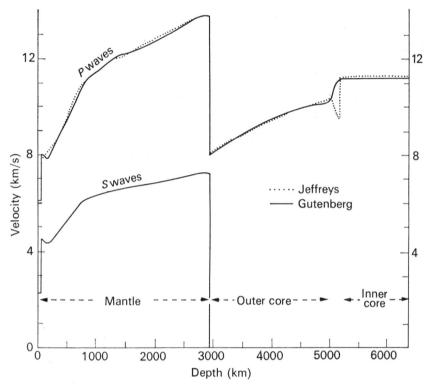

**Fig. 1.5** The broad velocity-depth distributions of Gutenberg ($P$ and $S$) and Jeffreys ($P$ only shown). Adapted from GUTENBERG (1959), *Physics of the Earth's interior*, p. 15, Academic Press.

**Table 1.1** The layers within the Earth. The layering is based on the seismic velocity-depth distribution (Figure 1.5). It is adapted from the layering given by BULLEN (1963). The values placed on the depth of boundaries by Bullen have been rounded for this table.

|  | Region | Depth range (km) | |
|---|---|---|---|
| CRUST | A | 0–33 (variable thickness) | |
| Mohorovičić discontinuity | | | |
| | B | 33–400 | upper mantle |
| MANTLE | C | 400–1000 | transition zone |
| | D | 1000–2900 | lower mantle |
| Gutenberg discontinuity | | | |
| | E | 2900–4980 | outer core |
| CORE | F | 4980–5120 | transition zone |
| | G | 5120–6370 | inner core |

Gutenberg's distribution does not differ greatly from that of Jeffreys except in the topmost mantle where Gutenberg has used evidence from amplitudes to support the idea of a low velocity zone. The main features of these velocity-depth distributions are still thought to be correct, although the discoveries of modern seismology have greatly amplified the details of the upper 1000 km. Within the mantle, a rapid increase in velocity with depth occurs between depths of about 400 and 1000 km. This is known as the *mantle transition zone*, and it is overlain by the *upper mantle* and underlain by the *lower mantle*. Within the core, a rapid increase in *P* velocity with depth at about 5000 km was discovered by MISS LEHMANN (1936) from observations of weak *P* arrivals in the shadow zone between 110° and 143° angular distance from epicentre. These could not be properly explained by diffraction at the core-mantle boundary, but required a strong increase of *P* velocity within the core. This is also known as the *core transition zone*, and it separates the low velocity *outer core* from the high velocity *inner core*.

On the basis of the velocity-depth distribution within the Earth, Bullen has subdivided the Earth into seven concentric shells which correspond to the above subdivisions, each being given a letter for identification. Bullen's subdivisions are shown in Table 1.1 and Figure 1.6. These subdivisions form the basis for most discussions of the physical and chemical properties of the Earth's interior.

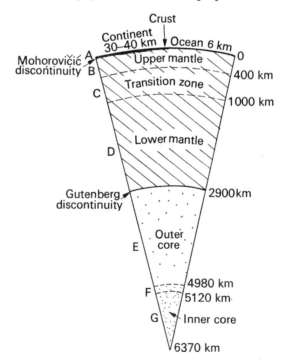

**Fig. 1.6** The layering within the Earth (refer also to Table 1.1).

## 1.5 Chemical composition of the Earth

The mantle and core together form about 99% of the Earth's volume. They are not accessible to chemical analysis. The rocks of the crust do not represent the mean composition of the Earth since they have much too low a density, even allowing for increase in density on compression. Thus indirect methods must be used to investigate the broad chemical composition of the interior. One useful guide is to draw analogy with meteorite compositions. Another method is to compare the observed physical properties at different depths with experimentally or theoretically determined properties of various likely materials at high pressure. Yet another method is to use igneous rocks differentiated from the mantle, but this is only applicable to the upper mantle. More detailed considerations on the composition of the crust, mantle and core are deferred to the following chapters, and here we review evidence for the broad composition of mantle and core.

Meteorites are small bodies in orbit round the Sun which sometimes fall onto the Earth. It is thought that they were formed by break-up of small planetary bodies, possibly asteroids. They provide the best samples we have of the material which forms the inner planets Mercury, Venus, Earth, Mars and the asteroids. Therefore they may have some relevance to the problem of the Earth's composition.

There are two main groups of meteorites. (1) The *irons*, or siderites, are mainly composed of an iron-nickel alloy (90% iron). They amount to about 10% of observed falls. (2) The *stones*, or aerolites, are mainly composed of silicates and are subdivided into:

(i) chondrites, with small rounded grains known as chondrules;
(ii) achondrites, which lack chondrules.

90% of the stones are chondrites. They have a fairly uniform composition resembling an ultrabasic rock. The average mineral composition is

$$46\% \text{ olivine: } (Mg, Fe)_2SiO_4,$$
$$25\% \text{ pyroxene: e.g. } (Mg, Fe)SiO_3 \text{ or } (Ca, Mg, Fe, Al)_2(Al, Si)_2O_6,$$
$$11\% \text{ plagioclase: } NaAlSi_3O_8\text{–}CaAl_2Si_2O_8$$

and    $$12\% \text{ nickel-iron phase.}$$

Achondrites are poorer in olivine and nickel-iron phase and they have a bulk composition closer to basalt. Intermediate between the irons and the stones are the siderolites. The glassy, silica-rich tektites form a further small group of doubtful affinities.

The existence of the two main types of meteorite suggests that there had been some separation of the silicate and nickel-iron phases in the parent body. Since the nickel-iron phase is denser, it would tend to segregate towards the centre of the body. By analogy, it may be suggested that the Earth has a nickel-iron core and a mantle similar in composition to the chondrites. However, there are difficulties in pressing the analogy too far by suggesting identical chemical composition for the parent meteorite body or bodies and the inner planets. For instance, the relative abundance of iron and stone meteorites does not correspond to the ratio of the Earth's core to mantle; also GAST (1960) has pointed out that igneous rocks from the upper mantle are depleted in potassium, rubidium and caesium relative to the chondrites. In fact, the densities of the inner planets are not consistent with their having the same chemical compositions. Variations in the relative proportions of the nickel-iron and silicate phases are to be expected both between planets and meteorites and also from planet to planet. The silicate phase itself probably also varies, certainly between meteorites, the Earth and the Moon.

When examined in detail, all hypotheses which attempt to equate the composition of the Earth as a whole, or the mantle, with one or other group of meteorites fall down on several counts. It is best to use the evidence from the composition of meteorites as an indication that two main phases may be present in the Earth's interior, nickel-iron in the core and some form of ultrabasic silicate in the mantle.

## 1.6 The age of the Earth

In the late nineteenth century it was widely accepted that the Earth was about 20–80 my old. This was Lord Kelvin's estimate, and was based on the premise that the Earth's outward heat flow represents the cooling of an initially hot body. However, the discovery of radioactivity at the turn of the century put a new complexion on the problem of the age of the Earth in two ways. Firstly, it provided a method for dating rocks which soon showed that many rocks are older than 80 my. Secondly, the energy given off during radioactive decay of uranium, thorium and potassium provides sufficient heat to explain the outward heat flow without need for the cooling hypothesis of Kelvin.

The rate of decay of a radioactive isotope is proportional to the number of atoms present. Expressing this algebraically, we obtain

$$\frac{dN}{dt} = -\lambda N,$$

where $N$ = number of atoms of parent isotope,
  $t$ = time
and   $\lambda$ = decay constant.

The time taken for half the original number of atoms to decay is called the 'half-life' or $t_{\frac{1}{2}}$. It can be shown that $t_{\frac{1}{2}} = (\ln 2)/\lambda$ where the logarithm is to the base $e$. The common radioactive isotopes with sufficiently long half-lives to be useful in dating rocks are shown in Table 1.2. Rubidium and potassium decay to their respective daughter isotopes by a single step, but uranium and thorium

**Table 1.2**   Radioactive isotopes commonly used for dating rocks. The uranium and thorium isotopes decay to the lead isotopes by a series of stages, but the half-lives are controlled by the first stage of decay since the later stages occupy negligible time by comparison. This table is based mainly on data given by HAMILTON (1965).

| Parent isotope | | Daughter isotope | Decay constant ($\times 10^{-3}$ my$^{-1}$) | Half-life ($\times 10^3$ my) | Remarks |
|---|---|---|---|---|---|
| $U^{238}$ | | $Pb^{206}$ | 0·154 | 4·51 | decay series |
| $U^{235}$ | | $Pb^{207}$ | 0·971 | 0·71 | decay series |
| $Th^{232}$ | | $Pb^{208}$ | 0·0499 | 13·9 | decay series |
| $Rb^{87}$ | | $Sr^{87}$ | 0·0139 | 50·0 | beta decay |
| $K^{40}$ | (89·05%) | $Ca^{40}$ | 0·472 | 1·26 | beta decay |
| $K^{40}$ | (10·95%) | $A^{40}$ | 0·0512 | | capture of orbital electron |

involve more complicated decay series. If, at time of crystallization, a mineral contains $N$ atoms of the parent isotope, then at some later time $t$ there will remain $N_r$ atoms of the parent isotope and $N_s$ atoms of the daughter isotope will have been produced. These are related by the equation

$$t = \frac{t_{\frac{1}{2}}}{0.693} \ln(1 + N_s/N_r)$$

where $0·693 = \ln 2$. By measuring $N_s$ and $N_r$ and using the experimentally determined value of $t_{\frac{1}{2}}$, the age $t$ is determined. It may, of course, be necessary to correct for initial presence of the daughter isotope.

The uranium and thorium series used to be widely used for dating rocks, but nowadays the potassium and rubidium methods predominate because they can be applied to a much wider range of rocks. These methods have been used to construct a time scale for the geological column. The base of the Cambrian, for instance, has been shown to be about 570 my ago (HARLAND, SMITH and WILCOCK, 1964). The methods have also helped greatly in unravelling the stratigraphy of Precambrian rocks. The oldest Precambrian rocks dated yield ages of about 3400 my, showing that the Precambrian lasted more than five times as long as the period from the base of the Cambrian to the present.

The Earth must be older than 3400 my, the age of the oldest known rocks. It must be younger than the time needed for all the $Pb^{207}$ within the Earth to form by decay of $U^{235}$, which according to geochemical estimates of terrestrial abundances is about 5500 my (HOLMES, 1965). A more exact estimate of the Earth's age can be made by using lead isotopes to date the primary crystallization of the material now forming the mantle. This may yield a true age for the Earth, but it could give an earlier or later date depending on the history of the silicate material which now forms the mantle. The method is as follows.

Common lead consists of four isotopes with mass number 204, 206, 207 and 208. It is a mixture of primeval lead ($Pb_p$) representing the isotopic composition when primary crystallization of the mantle occurred, and radiogenic lead ($Pb_r$) formed subsequently by radioactive decay of uranium and thorium. The decay series $U^{238} \rightarrow Pb^{206}$ and $U^{235} \rightarrow Pb^{207}$ can together be used to estimate the primary age of the source, provided there has been no chemical fractionation of uranium and lead between the times of primary crystallization and of withdrawal of lead from the source at a date not long ago compared with the Earth's age.

$U^{235}$ and $U^{238}$ are probably inseparable in nature. The present-day ratio $U^{238}/U^{235}$ is 137·8. Because $U^{235}$ decays more than six times as fast as $U^{238}$ does, the ratio has progressively increased with the passage of time. Its value at a given time in the past can be computed from the present ratio using the decay constants. Consequently the ratio $r = (Pb^{207}/Pb^{206})$ of daughter lead isotopes produced by decay of uranium between any fixed date in the past and the present can be computed, and is shown in Figure 1.7. If $r$ can be determined experimentally on samples of lead recently drawn from the mantle, such as lead in basalts or in galena deposits of Tertiary or recent age, then the age of the primary crystallization of the source can be read off Figure 1.7. The practical difficulty is that $Pb^{206}$ and $Pb^{207}$ were initially present in primeval lead. If we knew the isotopic composition of primeval lead, then the radiogenic fraction could be estimated and the age obtained. One way of doing this is to assume that the primeval lead of the Earth is the same as that of meteorites. Iron meteorites contain negligible uranium and thorium, and the isotopic composition of Pb in them has not changed significantly since their formation. Using estimates of the isotopic composition of primeval lead based on iron meteorites leads to a value of about 4550 my for the age of the Earth (HAMILTON, 1965).

The above method assumes that the primeval lead of the Earth and meteorites is identical, which may not be true. There is another method of estimating the age of the mantle which does not need this assumption (HOLMES, 1965, chapter 13). The ratio $r$ can be re-expressed in terms of the observed and primeval abundances of $Pb^{206}$ and $Pb^{207}$ as follows:

$$r = \frac{Pb_r^{207}}{Pb_r^{206}} = \frac{Pb^{207} - Pb_p^{207}}{Pb^{206} - Pb_p^{206}} = f(t)$$

where $r = f(t)$ is the curve shown in Figure 1.7. In a given sample, the abundances of the radiogenic lead isotopes relative to $Pb^{204}$ depend on the relative abundances of lead and uranium and thorium in the immediate vicinity of the source, which is likely to vary from place to place in the mantle. We can make good use of this variation. The method is to plot $Pb^{207}/Pb^{204}$ against $Pb^{206}/Pb^{204}$ for a variety of rocks containing lead recently brought up from the primary mantle source. According to the above equation, all these points should lie on a straight line possessing a gradient equal to $f(t)$. Thus $f(t)$ can be estimated and the age $t$ can be read off Figure 1.7. Holmes obtained an age of about 4550 my using this method, although in general the method based on meteoritic lead has been found to be more satisfactory.

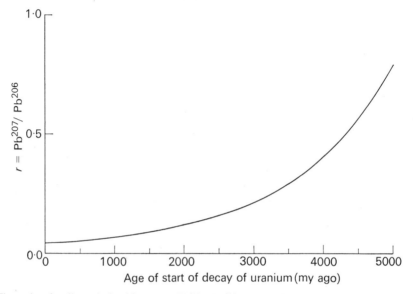

**Fig. 1.7**   The ratio of radiogenic lead isotopes $Pb^{207}$ : $Pb^{206}$ produced by decay of pure uranium between a given time in the past and the present. This graph makes it possible to compute the date at which the pure uranium started to decay from the lead isotope ratio provided a correction can be applied for the lead isotopes initially present.

The methods of radioactive age dating have also been applied to meteorites. The results suggest that they crystallized about 4550 my ago (HAMILTON, 1965, Chapter 11). This suggests that the meteorite parent bodies formed at about the same time as the Earth's mantle was formed. We anticipate that the other planets formed at about the same time.

Important dates relating to the geological time scale and to the origin of the Earth are shown in Table 1.3.

## 1.7 The origin of the Earth

To put the origin of the Earth in its proper setting, a short description of its place in the history of the universe will be given. The mean distance of the Earth from the Sun is $149.6 \times 10^6$ km and the average velocity of the Earth in its orbit is $29.8$ km/s (GUTENBERG, 1959). Some numerical facts about the Sun and its planets are given in Table 1.4.

The Sun, with its planetary system, is situated in the outer part of a lens-shaped disc of stars and interstellar gas and dust which is our galaxy, visible to us as the Milky Way. The Sun is about 27 000 light years from the centre of the galaxy (one light year is about $10^{13}$ km) and it is rotating about the centre with a velocity of about 230 km/s. It takes 220 my to complete one orbit. Our galaxy forms only a minute part of the whole universe. There are a multitude of other galaxies which have been observed up to 4500 million light years away. Light now reaching us from the most distant galaxies started on its way at about the time when the solar system was being formed!

**Table 1.3** The geological time-scale. Ages Cambrian to recent taken from HARLAND, SMITH and WILCOCK (1964).

| Eras | Periods | Beginning of periods (my) |
|---|---|---|
| QUATERNARY | Recent | 0·01 |
| | Pleistocene | 1·5–2·0 |
| TERTIARY | Pliocene | 7 |
| | Miocene | 26 |
| | Oligocene | 38 |
| | Eocene | 54 |
| | Palaeocene | 65 |
| MESOZOIC | Cretaceous | 136 |
| | Jurassic | 190–195 |
| | Triassic | 225 |
| PALAEOZOIC | Permian | 280 |
| | Carboniferous | 345 |
| | Devonian | 395 |
| | Silurian | 430–440 |
| | Ordovician | 500 |
| | Cambrian | 570 |
| PRECAMBRIAN | oldest known rock | 3400 |
| | origin of Earth | 4550 |
| | origin of meteorites | 4550 |
| | solar nebula condenses* | 4550–4750 |
| | origin of galaxy* | 8500–15 000(?) |
| | origin of universe* | 10 000–15 000(?) |

* see section **1.7**.

**Table 1.4** Table of planetary constants based on tables in the *British Astronomical Association Handbook,* 1970.

| | Mean distance from the Sun* (×10⁶ km) | Eccentricity of orbit* | Sidereal period* (days) | Inclination to ecliptic* | Equatorial radius (km) | Mass (Earth =1) | Density (g/cm³) | Sidereal period of axial rotation (days) |
|---|---|---|---|---|---|---|---|---|
| Sun | | | | | 696 000 | 332 958 | 1·409 | 25·380† |
| Moon | 0·384 400 (from Earth) | 0·0549 | 27·321 661 | 5°08′43·4″ | 1738 | 0·0123 | 3·342 | 27·322 |
| Mercury | 57·91 | 0·2056 | 87·969 | 7 00 15·0 | 2420 | 0·054 | 5·41 | 59 |
| Venus | 108·21 | 0·0068 | 224·701 | 3 23 39·6 | 6150 | 0·8150 | 4·99 | 244·3‡ |
| Earth | 149·60 | 0·0167 | 365·256 | 0 00 00·0 | 6378 | 1·0000 | 5·517 | 0·997 |
| Mars | 227·94 | 0·0934 | 686·980 | 1 50 59·5 | 3395 | 0·107 | 3·94 | 1·026 |
| Jupiter | 778·34 | 0·0485 | 4332·59 | 1 18 17·3 | 71 400 | 317·89 | 1·330 | 0·410† |
| Saturn | 1427·01 | 0·0556 | 10 759·20 | 2 29 22·1 | 59 650 | 95·14 | 0·706 | 0·426 |
| Uranus | 2869·6 | 0·0472 | 30 685·0 | 0 46 23·2 | 23 550 | 14·52 | 1·70 | 0·451‡ |
| Neptune | 4496·7 | 0·0086 | 60 190·0 | 1 46 22·2 | 22 400 | 17·46 | 2·26 | 0·625 |
| Pluto | 5900 | 0·25 | 91 000 | 17 08 24 | 2950 | 0·10 | 5·5 (?) | 6·39 |

* Some of the orbital elements are affected by long period variations, which are most noticeable for the outermost planets. Except for Pluto, these are given for Epoch January 0·5, 1969 E.T.

† At equator (period varies with latitude).

‡ Retrograde.

The spectral lines from the distant galaxies are shifted towards the red end of the spectrum. This is usually interpreted as a Doppler shift of wavelength caused by movement of the sources away from us with velocities proportional to distance. Taken at face value, this suggests that the universe is expanding. Extrapolating back in time, expansion from a concentrated nucleus started over 10 000 my ago. The universe may have been spontaneously created 10 000–15 000 my ago; or it may undergo a cycle of expansion and contraction as Dicke has suggested. There is further support for a 'big bang' about 10 000 my ago or earlier coming from the claim of radioastronomers to have detected the microwave radiation emitted just after the event.

The cosmic abundances of uranium isotopes, rhenium, and their decay products as estimated by chemical analyses of meteorites, suggest that the galaxy formed at least 5000 my before the formation of the solar system, and possibly much earlier (CLAYTON, 1964). There are also many stars in the galaxy having a more primitive, hydrogen-rich composition than the Sun. The available evidence does suggest that the galaxy was in existence long before the solar system began to form. One idea is that it formed by gravitational collapse of a turbulent body of gas, possibly soon after the initial 'big bang'.

Astronomers think that stars, similar to the Sun, are forming at the present time in the galaxy (SPITZER, 1963; KLEINMANN and LOW, 1967; HARTMANN, 1967). They form from clouds of interstellar gas and dust which have become sufficiently dense to be gravitationally unstable. The interstellar material is partly hydrogen and helium dating back to the formation of the galaxy, and partly helium and heavier elements synthesized by nuclear reactions in the interior of stars and caused by other more violent events such as supernovae explosions. Gravitational collapse of the cloud at first forms a cluster of protostars, and later the protostars themselves collapse to become young stars. As collapse occurs, gravitational energy released heats up the interior of the star and causes it to radiate and become luminous. Eventually the internal temperature becomes high enough for nuclear reactions to commence and the star ceases to contract as it joins the 'main sequence'. The Sun is a rather typical main sequence star which underwent its contraction from interstellar material somewhat over 4550 my ago.

If modern opinion is correct (JASTROW and CAMERON, 1963), the planets and their satellites formed at the same time as the Sun by condensation of the solar nebula of gas and dust. This is a modified version of the old condensation hypothesis which goes back to Descartes and Kant but which was out of favour during the first half of the present century. The rival group of theories attributes the formation of the planets to a catastrophic event, such as a near approach of another star to the Sun or the explosion of a nearby supernova; it is supposed that this event caused a filament of gas to be drawn out of the Sun and to condense to form the planets. An excellent historical review of theories for the origin of the solar system is given by TER HAAR and CAMERON (1963).

All existing theories meet some unsolved problems. But the catastrophic theories are confronted by some insuperable difficulties. One of these, pointed out by SPITZER (1939), is that a filament large and hot enough to form the planets would dissipate itself into space in a matter of an hour, long before it could cool enough to begin condensation. All that a 'catastrophe' could do would be to form a solar nebula, which is anyway the starting point for the condensation or nebular hypothesis. Hence the nebular hypothesis has again come into favour. It fits well into modern ideas of star formation and it can also explain many of the regularities of the solar system. There are still problems concerning the distribution of angular momentum, the chemical differences between the terrestrial and giant planets, and the mechanism by which condensation took place (GOLD, 1963). But at least we see possible ways of meeting these problems.

Let us look at some of the facts about the solar system which need explaining. Foremost is the regular pattern of rotation and the distribution of angular momentum. All planets except Pluto occupy nearly circular orbits which lie close to a single plane and they move round the Sun in the direction that the Sun itself rotates. Most satellites orbit in the same direction as their primary planet rotates and in the equatorial plane, although there are exceptions which may be attributed to orbital capture or tidal friction. The planets are spaced at approximately regular intervals from the Sun as expressed by the modified Titius-Bode law. The mean distance of the $n$th planet from the Sun, counting the asteroids together as a single planet, is given by $r_n = r_o m^n$, where $m = 1\cdot89$. The physical significance of the law is not yet understood.

The distribution of angular momentum in the solar system has created difficulty for all theories. 98% of the angular momentum is possessed by the planets and the Sun itself rotates slowly with a period of 24·65 days. On the other hand, most of the mass of the system is contained in the Sun. The problem is to explain how the angular momentum has been transferred from the central body to the outer parts of the system. It was this difficulty which led to the abandonment of the nebular hypothesis of Kant early this century, because the angular momentum of a contracting disc of rotating gas and dust would remain firmly anchored to the main mass condensing to form the Sun itself. The catastrophic theories were put forward to meet this difficulty, but in fact they failed to do it satisfactorily. With the revival of the nebular hypothesis, it is now recognized that interaction between the solar magnetic field and an ionized nebula (ALFVÉN, 1967), or the effects of turbulence in the nebula (VON WEIZSÄCKER, 1944), could cause outward transfer of angular momentum in a contracting solar nebula.

Until UREY (1952) pointed out its importance, chemical evidence bearing on the origin of the solar system was largely ignored. Three groups of elements form the major constituents of different parts of the solar system. These are

Group I      H, He (about 98% of the Sun's mass),
Group II    C, N, O (about 1·5% of the Sun's mass), and
Group III   Mg, Fe, Si (about 0·25% of the Sun's mass).

Mg, Fe and Si form the major constituents of the high density inner (terrestrial) planets—Mercury, Venus, Earth, Mars and the asteroids. Jupiter and Saturn, although much larger than the terrestrial planets, have much lower densities (Table 1.4) showing that they are formed mainly of H and He; possibly their overall composition does not differ much from that of the Sun or of the primitive solar nebula. Uranus and Neptune, having intermediate densities, predominantly consist of Group II elements in the form of solid methane, ammonia and ice. Severe chemical differentiation must have occurred within the solar nebula as the planets were formed. In the vicinity of the terrestrial planets, the less volatile Group III elements must have condensed out of the nebula as it was being pushed outwards by magnetic or other forces. There must have also been a very substantial escape of hydrogen and helium into outer space from the vicinity of Uranus and Neptune, amounting to over 90% of the original mass of the nebula. The mechanism by which this 'blow-off' occurred is not clear.

It was stated above that the study of the abundances of certain isotopes in meteorites yields an estimate of the age of the galaxy. The study of the isotopic ratios of xenon in meteorites gives us an estimate of another important time interval, that between completion of synthesis of heavy elements (in supernovae explosions?) and the formation of the parent meteorite bodies. $X^{129}$ is the decay product of $I^{129}$ which has a half-life of 17 my; $Pu^{244}$ produces $X^{136}$ by fission. SABU and KURODA (1967) used these methods to show that the interval may be between 200 and 300 my. Some other scientists (e.g. LEVIN, 1969) consider that 200 my is an upper limit for the interval. These results suggest that a supernova explosion may have occurred within the vicinity of the future solar nebula less than 200 my before the solar system formed. We may speculate that this explosion may have triggered the initial condensation of the solar nebula.

Some meteorites also show evidence of metamorphism and differentiation, suggesting that their parent bodies were heated before disruption. It is generally supposed that such heating must be caused by decay of the relatively short-lived isotope $Al^{26}$ which has a half-life of $0.73$ my (p. 188). Thus an additional nucleosynthesis which produced light elements such as $Al^{26}$ must have occurred a few million years at most before the meteorite bodies were formed. This probably occurred within the solar nebula itself through irradiation by high energy particles emitted by the condensing nucleus of the nebula. This is particularly clear evidence that the condensation of the solar nebula and the formation of the planetary bodies occurred at the same time.

Let us conclude by summarizing the stages through which the solar system may have evolved. The ideas are mostly taken from the references cited above. The first five stages may have been concurrent.

1. As the rotating Sun contracted, its angular velocity progressively increased. It developed a disc of rotating gas and dust in its equatorial plane. Possibly the nebula formed as a result of material being thrown off the Sun's equator when centrifugal force exceeded gravity (as originally suggested by Laplace), or the nebula may have formed through some other process.

2. Angular momentum was transferred from the Sun to the nebula, slowing down the Sun's rotation and pushing out the nebula towards the future position of the major planets. This may have occurred through interaction of the solar magnetic field (about 1 gauss?) with the ionized part of the nebula, or through turbulent convection in the nebula.

3. The outward transfer of angular momentum involves the loss of rotational energy. This may have occurred by emission of high energy particles from magnetic flares, causing irradiation of the nebula. Some light elements, such as lithium, and some short-lived radioactive isotopes, such as $Al^{26}$, may have been produced by this process.

4. Chemical differentiation of the cloud occurred, with most of the original hydrogen and helium of the nebula being lost to outer space from the vicinity of the outer planets, and silicon, iron and magnesium dropping out by condensation in the vicinity of the future terrestrial planets.

5. As the nebula cooled, it condensed to form grains and solid particles orbiting the Sun in elliptical orbits under the control of the Sun's gravitational attraction.

6. Grains and particles in neighbouring orbits sometimes collided and coalesced, progressively growing to form larger bodies. The mechanism whereby the particles stuck together during the early stages of accretion is problematical. As the bodies reached 1 km or more in size, the process of collision and sticking together would be assisted by gravitational attraction. Eventually, bodies of planetary, lunar and asteroidal size would be produced. Most of the gas and dust of the original nebula would be in these bodies or dispersed into space.

7. During the process of accretion, some angular momentum would be transferred from the nebular rotation to the newly formed planets and satellite systems, by a process not yet understood.

8. During early stages of accretion, strong heating within the small bodies may have occurred through decay of short-lived isotopes such as $Al^{26}$, which had been produced by irradiation of the nebula. This would allow the separation of the nickel-iron and silicate phases and other thermal effects to occur in the parent meteorite bodies. Later, some of these bodies broke up to form meteorites.

9. The main process of forming the solar system was complete by about 4550 my ago, and the pattern of the system has not greatly changed since then. There may have been some capture of

satellites by planets (the Moon may have been captured by the Earth), and some slowing down of planetary rotation by tidal friction especially affecting Mercury, Venus and Earth.

## 1.8 Past History of the Earth–Moon system

The Earth and its satellite Moon form a system which is to some extent unique in the solar system. This is because the ratio of the mass of the Moon to that of the Earth is 1/81·3 which is exceptionaly high for a satellite. Over 80% of the angular momentum of the system is tied up in the orbital motion of the Moon. In all the other satellite systems most of the total angular momentum is possessed by the rotating planet itself.

When we trace the Moon's orbital history back into the geological past, we find that the Moon may have been uncomfortably close to the Earth between 1000 and 2000 my ago. The history of the Moon's orbit has been inferred (1) over the last 200 years by direct observation, (2) between 0 and 1000 B.C. from ancient records of eclipses, and (3) back to the Lower Palaeozoic by using fossil coral 'clocks'.

The Moon's past position, expressed in terms of latitude and longitude, can be accurately computed from the present orbit provided the Earth and Moon are perfectly rigid bodies. The telescope observations of the Moon's actual position over the past two centuries do not agree with the theoretically computed positions (Figure 1.8). The Moon appears to move progressively faster in its orbit by about 10″ of longitude per century; this is known as the *secular acceleration of the Moon.* Superimposed on this steady change, there are also irregular fluctuations in the observed longitude of the Moon.

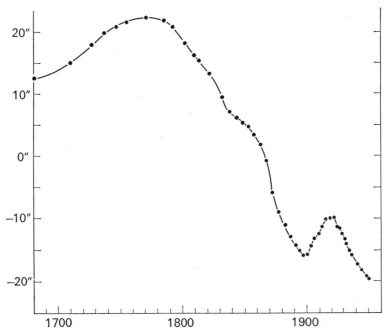

**Fig. 1.8** The discrepancy between the observed and calculated longitude of the Moon, showing irregular fluctuations superimposed on the secular acceleration. Redrawn from MUNK and MACDONALD (1960), *The rotation of the Earth*, p. 180, Cambridge University Press.

Variations in the apparent angular velocity of the Moon as viewed from the rotating Earth could be caused *either* by variations in the Moon's period of orbital rotation *or* by variations in the Earth's rate of rotation *or* by a combination of both. These two possible causes can be separated from each other over the last 200 years by using past observations of the celestial positions of the Sun, Mercury and Venus. The orbits of these bodies are unlikely to have been significantly disturbed over the past 200 years and this is borne out by the internal consistency among the observations. These observations show that the secular acceleration is partly caused by a slowing down of the Earth's rotation and partly by changes in the Moon's angular velocity, and that the irregular fluctuations of the Moon's motion are entirely caused by variations in the Earth's rate of rotation.

About half of the observed secular acceleration of the Moon, amounting to 5″ of longitude per (century)$^2$, is caused by perturbation of the Earth's orbit by the gravitational attraction of other planets which has the secondary effect of accelerating the Moon in its orbit; Laplace originally attributed the whole of the secular acceleration to this effect but Adams later showed significant terms had been neglected in Laplace's calculation. The remaining part of the secular acceleration is the combined effect of slowing down of the Earth's rate of rotation and of a lesser slowing down of the Moon's rate of orbital rotation. All the slowing down of the Moon and most of that of the Earth are caused by tidal interaction between the two bodies. Solar tides cause a further slowing down of the Earth's rate of rotation. The non-seasonal irregular fluctuations of a few years' period or more are believed to be caused by interchanges of angular momentum between the core and mantle resulting from electromagnetic coupling between them (p. 174).

Tidal interaction between the Earth and Moon (p. 155) causes a progressive loss of rotational energy from the system. The Moon has already been brought to a 'standstill' so that the same side always faces the Earth. However, the total angular momentum of the system must be conserved. The loss of angular momentum as the Earth slows down is balanced by an equal increase in the angular momentum of the Moon's orbital motion, which means that the Moon progressively recedes from the Earth. The rotational energy lost by the Earth is partly used to increase the orbital energy of the Moon but is mainly dissipated as heat by tidal friction in the shallow seas and in the body of the solid Earth. The apportionment of the energy loss between ocean tides and Earth tides is not accurately known but is probably in a ratio between 1:1 and 10:1.

Tidal friction works by exerting a couple on the Earth, which slows down the rate of rotation. The slight bulge on the Earth caused by the tides also exerts a couple on the Moon which increases its velocity in orbit, thereby causing it to recede from the Earth, which increases its period of revolution (Figure 1.9).

The rate of increase of the Moon's period of revolution caused by tidal interaction with the Earth is accurately known from the astronomical observations over the past 200 years. The rate of slowing down of the Earth caused by lunar tidal friction can be accurately calculated by the principle of conservation of momentum without reference to the mechanism of tidal friction. The result is that the length of day is increased by $1·81 \times 10^{-3}$ seconds per century through lunar tidal friction. Tidal interaction between the Sun and the Earth contributes a further increase in the length of day of between 0·35 and $0·53 \times 10^{-3}$ seconds per century; the exact value is not known because the Sun's response cannot be measured, and the estimate varies depending on the assumed mechanism of tidal friction. Thus the total tidal slowing down of the Earth's rotation over the past two centuries amounts to between 2·16 and $2·34 \times 10^{-3}$ seconds per century.

Going further back in time, the average secular acceleration of the Moon can be estimated over the last 2000–3000 years from records of several ancient eclipses between 0 and 1000 B.C. There is some doubt about the reliability, but DICKE (1966) found internal consistency which suggested to him that the irregular fluctuations in the Earth's rotation average out over long periods. He estimated that the length of day has been increasing by $1.55 \times 10^{-3}$ seconds per century, which is about 25% lower than the estimate for the last 200 years. One explanation of the discrepancy is that tidal friction has been more active over the last 200 years than over the last 3000 years. Alternatively, a slight speeding up of the rate of rotation caused by a decrease in the Earth's moment of inertia is superimposed on the tidal deceleration. The melting of the Pleistocene icecaps could contribute a small part of this decrease in moment of inertia or as Runcorn has suggested there may be a progressive downward concentration of iron within the Earth towards the core. Another suggestion put forward by Dicke is that the moment of inertia is decreased as the universal gravitational constant $G$ becomes progressively smaller with time (p. 271).

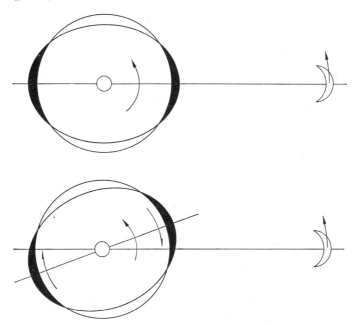

**Fig. 1.9**   How tidal friction works. The upper diagram shows the tidal bulge on the Earth which would be produced by the Moon if there is no tidal friction. This is produced because the Moon's gravitational attraction on the Earth only exactly balances the centrifugal force at the centre. The lower diagram shows how tidal friction delays the high tide. This produces a couple on the Earth, slowing down the axial rotation, and a force on the Moon, speeding up its velocity in orbit and causing it to recede. Redrawn from MUNK and MACDONALD (1960), *The rotation of the Earth*, p. 199, Cambridge University Press.

The story of the Earth's rotation and the Moon's orbit can be carried much further back into the past by using fossil 'clocks'. Some fossil rugose corals of Palaeozoic age show a characteristic banding on the outer skin (epitheca). WELLS (1963) was able to recognize tentatively both a daily and annual banding in corals of Middle Devonian age (Figure 1.10). He calculated that there were $400 \pm 7$ days in the Middle Devonian year, which was about 375 my ago. This gives an average increase in the length of day between then and now of $2.4 \times 10^{-3}$ seconds per century, which is in excellent agreement with the estimates of tidal slowing down over the last two centuries. Later SCRUTTON (1964) recognized apparent monthly banding in Middle Devonian corals and he was able to suggest that the Devonian year was divided into 13 lunar months of 30·5 days each. More recently the method of fossil clocks has been applied to other periods in the past.

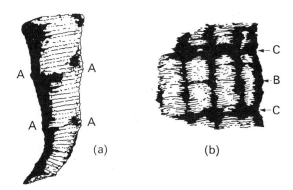

**Fig. 1.10** Diagrammatical representation of a fossil coral 'clock'; (a) shows a coral with three complete annual growth cycles each divided into thirteen monthly bands; (b) shows an enlargement of a single monthly band subdivided into thirty daily growth ridges. Reproduced from SCRUTTON (1967), *International Dictionary of Geophysics*, Vol 1, p. 1, Pergamon Press.

These estimates of the length of day and lunar month 375 my ago are particularly interesting because they enable the slowing down of the Earth by tidal friction to be separated from changes in the Earth's rotation rate caused by other processes (RUNCORN, 1964). The length of the lunar month makes it possible to estimate the angular momentum of the Moon's orbital motion. Because angular momentum is conserved, the slowing down of the Earth's rotation caused by lunar tidal friction between then and now can be estimated. Runcorn suggested that any residual effect may be the result of a change in the moment of inertia of the Earth caused by growth of the dense core. In fact he found that nearly all the change in length of day can be attributed to tidal interaction and that the moment of inertia in the Devonian does not differ significantly from the present value.

The above results suggest that the lunar month was about 4% shorter in the Devonian than now. Kepler's third law of planetary motion shows that the Moon's semi-major axis must have been about 6% less than now. Tidal friction probably depends on the sixth power of the distance between the interacting bodies. Consequently it would have been 40% more effective then than now, provided the dissipation of energy was occurring in similar conditions to now. Thus the rate at which the Moon has receded from the Earth becomes greater the further back in geological time one goes. Extrapolating back into the Precambrian, the Moon must have been very close to the Earth at some time between 1000 and 2000 my ago unless tidal friction was much less effective before than after the Devonian (MACDONALD, 1966) or the Moon was captured later. It could not have been closer than about three Earth radii because otherwise it would have broken up and formed a ring round the Earth, analagous to Saturn's rings. Thus some sort of major event in the Earth-Moon system with close proximity of both bodies may have occurred about 1400–1600 my ago, as originally suggested by GERSTENKORN (1955).

One might expect to find some evidence for this event in the Precambrian geological record. MUNK (1968) graphically describes it as follows: 'A heavy hot atmosphere over a darkened Earth. Giant tides on a $5^h$ day, with steaming tidal bores following the Moon on a $7^h$ polar orbit. Mr.

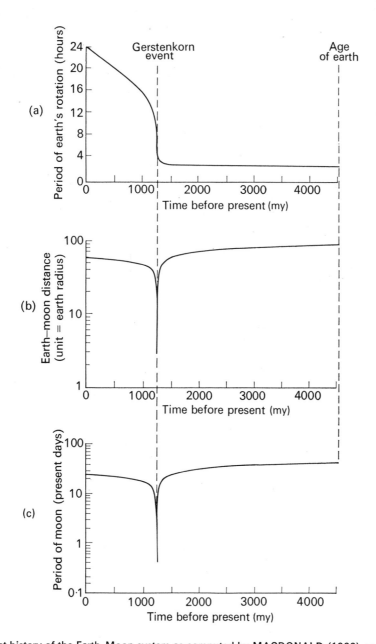

**Fig. 1.11**    Past history of the Earth-Moon system as computed by MACDONALD (1966) on the assumption that the tides lag 2·5° behind their equilibrium position as a result of tidal friction (Figure 1.9)—the present day lag is about 2·25°. The diagram shows the past history of (a) the Earth's period of rotation, (b) the Earth-Moon distance, and (c) the Moon's period of orbital rotation. Redrawn from MACDONALD (1966), *The Earth-Moon system,* pp. 187, 182 and 186, Plenum Press.

Gerstenkorn may not appreciate having his name attached to this era.' The Gerstenkorn event probably lasted about 1000 years and before it the Moon may have been approaching the Earth as its retrograde orbit decreased in size (Figure 1.11), or may have been recently captured (see below).

## 1.9 The Moon

### *Early results of manned exploration of the lunar surface*

The Moon is our nearest planetary neighbour and speculations about its origin have excited interest for a long time. This interest has been intensified by the outstanding success of the early manned exploration of the lunar surface starting with the Apollo 11 mission. Already, the scientific results of the Apollo 11 and 12 missions have greatly advanced our understanding of the Moon, although the problem of its origin still remains to be solved.

Apollo 11 landed on the Mare Tranquillitatis (Sea of Tranquillity) on July 20, 1969, and Apollo 12 landed on Oceanus Procellarum (Ocean of Storms) on November 19, 1969. These are both maria regions which typify only about 20% of the Moon's surface, but later missions are planned to land on or near the more common cratered highland parts of the Moon.

The astronauts Armstrong and Aldrin collected 7·5 kg of rock fragments of over 1 cm in diameter and 12·5 kg of finer material from the Moon's surface during the Apollo 11 mission. A preliminary investigation of the samples was quickly completed and published (Lunar Sample Preliminary Examination Team, 1969). The rock fragments can be divided into two main groups: (1) fine and medium grained igneous rocks; and (2) breccias of igneous fragments. The igneous rocks resemble basalt but differ in detail from any known terrestrial igneous rocks. Their mineralogy is broadly that of olivine-bearing, titaniferous basalts and microgabbros. The mean chemical composition of eight igneous rock fragments, expressed in terms of weight percent of oxides, is as follows:

| | | | |
|---|---|---|---|
| $SiO_2$ | 41 | $Na_2O$ | 0·5 |
| $Al_2O_3$ | 11 | $K_2O$ | 0·1 |
| $TiO_2$ | 10 | MnO | 0·4 |
| FeO | 17 | $Cr_2O_3$ | 0·6* |
| MgO | 8 | $Zr_2O$ | 0·1* |
| CaO | 10 | | |

* More recent analyses give lower estimates.

The silica percentage is lower than would be expected for terrestrial basalt. The rocks are also distinguished by particularly high content of Fe, Ti, Zr, Y and Cr, and low content of the alkalis Na, K and Rb. Volatile elements such as Cu, Pb and Zn are extremely low in abundance, and no hydrous minerals are present. There is not much variation in composition between the individual fragments analysed, and the breccias and fine material have approximately the same composition, apart from higher nickel content, as the igneous fragments. The igneous rocks have densities ranging from 3·1 to 3·5 g/cm³ which is higher than would normally be expected for basic igneous rocks and is attributable to the high iron content. Some of the fragments are higher in density than the mean density of the Moon. Most of the specimens show evidence of intense shock metamorphism in localized regions, suggesting a long history of bombardment by micro-meteorites. Chondritic fragments in the regolith fine-grained material are estimated as less than 2% by weight.

The results of more thorough investigation of the Apollo 11 samples by many scientists are reported in *Science*, **167**, **3918**, 447–784, including an introductory summary by the Lunar Sample Analysis Planning Team (1970). More detailed accounts of the work are to appear in *Geochimica et Cosmochimica Acta* (1970). Particularly important results, apart from showing the predominance of volcanic rocks, are the demonstration that the fine material and breccia has an age of 4600 my as determined by Rb-Sr and U-Th-Pb methods, and that the igneous rocks have an age of crystallization of 3700 my. Preliminary analysis of Apollo 12 samples (Lunar Sample Preliminary Examination Team, 1970) shows similar types of rock to the Apollo 11 samples, although there is a greater variety in texture and composition and breccias are less abundant; the titanium content is much lower. Preliminary age dating using the K-Ar method yields an estimate of 2300 my for the igneous fragments.

These results show that the Moon formed at the same time as the Earth and the meteorites. Subsequently, the igneous rocks which have been sampled have formed by crystallization of magma which formed either by melting of the Moon's interior relatively early in its history or by meteoritic impact melting. The peculiar rock compositions show that the parent magma must have been derived by partial melting of silicate material possessing considerable geochemical differences from the Earth's mantle and from chondritic meteorites.

Another important scientific accomplishment of the Apollo 11 and 12 missions was the placing of highly sensitive seismograph stations on the Moon's surface (LATHAM and others, 1970). The seismographs have already shown that the Moon is exceptionally quiet seismically. Seismic signals recorded on the long-period, vertical component seismometer, resulting from the impact of the lunar module 75·9 km away and from two natural events (meteorite impacts?), are shown in Figure 1.12. These are similar in character to each other, and quite unlike terrestrial seismic records of earthquakes.

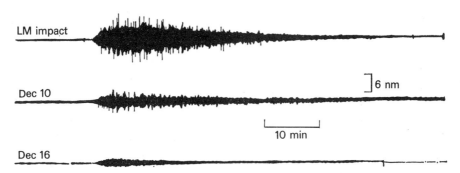

**Fig. 1.12**   Seismic signals received on the long-period vertical component seismometer from the Lunar Module impact on November 20 and from natural events on December 10 and 16, 1969. Redrawn from LATHAM and others (1970), *Science, N.Y.,* **167**, 456.

The records show a relatively slow build-up of seismic energy over a period of about 5–10 minutes, followed by a slow decay, the whole event continuing for up to one hour. The records may be explained along the following lines: (1) the seismic energy is trapped mainly in a low velocity, near-surface layer of the Moon; (2) the elastic waves are strongly scattered within this layer; and (3) the non-elastic attenuation of seismic energy is an order of magnitude weaker than it is in the Earth's crust (p. 247). We look forward with great interest to the development of lunar seismology, as further stations are placed on the Moon's surface and as the number of observed artificial and natural events increases. It may be one of the main keys to unravelling the internal structure of the Moon.

The Apollo 11 astronauts also placed a laser ranging retro-reflector on the Moon's surface (ALLEY and others, 1970). This enables ruby laser pulses from the Earth to be reflected back and timed, yielding an estimate of the distance between points on the Earth and the Moon to a precision which is already better than 1 m and will probably soon be improved to 15 cm. This will eventually make it possible to measure the Moon's orbital motion and the Earth's anomalies of rotation with greatly improved accuracy; it will also be possible to use the method to measure relative motion between crustal plates on the Earth's surface. A further experiment which is programmed for future missions is the measurement of lunar heat flow—the results will be of importance in discussing temperatures and possible processes within the Moon and its radioactive material content.

## Origin of the Moon

Let us now turn to a brief review of theories of origin of the Moon. For a fuller account, the reader is referred to a recent paper by WADE (1969). The three main theories are:

(1) *The Moon formed by fission from the Earth*, early in its history. Darwin's original version of the theory attributed the Moon's separation to the build-up of a very large solar tide when the Earth rotated with the same period as its free vibration. The fatal objection pointed out by Jeffreys is that friction would prevent separation of the tidal protuberance. Modern versions of the theory appeal to rotational instability of the Earth, but they are not widely accepted because among other difficulties they imply excessively high initial angular momentum for the Earth.

A more viable version of the theory is that the Earth, Mars and the Moon all formed by fragmentation of a larger condensing protoplanet in the solar system. LYTTLETON (1960) has shown that rotational instability of such a protoplanet would result in a separation of two major fragments with a mass ratio of at least 8 : 1 (the Earth and Mars). A stream of material drawn out between the two separating planets would condense to form one or more satellites of the larger planet.

(2) *The Moon and Earth formed together as a binary system* within the solar nebula, the Moon being formed by accretion of the part of the dust cloud between distances of 10 and 30 Earth radii from the centre (GOLDREICH, 1966). This theory fails to account for the apparent close approach of the Moon about 1400 my ago and requires a large amount of angular momentum in the original binary system.

(3) *The Moon was captured by the Earth*, either early in its history or at a later date. GERSTENKORN (1955) originally suggested that the Moon was captured in a retrograde orbit about 3000 my ago. Tidal friction would cause the retrograde satellite to approach the Earth. When the Moon's orbit came within a few radii of the Earth, the Gerstenkorn event described above (p. 22) would occur. Excessive tidal friction would cause the orbit to be deflected at first over the poles and then into a prograde orbit, as the Earth's period of rotation increased substantially (Figure 1.11). Tidal friction would then cause the Moon to recede progressively to its present orbit. The main difficulty of this theory is that the Earth's rotational period before the event would need to be about 2·6 hours to provide sufficient tidal dissipation of energy during the close approach of the Moon. GOLDREICH (1966) pointed out that Gerstenkorn and MACDONALD (1966) had neglected the influence of the Sun on the orbits. GERSTENKORN (1969, 1970) has now taken the perturbations of the Sun into account in new calculations, which suggest that the Moon was captured about 1400 my ago *either* in an orbit of inclination about 90° at distance of a few Earth radii, *or* in a prograde orbit of large eccentricity. These recent calculations suggest that the Moon approached the Earth at distance with a relative velocity of less than 1·5 km/s. The problem here is to understand how the Moon could have remained parked in such an orbit from 4600 to 1400 my ago without suffering earlier collision or capture. Yet another version of the capture theory suggested by MACDONALD (1966) is that the Earth captured six or more satellites,

one of them a quarter of the Moon's size, early in its history. At a later date in the Earth's history, it is supposed that these small moons coalesced to form the Moon without need for a Gerstenkorn event.

How does the new evidence obtained during the Apollo 11 and 12 missions bear on the problem of origin of the Moon? *Firstly*, it suggests that the Moon must be of similar age to the Earth, and that it must have existed long before the supposed Gerstenkorn event. *Secondly*, if the lunar igneous rocks formed by melting of material within the Moon, then the composition of the Moon and the Earth's mantle are chemically distinct from each other. Of all the theories, the one which seems to show best promise is the capture theory in one form or another. However, we look forward to the results of later Apollo missions which are sure to produce further evidence on the problem of origin of our nearest neighbour, and indirectly on the origin of the Earth itself.

# 2 The continental crust

## 2.1 Introduction

A century or more ago, it was believed that the Earth consisted of a thin rigid crust overlying a hot fluid substratum which provided magma to feed volcanoes. This concept was being questioned before 1900. It was finally abandoned when early seismological research showed that the Earth is normally solid down to 2900 km depth. It is now known that magma is not drawn from a permanently fluid region within the Earth but forms by local fusion of the normally solid rocks of the upper mantle and crust.

The crust is nowadays almost always defined as the region above the Mohorovičić discontinuity (or Moho). This discontinuity has been found to be almost universally present beneath continents and oceans. It is marked by a discontinuous or rapid downward increase in $P$ velocity to above 7·6 km/s (JAMES and STEINHART, 1966). Defined in this way, the crust forms less than 1% of the Earth by volume and less than 0·5% by mass. But it is the only major subdivision as yet directly accessible to man and its importance is out of all proportion to its size. The topmost part of the continental crust is the most fully investigated part of the Earth and study of it has provided most of the evidence we have of the past history of the Earth. In contrast, surprisingly little is known of the structure of the lower part of the continental crust.

Two important facts about the structure of the uppermost part of the Earth were known before the Moho had been discovered. *Firstly*, it had been recognized that the mean density of the Earth is substantially greater than that of rocks at or near the surface, suggesting the existence of a low density layer near the surface. *Secondly*, measurements of the local variations in the vertical direction near mountain ranges led to the discovery of the theory of isostasy during the eighteenth and nineteenth centuries. This shows that there are large lateral variations in density within the upper layers of the Earth and that the near surface layer of relatively strong and brittle rocks now called the *lithosphere* must be underlain by a weaker substratum which deforms by flow, now called the *asthenosphere*. Occasionally the term 'crust' has been used for the lithosphere, but its lower boundary does not in general coincide with the Moho and is gradational rather than sharp.

### Geological structure of the uppermost crust

Over a century of geological investigation has given us a detailed knowledge of the surface rocks forming the continents. Typically a variable thickness of partly consolidated sedimentary rocks overlies a strongly folded and metamorphosed basement, or alternatively the basement itself crops out at the surface. Most of the sedimentary rocks have been formed by erosion of pre-existing sedimentary rocks. Over large areas, such as parts of the Precambrian shields, unmetamorphosed sediments are absent. At the other extreme, local accumulations of sediment may exceed 10 km in geosynclines and deep basins. According to POLDERVAART (1955), the average thickness of sediments in regions of young fold belts is about 5 km and in continental shield areas it is 0·5 km. Lava flows and minor igneous intrusions commonly occur in sedimentary sequences, but penetration by large igneous intrusions is relatively rare.

The underlying rocks of the basement are metamorphosed sedimentary and igneous rocks which are locally penetrated by large igneous intrusions, especially granites and granodiorites.

It is convenient to subdivide the continental regions into structural provinces based on the history of deformation over the last hundred million years or more. The main subdivisions are:

(1) *Stable regions*, sometimes called *cratons*, which show little evidence of vertical or horizontal movement apart from broad warping and a few minor faults. Included in this category are the *Precambrian shields* which are gently arched regions of large areal extent where Precambrian rocks are found at the surface, and *platforms* where the basement rocks are overlain by a thin cover of flat-lying sediments.

(2) *Semi-mobile regions*, which are characterized by relatively strong differential vertical movement including the formation of sedimentary basins. Great Britain has been a semi-mobile region since the end of the Palaeozoic mountain-building movements.

(3) *Mobile belts*, or young mountain ranges, which have been strongly deformed with indication of powerful vertical and horizontal movement. The two main mobile belts are the circum-Pacific belt of mountain chains and island arcs which forms a ring round the Pacific Ocean and the Alpine–Himalayan mountain belt. Metamorphism and emplacement of large granite batholiths occur in mobile belts. Most parts of the continental crust have been mobile belts at some time during the Precambrian or later.

## 2.2 Earthquake seismology and the discovery of the crust

Earthquake seismology laid the foundation for the modern study of the crust. As a method of investigating the thickness and internal structure of the crust, it has now been largely superseded by refraction and reflection studies using artificial explosions.

The starting point was Mohorovičić's discovery of the discontinuity at the base of the continental crust, now called the Moho. He recognized two $P$ and two $S$ pulses on seismograph records of the Croatia earthquake of October 8, 1909, at observatories within a few hundred kilometres of the epicentre. Near the epicentre, the slower travelling $P$ and $S$ pulses which are now usually called $P_g$ and $S_g$ were prominent and arrived first. They progressively died out. Beyond 200 km $P_g$ and $S_g$ were overtaken by $P_n$ and $S_n$ pulses travelling with a higher apparent velocity, which could be identified as the normal $P$ and $S$ phases observed up to an angular distance of 103° from the epicentre. Mohorovičić interpreted $P_g$ and $S_g$ as the waves transmitted direct from the focus to the seismograph station through an upper low velocity layer which is the continental crust. The faster $P_n$ and $S_n$ arrivals were interpreted as so-called head waves which had travelled for most of their path in a higher velocity underlying medium, which is the mantle. The principle, applied to a surface source, is shown in Figure 2.1. MOHOROVIČIĆ (1901) obtained the following results:

$$P_g = 5 \cdot 6 \text{ km/s},$$
$$P_n = 7 \cdot 9 \text{ km/s},$$
and crustal thickness $= 54$ km.

The next important step was made by CONRAD (1925) who recognized two new body wave phases $P*$ and $S*$ with intermediate velocities of $6 \cdot 29$ and $3 \cdot 57$ km/s respectively by studying records of the Tauern earthquake of November 28, 1923. On this evidence, he subdivided the crust into an upper layer which transmits $P_g$ and $S_g$ overlying an intermediate layer giving rise to the refracted arrivals $P*$ and $S*$. The intervening interface is called the *Conrad discontinuity*.

From a quite different standpoint, Daly had suggested earlier that beneath the continents there is an upper silicon-aluminium-rich layer (the SIAL) providing the source of granite magma overlying a silicon-magnesium-rich layer (the SIMA) which is the source of basalt magma. Earthquake seismologists took over Daly's model because they found that the seismic velocities in granite and basalt

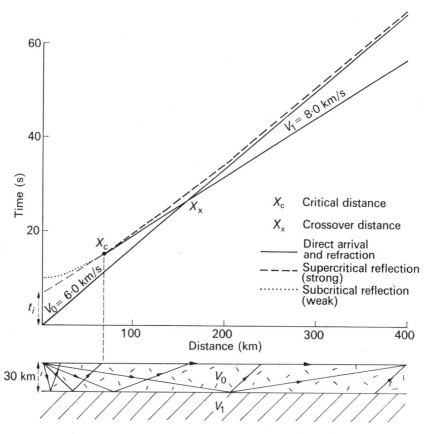

**Fig. 2.1** Time-distance graph for direct, refracted and reflected arrivals for an explosion source at the surface of a uniform horizontal layer overlying a higher velocity half-space. The velocities are equal to the reciprocal gradients of the respective first-arrival segments, and the depth of the interface is given by the following relationships

$$d = \frac{t_i}{2}\sqrt{\frac{V_0 V_1}{(V_1^2 - V_0^2)}}, \quad d = \frac{X_x}{2}\sqrt{\left(\frac{V_1 - V_0}{V_1 + V_0}\right)} \quad \text{or} \quad d = \frac{X_c}{2}\frac{\sqrt{(V_1^2 - V_0^2)}}{V_0}.$$

were closely similar to the observed velocities of the upper and intermediate layers respectively. They called these two layers the 'granitic layer' and the 'basaltic layer'. It will be shown later in the chapter that the upper crust is not granitic and that the lower crust is probably not basaltic, nor is the twofold subdivision found everywhere. It is now preferable to use the non-committal terms 'upper crustal layer' and 'lower crustal layer' wherever a twofold subdivision of the continental crust is made.

Until the advent of crustal explosion seismology about 1950, earthquake body wave studies were widely used to investigate the structure and thickness of the continental crust. The dispersion of earthquake surface waves was used to investigate broad structure across wide tracts of country (PRESS and EWING, 1955) and latterly regional variations in crustal structure (EWING and PRESS, 1959). Although explosion seismology has been found to be a much more effective method for probing crustal structure, earthquake studies have not been entirely superseded. They do still hold one advantage, which is that earthquakes occur at different depths in the crust and below it, and thereby can provide complementary information on lower crustal structure unattainable by surface sources; the use of seismological array stations may enable this advantage to be more fully exploited.

## 2.3 Explosion seismology and the structure of the crust

### The refraction method

Nearly all our knowledge of the structure of the continental crust has been obtained since about 1950 by explosion seismology. The advantages of using artificial explosions rather than earthquakes are that the time and position of the shot are accurately known. Experiments can be planned in relation to geological structure and regions without earthquakes can be investigated without difficulty.

Most applications of explosion seismology to investigation of crustal layering and thickness make use of the refraction method of seismic prospecting (e.g. DOBRIN, 1960). The basic method is to determine the travel-time of $P$ waves between shot points and seismic recorders at varying distance apart up to several times the depth of penetration required, usually along straight line profiles. Figure 2.1 shows the simplest possible situation, with a horizontal layer of velocity $V_0$ overlying a higher velocity substratum ($V_1$). From the shot point to the crossover point $X_x$ the direct wave arrives first. Beyond the critical distance $X_c$ the refracted head wave from the underlying layer reaches the surface and it becomes the first arrival beyond $X_x$. A wave reflected from the interface also reaches the surface; this is a relatively weak arrival below the critical distance, but its amplitude increases strongly near $X_c$. Beyond $X_c$ the supercritical reflection (as it is called) may be the largest amplitude arrival on the record. In this simple horizontal two-layer case, the velocities are given by the reciprocal gradients of the first arrivals on the time-distance graph and the depth to the interface can be calculated from $t_i$ (the intercept time), $X_x$ or $X_c$.

The dip of a plane interface can only be determined by shooting in both directions. The velocities and the dip and depth to the interface are computed from the two travel-time graphs as shown in Figure 2.2. The same procedure can be extended to the multi-layer case provided the velocity increases downwards at each interface.

In practice, seismic refraction lines need to be about 200–300 km long to determine continental crustal structure. It is nowadays standard practice to reverse the lines. The normal method of interpreting is to plot a reversed time-distance graph and to fit straight-line segments to the first arrivals by least squares. The inverse gradients of the segments and their intercepts on the time axis are measured. Interpretation proceeds on the assumption that the underlying structure consists of layers of uniform velocity separated by plane interfaces (which may dip).

More sophisticated methods of interpretation of refraction surveys can take into account the deviations of the segments of the time-distance graph from straight lines. For instance, the method of HAGEDOORN (1959) enables the shape of a refractor to be determined along a reversed profile. A generalization of this approach to surveys where the shots and recorders are not along a line is known as the 'time-term' method and it has proved useful in some crustal structure investigations (WILLMORE and BANCROFT, 1960). In theory, the time-term method enables the subsurface shape of one or more refractors to be mapped provided the overlying velocity distribution is known and one shot point and recording point are common.

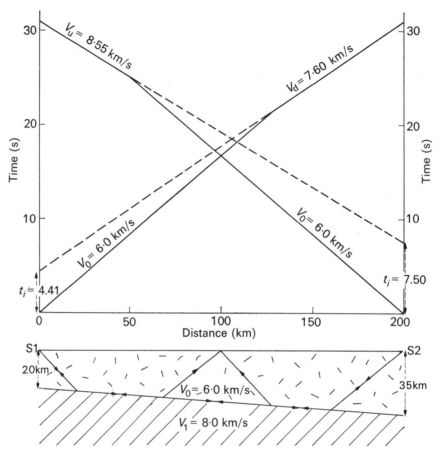

**Fig. 2.2** Time-distance graph for reversed profiles along the surface of a dipping layer (velocity= $V_0$) overlying a substratum (velocity= $V_1$). $V_u$ and $V_d$ are the apparent velocities of the refracted head wave shooting up-dip and down-dip respectively; these apparent velocities are equal to the reciprocal gradients of the corresponding segments on the graph. The dip of the interface is

$$\theta = \tfrac{1}{2}\left\{\sin^{-1}\left(\frac{V_0}{V_d}\right) - \sin^{-1}\left(\frac{V_0}{V_u}\right)\right\}.$$

The true velocity of the underlying layer is

$$V_1 = 2\,V_u V_d \cos\theta / (V_u + V_d).$$

The depth of the interface at $S_1$ (or $S_2$) is

$$d = \frac{t_i}{2}\,\frac{V_0 V_1 \sec\theta}{\sqrt{(V_1^2 - V_0^2)}}.$$

There are two important limitations on the scope of the seismic refraction method for determining crustal structure. *Firstly*, the method is unsuitable for pinpointing detailed structure, and at best gives a simple layered model which may be a great simplification of a more complicated structure.

*Secondly*, the presence of low velocity layers cannot be detected and if they do occur they lead to erroneous depth estimates for the underlying interfaces. Similarly, layers which are thin or represent small increases in velocity with depth may not give rise to first arrivals and may therefore be missed, causing wrong estimates of the depth to interfaces below. The method also assumes that there is no lateral variation of velocity within the layers.

The second limitation applies as follows. Commonly in crustal structure refraction surveys only the direct arrival $P_g$ and the Moho refraction $P_n$ are observed as first arrivals. Such a result would probably be interpreted in terms of a single-layered crust of uniform velocity (as in Figure 2.1). But such an interpretation is not a unique solution because widely different velocity-depth distributions in the middle and lower crust could give rise to exactly the same first arrival segments. This ambiguity can in fact be reduced to some extent by making use of the supercritical reflections from the Moho and other interfaces, as described below.

At a discontinuity marked by an increase in velocity, the amplitude of the reflected ray is relatively small if the angle of incidence is less than the critical angle. In contrast, for angles of incidence above the critical angle, large amplitude reflections occur because of total reflection (Figure 2.3). Such large amplitude arrivals are commonly observed in crustal structure surveys and may occur through reflection at the Moho ($P_mP$) and other interfaces in the crust such as the Conrad discontinuity ($P_IP$). According to ray theory, the maximum amplitude occurs at the critical distance $X_c$, at which point the time-distance segment of the refracted arrival is tangent to the reflected arrival. However, ·ČERVENÝ (1966) has shown that ray theory is an oversimplification and that the curved wavefront causes the maximum amplitude to occur somewhat beyond this point.

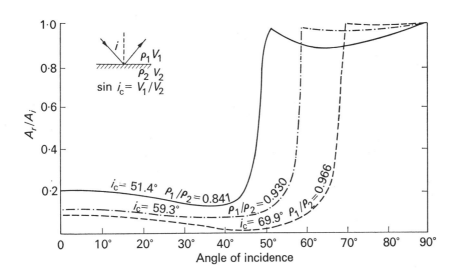

**Fig. 2.3** The ratio of amplitude of the reflected and incident plane wave at a boundary marking an abrupt increase in velocity, plotted as a function of angle of incidence. Redrawn from STEINHART and MEYER (1961), *Explosion studies of continental structure*, p. 97, Carnegie Institution of Washington.

Time-distance curves for the reflected and refracted arrivals for a single-layered and two-layered crust are shown in Figures 2.1 and 2.4 respectively. The amplitude-distance curves for the head waves $P_g$, $P^*$ and $P_n$, and for the reflected arrivals $P_IP$ and $P_mP$ (using ray theory) for a two-layered crust

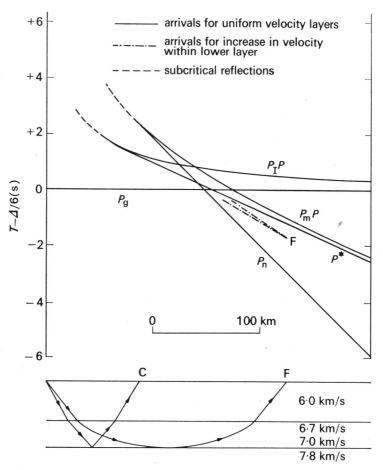

**Fig. 2.4** Reduced time-distance graph showing refractions and reflections for a two-layered crust with (a) uniform velocity layers, and (b) increase in velocity with depth in the lower crustal layer (giving rise to the segments meeting at F). The increase in velocity with depth changes the curve considerably from the uniform case, $P_mP$ arriving earlier but not being observed beyond F.

Reduced time-distance graphs like this, where (travel-time minus distance × velocity) is plotted against distance, are commonly used in presenting crustal refraction results because they allow the time-scale to be expanded without letting the graph become unmanageable. Redrawn from EATON (1963), *J. geophys. Res.*, **68**, 5795.

overlain by a thin veneer of sediments are shown in Figure 2.5; these show that the reflected arrivals beyond the critical distance have larger amplitudes than the head waves by one to two orders of magnitude.

The use of these supercritical reflections can greatly enhance the interpretation of crustal refraction surveys in two main ways: (1) the reflections make it possible to recognize with conviction significant interfaces within the crust which do not give rise to first arrivals (e.g. Figure 2.4); and (2) by using the reflection time-distance curve in addition to the first arrival segments, a realistic estimate of the mean velocity of the crust and of the true thickness can be made, thereby overcoming some of the ambiguity of using first arrivals only.

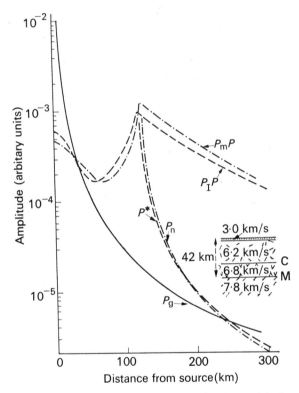

**Fig. 2.5**    Theoretical amplitudes of refracted and reflected rays for the model of crustal structure as shown. Redrawn from BERRY and WEST (1966), *The Earth beneath the continents*, p. 474, American Geophysical Union.

This section on the technique of crustal refraction survey is concluded by an example which demonstrates the use of first arrivals and supercritical reflections to determine the structure and thickness of the continental crust. The example chosen is a project carried out in late 1966 to determine the crustal structure beneath the composite granite batholith of south-west England (BOTT, HOLDER, LONG and LUCAS, 1970); the granite batholith itself is known from gravity surveys to extend to a depth of about 10 km (p. 53). Recording stations were set up on each of the main granite outcrops shown in Figure 2.15 and also on the Scilly Isles. Twenty 300-lb depth charges were exploded by H.M.S. *Hecla* at about 10 km interval along a line extending approximately south-westwards from Land's End. Unfortunately no recording station could be placed at the far end of the line. However, the line was effectively reversed because the observed travel-times for several shots at a single recording station yielded an estimate of the apparent $P_n$ velocity in the direction shot to station, while the observed travel-times for a single shot at different recording stations yielded an estimate of the apparent velocity of $P_n$ in the opposite direction. This is not an ideal method of reversing the line, but was all that could be done under the circumstances. It showed that the well-defined Moho is practically horizontal along the central part of the line.

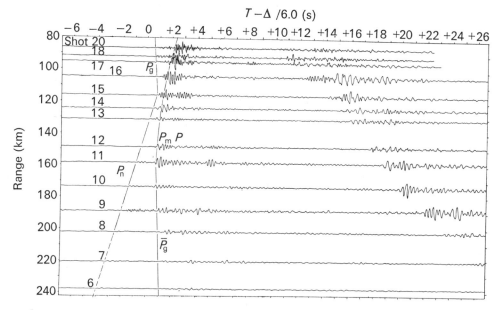

**Fig. 2.6**  Recordings of shots 6 to 20 of the south-west England crustal refraction project at a single seismometer of the Bodmin Moor station. The records have been presented in 'stacked' form so that the arrivals lie on a reduced time-distance graph. The shot-receiver distance $\Delta$ in km is plotted against $T-\Delta/6\cdot0$, where $T$ is the travel-time. This line was observed along AA (Figure 2.12(a)) and its extension towards the WSW.

The following phases can be recognized :

$P_g$—the direct wave travelling through the uppermost crust
$P_n$—the refracted head-wave from the Moho
$P_mP$—the supercritical reflection from the Moho
$\bar{P}_g$—a channel wave trapped in the upper crust by the velocity increase below
$S$ arrivals are also visible later on the records.

After BOTT, HOLDER, LONG and LUCAS (1970), *Mechanism of igneous intrusion, Geol. J. Spec. Issue* No. 2, p. 97.

Recordings of shots 6 to 20 by one of the seismometers at Bodmin Moor are stacked together in the form of a reduced time-distance graph in Figure 2.6. The first arrivals fall on two straight line segments corresponding to the direct wave $P_g$ (5·77 km/s) and the refracted head wave from the Moho $P_n$ (7·93 km/s). These first arrivals give a crustal thickness of 23 km, assuming it to be a single layer of uniform velocity. However, a clearly defined supercritical reflection interpreted as $P_mP$ occurs on Figure 2.6 and was also observed at the other stations. This can be used together with the first arrival data to show that the mean crustal velocity is 6·15 ± 0·15 km/s which is significantly higher than $P_g$, indicating that the velocity increases with depth in the crust. Knowledge of the mean crustal velocity makes it possible to obtain a more realistic estimate of crustal thickness of 27 ± 2 km. As there is no evidence for refractions or reflections from sharp discontinuities within the crust, our preferred crustal model shown in Figure 2.7 incorporates a uniform upper crustal layer formed by the granite batholith underlain by a progressive increase in velocity with depth through the lower crust. The Moho itself is well defined.

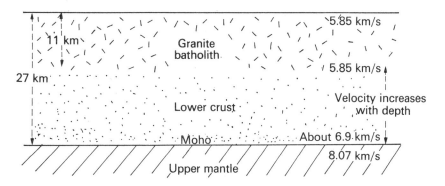

**Fig. 2.7**   Model of crustal structure beneath the granite batholith of south-west England (line AA Figure 2.12(a), and Figure 2.15) based on the interpretation of the crustal structure seismic experiment (Figure 2.6) given by BOTT, HOLDER, LONG and LUCAS (1970), *Mechanism of ingeous intrusion, Geol. J. Spec. Issue*, No. 2, and incorporating the granitic nature of the uppermost 11 km of the crust here as deduced by gravity survey (Figure 2.16).

## The normal incidence reflection method

Another possible seismic method of investigating the depth to the Moho, and to intermediate discontinuities within the crust, is to look for reflections of waves which initially travel downwards in a nearly vertical direction. Such vertical-incidence reflections can only occur where the discontinuity is sharp in relation to the wavelength; in theory, this should make it possible to determine whether the discontinuities within the crust and at its base are sharp or gradational. Vertical-incidence reflections are less easy to recognize than the wide-angle reflections discussed above because of their relatively small amplitude in relation to the seismic background 'noise' produced by other phases. Their detection therefore requires sophisticated field and interpretation techniques. Such techniques, however, are available in the seismic reflection prospecting method used in particular in exploration for oil.

Many alleged reflections have been reported from the Moho, but not all of these are convincing. The small-amplitude reflections which are sought could easily be confused with multiple reflections from shallower boundaries or with non-vertical reflections, unless modern processing techniques are used. One of the more convincing surveys has been carried out by DIX (1965) in the Mohave Desert, California, using modern methods of survey and interpretation. He obtained reflections from the depth at which the Moho was expected as estimated from refraction surveys. Dix did not obtain a single reflection, but a cluster of events spread over about 0·8 seconds, suggesting that the Moho there may be spread over about 2 km as a series of discontinuities.

Perhaps the most successful attempts to use near-vertical reflections are by German seismologists (e.g. LIEBSCHER, 1964; DOHR and FUCHS, 1967), who have made statistical studies of late reflections obtained in commercial prospecting. Reflections are read from a large number of records in a region and the numbers of reflections in given depth intervals are plotted as frequency polygons. Convincing peaks in the number of reflections occur for some (but not all) of the regions studied. In Figure 2.8 the reflections occurring between about 10 and 11 seconds are interpreted as the Moho, and those between 7 and 8 seconds as the Conrad. The prominent reflections at about 4 seconds are attributed to a discontinuity named the *Fortsch discontinuity* by German seismologists.

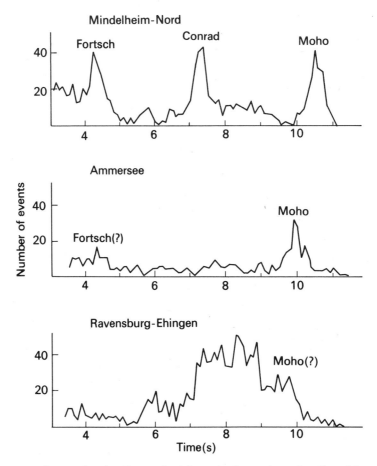

**Fig. 2.8**  Frequency polygons showing the number of events observed as a function of two-way travel-time in normal-incidence seismic reflection surveys in south Germany. The peaks are interpreted as major discontinuities within the crust (Fortsch and Conrad) and at the base (Moho). Based on LIEBSCHER (1964) and redrawn from JAMES and STEINHART (1966), *The Earth beneath the continents*, p. 314, American Geophysical Union.

*The thickness of the continental crust*

Extensive investigations of crustal structure by explosion seismology have been made in U.S.A., Russia and Japan, and quite large programmes have also been done in many other countries. In the United States, pioneer work was done from the Carnegie Institution at Washington (TATEL and TUVE, 1955) and by Wisconsin University (STEINHART and MEYER, 1961). Between 1960 and 1963 refraction surveys on an unprecedented scale were made by the U.S. Geological Survey in connection with the underground nuclear test detection programme VELA UNIFORM (PAKISER, 1963). Since that time, emphasis has been on individual large-scale projects such as the Lake Superior experiments of 1963 and 1964 described in *The Earth beneath the continents* (STEINHART and SMITH, 1966). The Russians have used a technique which they call Deep Seismic Sounding (DSS), in which seismic detectors are spread at 100–200 m intervals along as much of the line as possible, making it possible for them to trace individual waves from detector to detector, and thereby recognize late arrivals such as reflections with confidence (KOSMINSKAYA, BELYAEVSKY and VOLVOVSKY, 1969).

STEINHART and MEYER (1961) and JAMES and STEINHART (1966) have given excellent reviews of the state of knowledge at the time they were written, and a tabulation of results has been given by McCONNELL, GUPTA and WILSON (1966). All we can do here is to summarize some general conclusions and to present a few examples.

A most important result is that the Moho appears to be present universally beneath the continents. It is normally found between depths of 20 and 50 km but locally it may be deeper as, for instance, beneath young fold mountain ranges. The average thickness of the continental crust is about 35 km. In general, the base of the crust is marked by an easily distinguishable velocity contrast, although in a few regions there is some ambiguity in recognizing it uniquely. Examples of the results of crustal thickness determinations are shown in Figures 2.7, 2.9 and 2.12.

We are not yet in a position to produce a reliable map of the Moho, except in a few special regions. But a sufficiently large number of refraction surveys have been done to show that there are regional variations in the depth to the Moho which correlate quite well with the boundaries between structural regions. An important example is the variation in crustal thickness across the western part of U.S.A. shown in section in Figure 2.9 (PAKISER, 1963). This shows that local thickening of the crust occurs beneath the Sierra Nevada mountains and that each of the main geological provinces is associated with a characteristic crustal thickness as follows:

| | |
|---|---|
| Coast Range, California | 25 km |
| Central Valley, California | 20 km |
| Basin and Range province | 25–30 km |
| Colorado province | 40 km |
| Great Plains | 45–50 km |

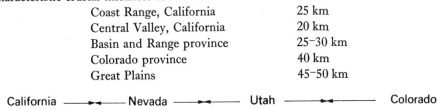

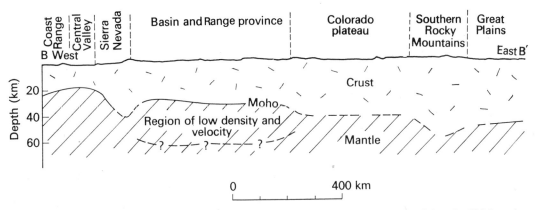

Fig. 2.9  Variations in crustal thickness from San Francisco, California, (B) to Lamar, Colorado, (B') based on crustal refraction surveys (BB' is shown in Figure 2.12(a)). Redrawn from PAKISER (1963), *J. geophys. Res.*, **68**, 5751.

A second example of local variation in crustal thickness is the crustal profile centred on the seismo-logical array station at Eskdalemuir in south Scotland, shown in Figure 2.12. The crust is estimated to be about 25 km thick beneath the north Irish Sea but it thickens slightly but significantly beneath south Scotland which is a different structural province. Regional variations in the sub-Moho $P$ velocity are discussed in Chapter 4 (p. 112).

*Seismological structure of the upper crust*

Seismic refraction surveys typically yield a velocity of 5·9–6·2 km/s for the direct wave $P_g$ travelling in the upper part of the crust. This is significantly higher than the value of 5·6 km/s yielded by the old earthquake studies of crustal structure. This discrepancy between earthquake and explosion seismology has been explained in the following ways:

(1) Lack of the exact time and focus of an earthquake may lead to serious errors in the estimate of $P_g$. For instance, WOOD and RICHTER (1931, 1933) using quarry blasts in California found $P_g$ to be 5·9–6·2 km/s when they knew the time of the blast, but estimated it to be 5·5 km/s when they did not know it.

(2) Earthquake studies of crustal structure have been concentrated in the active seismic belts where crustal structure may not be typical.

(3) Earthquake waves may travel in a low velocity channel within the continental crust while the direct wave in explosion studies travels in the higher velocity layer above (GUTENBERG, 1954a).

Most seismologists favour (1) or possibly (2) as the explanation of the discrepancy. Gutenberg, however, brought forward further arguments for (3). He called attention to the horizontally polarized shear waves of 4 second period and 3·5 km/s velocity which are commonly observed in trains of earthquake waves which travel entirely along continental paths, and are known as $L_g$. These are channel waves propagated in the crust. It is thought that they propagate by total internal reflection at the upper and lower boundaries of the channel. Gutenberg considered that $L_g$ is propagated in a low velocity channel within the middle crust. The more generally favoured interpretation is that the boundaries of the channel are the Earth's free surface above and the Moho below.

Recently, MUELLER and LANDISMAN (1966) have revived the idea of a low velocity layer within the upper crust. They base this on their interpretation of a strong amplitude supercritically reflected phase which they name $P_c$, which occurs in refraction experiments in Germany and elsewhere starting at a distance of about 50–60 km from the shot (this being the emergence of the critical reflection). On the seismograms, $P_c$ follows $P_g$ by about one second. They argue that a discontinuity at a depth of 10 km marking a downward increase in velocity can explain both the reflections at about 4 seconds observed by Liebscher (Figure 2.8) and the supercritically reflected phase $P_c$. In order to reconcile both sets of observations, and also the almost constant time difference between $P_g$ and $P_c$, they consider that the discontinuity must mark the base of a low velocity channel a few kilometres thick. The model of crustal structure incorporating this idea is shown in Figure 2.10. In it the velocity decreases at a depth of about 6 km and then it abruptly increases again at the Fortsch discontinuity at 10 km depth to a value at least 0·2 km/s higher than above the channel. It is yet too early to say whether this revived hypothesis of the low velocity channel in the upper crust will gain wide acceptance.

*Structure of the lower crust*

Present-day ideas on the lower crust and its layering are much less definite than those of twenty years ago. Following Conrad's recognition of $P^*$ and $S^*$, these phases became widely recognized in near-earthquake studies and they were attributed to the refracted head waves from the Conrad discontinuity which was interpreted as the boundary between the upper and lower crust. As late as 1950–60, earthquake seismologists generally believed that the Conrad discontinuity is universally present in the continental crust (GUTENBERG, 1959; BYERLY, 1956) although a minority of them such as JEFFREYS (1959) thought that the evidence was far from clear-cut.

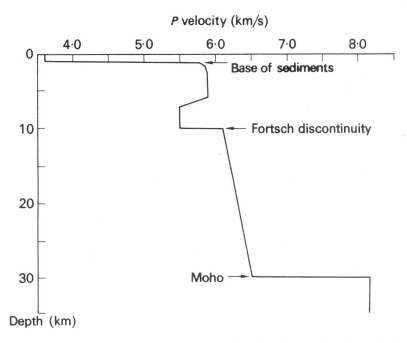

**Fig. 2.10**  Model of *P* velocity distribution within the crust, showing the postulated low velocity layer in the upper crust. Redrawn from MUELLER and LANDISMAN (1966), *Geophys. J. R. astr. Soc.*, **10**, 530.

Explosion seismologists have been much more sceptical about the widespread existence of the Conrad discontinuity. At the extreme, TATEL and TUVE (1955) failed to find any evidence from explosion seismology in widely separated regions of U.S.A. for layering within the crust. However, they did show that an increase in velocity with depth was required to reconcile the travel-times of refracted and critically reflected arrivals from the Moho. Later work has confirmed the general increase in velocity with depth through the continental crust, but it has also shown that layering does occur in some regions.

There is a fundamental difficulty in convincingly recognizing $P^*$ in earthquake and explosion studies of crustal structure in that it seldom occurs as a first arrival. As a late arrival, it tends to be swamped by the much larger amplitude supercritical reflections from the Moho and other interfaces in the crust, which would arrive at much the same time. Because of this, phases which have been picked as $P^*$ in some studies would now be interpreted as the reflected phase $P_mP$. On the time distance graph, $P_mP$ is asymptotic to a line with inverse gradient equal to the highest $P$ velocity in the crust, so would be expected to give an apparent velocity within the range 6·5–7·0 km/s. Let us take an example. HALES and SACKS (1959) made a careful study of crustal structure in Eastern Transvaal using earth tremors which occur frequently in the deep gold mining district of Johannesburg and which were better located in position and time than most earthquakes. They observed a large amplitude $P$ wave with apparent velocity 6·7–7·2 km/s which was widely traced and always occurred as a second arrival (Figure 2.11). Hales and Sacks interpreted this arrival as the refracted wave $P*$ from the Conrad discontinuity at 25 km depth, the Moho being 37 km deep. The large amplitude of this phase suggests that it is probably $P_mP$, and that it indicates an increase in velocity with depth in the crust but does not reveal a Conrad discontinuity.

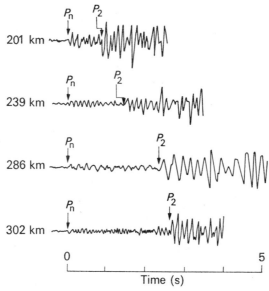

**Fig. 2.11**   Seismometer records of earth tremors near Johannesburg observed at the distances indicated along a line east-south-eastwards from Johannesburg, re-aligned so that $P_n$ arrivals are all in the same vertical line. Each record shows $P_n$ as a first arrival, followed by a strong amplitude second arrival $P_2$ which was originally interpreted as $P*$ but may be a supercritical reflection $P_mP$ from the Moho. Redrawn from HALES and SACKS (1959), *Geophys. J. R. astr. Soc.*, **2**, 24.

   With present techniques, there are two methods of convincingly recognizing a Conrad discontinuity where it exists. *Firstly*, it may give rise to a refracted first arrival where the lower crustal layer is relatively thick and the velocity contrast is relatively large. *Secondly*, even if there is no first arrival, the supercritical reflection $P_IP$ may show up the existence of a discontinuity within the crust with a downward increase in velocity. The criteria for recognizing $P_IP$ are (1) large amplitude, (2) the travel-time curve for it becomes parallel to that of $P_g$ at large distances, and touches that of $P*$ at the critical distance. It is to be hoped that other methods, such as use of normal-incidence reflections and common-depth point reflection surveys, may in future lead to further ways of probing the lower crustal structure.

   Some of the most reliable evidence on the subdivision of the crust comes from the intensive seismic surveys carried out in U.S.A. The continuous spreads of seismometers from shot point to the end of the line used by the United States Geological Survey have enabled unambiguous picking of phases including second and later arrivals. The closely spaced shots of the Lake Superior project give a similar confidence in the interpretation. The outcome is that a lower crustal layer appears to be present beneath several of the structural provinces of U.S.A. but not all of them. Where present, the layer varies greatly in character from region to region. It is not clear whether the upper boundary of it is sharp or gradational. These results support neither the scepticism of Tatel and Tuve, nor the existence of the universal Conrad discontinuity of early earthquake seismologists.

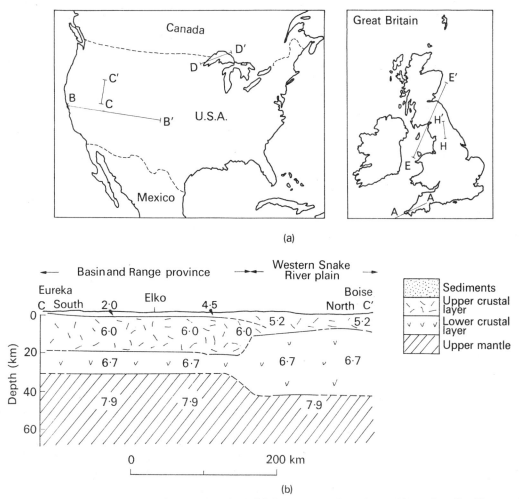

(a)

(b)

**Fig. 2.12**  Some different types of crustal 'layering'. (a) Key maps to show the positions of profiles illustrated in Figures 2.7, 2.9, 2.12 and 2.17. (b) Crustal structure along CC′ across the boundary between the Basin and Range province and the western Snake River plain, illustrating a lower crustal layer which changes thickness across a structural boundary. Redrawn from HILL and PAKISER (1966), *The Earth beneath the continents*, p. 410, American Geophysical Union. (Figure 2.12 is continued on p. 44)

Let us look at some examples (Figure 2.12). In eastern Colorado and eastern New Mexico there is evidence for a lower crustal layer about 20 km thick which in some places does give rise to first arrivals. A thin lower crustal layer which does not give first arrivals has been recognized beneath the Basin and Range province from the reflected phase $P_I P$ (HILL and PAKISER, 1966; EATON, 1963). Beneath the Snake River basalt plateau to the north a 6·7 km/s layer comes within a few kilometres of the surface and gives a first arrival. Beneath the highly anomalous Lake Superior region, 5–10 km of Keeweenawan sediments and volcanic rocks overlie a lower crustal layer which gives first arrivals of 6·8 km/s; here there is an unusually great variation in the interpreted depths to the Moho ranging from 25 to 60 km. Beneath the Coast Range of California there is no evidence for a lower crustal layer (HEALY, 1963).

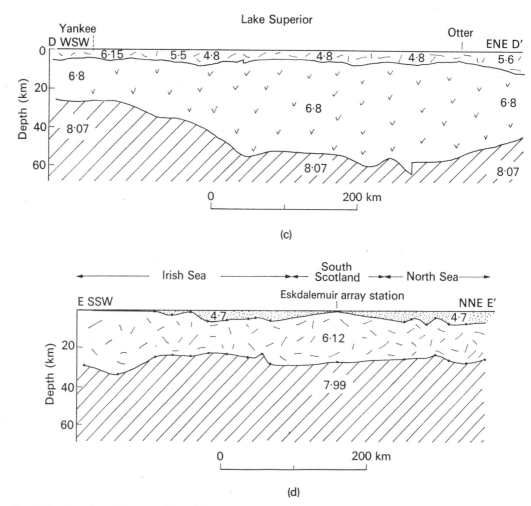

Fig. **2.12** (continued from p. 43). (c) Crustal structure along DD′ across Lake Superior, illustrating a particularly thick lower crustal layer. Redrawn from SMITH, STEINHART and ALDRICH (1966), *The Earth beneath the continents*, p. 194, American Geophysical Union. (d) Crustal structure along line EE′ from the Irish Sea to the North Sea across south Scotland, illustrating an apparently unlayered crust. Redrawn from AGGER and CARPENTER (1964), *Geophys. J. R. astr. Soc.*, **9**, 78.

Turning to Great Britain, the north-east orientated line through Eskdalemuir yielded no evidence for a lower crustal layer (AGGER and CARPENTER, 1964; Figure 2.12). A crustal refraction survey on the granite belt of south-west England (p. 36) showed an increase in velocity with depth but gave no evidence for any layering in the crust below the granites; the crustal thickness is about 27 km and the mean crustal velocity is 6·2 km/s, $P_g$ being 5·9 km/s. However, the phase $P*$ has been recognized in Eskdalemuir recordings of local earthquakes and recently BLUNDELL and PARKS (1969) have found evidence for a 7·3 km/s layer occurring in the lower crust beneath the south Irish Sea.

Thus the lower crustal layer appears to exist in several continental regions but its thickness and velocity vary greatly. In other regions it has not been detected. There is little support for the idea of a universal, reasonably homogeneous, lower crustal layer; lateral variations within the lower crust may be as great as the vertical inhomogeneity of the whole crust. But in general, there does appear to be a well substantiated increase in velocity of about 10% with depth and in some places this approximates a layered structure.

## 2.4 Gravity anomalies and crustal structure

### The discovery of isostasy

The original observations which led to the discovery of the principle of isostasy were made between 1735 and 1745 during the measurement of an arc of meridian in Peru by the French geodetie expedition under Bouguer's leadership. They recognized that the Andes would cause a horizontal attraction on the plumbline which would result in local variations in the vertical direction. On investigation, they found that the observed deflection of the vertical was much smaller than the value computed theoretically from the known topography of the Andes. Bouguer originally noted this discrepancy and a few years later Boscovitch postulated attenuation of matter beneath the mountains to explain it. Next century, similar results were found near the Himalayan mountain chain, and it is now known to be a fairly general phenomenon associated with the Earth's major surface features.

For both Andes and Himalaya, the underlying mass deficiency needed to explain the observed deflection of the vertical is approximately equal to the surface load represented by the mountain ranges. The term 'isostasy' was introduced by Dutton in 1889 to explain this phenomenon.

To elaborate a little, the principle of isostasy states that beneath the 'depth of compensation', pressures within the Earth are hydrostatic. This means that the weight of the overlying columns of unit cross-section must all be equal at and below the depth of compensation, allowance being made for a small correction for the Earth's curvature. If there is an excess load on the Earth's surface such as a mountain range or an ocean ridge or an icecap, then if isostatic equilibrium has been reached there must be an equivalent compensating mass deficiency beneath the surface feature but above the depth of compensation; and vice versa for deficient loads such as oceans.

Isostasy is merely the application of Archimedes' principle to the uppermost layers of the Earth. The existence of isostatic movements and of other types of vertical movement affecting the crust shows that lateral flow must be able to occur in the relatively weak region below the depth of compensation, which is commonly called the *asthenosphere*. In contrast, the overlying relatively strong *lithosphere* must reach isostatic equilibrium either by elastic bending or by a combination of fracture and flow.

The two main hypotheses of isostasy were both put forward in 1855. Each of these attempts to explain the shape of the underlying mass distribution which compensates the surface topography. These hypotheses (Figure 2.13) are as follows:

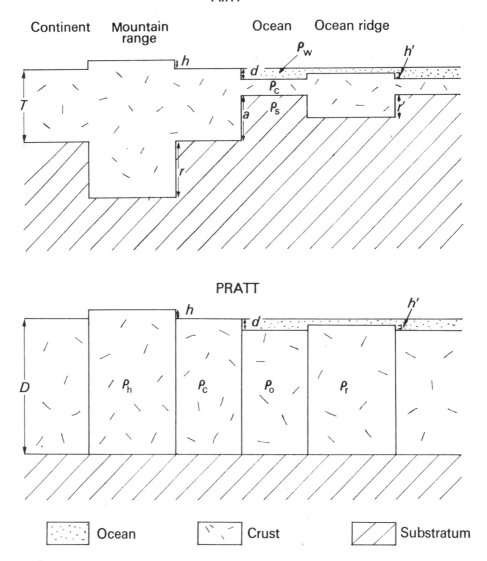

Fig. 2.13 Isostatic compensation according to the Pratt and Airy hypotheses, where $\rho_c$ = density of crust,
$\quad\rho_w$ = density of sea-water,
$\quad\rho_s$ = density of substratum,
$\quad\rho_h$ = density of crust beneath mountain height $h$ (Pratt),
$\quad\rho_o$ = density of crust beneath ocean depth $d$ (Pratt),
and $\quad\rho_r$ = density of crust beneath ocean ridge height $h'$ (Pratt).

(1) *Pratt's hypothesis* (1855) assumes that the density within the shell of the Earth above the depth of compensation varies laterally depending on the elevation of the overlying topography. The condition of isostasy requires that

$$\rho(h+D) = \text{constant},$$

where $D$ = depth of compensation,

$h$ = height of topography

and $\rho$ = the underlying density.

A small correction needs to be applied if the Earth's curvature is taken into account. According to Pratt's hypothesis, mountain ranges are underlain by anomalously low density rocks which extend downwards to the depth of compensation; oceans are underlain by relatively high density rocks. The American geodesist Hayford developed this hypothesis in the early part of this century; he arbitarily took the depth of compensation to be 113·7 km.

(2) *Airy's hypothesis* (1855) assumes that the uppermost shell of the Earth is a low density 'crust' overlying a higher density substratum. The 'crust' and substratum are each assumed to have uniform density. The relatively rigid 'crust' or lithosphere is assumed to float on the fluid substratum (i.e. the asthenosphere). In the original form of the hypothesis, the base of the low density crust is identical with the boundary between rigid lithosphere and weak asthenosphere, although the more realistic situation where the boundaries are distinct can be incorporated without difficulty. Compensation occurs as a result of variation in thickness of the low density crust; mountain ranges are considered to be underlain by a thicker crust than normal (a *root*) and oceans by a thinner crust than normal (an *antiroot*). The condition for isostasy, neglecting the Earth's curvature, is

$$r = h\rho_c/(\rho_s - \rho_c)$$

where $r$ = depth of root,

$h$ = height of topography,

$\rho_c$ = density of 'crust'

and $\rho_s$ = density of substratum.

There are several variations on Airy's hypothesis. For instance, Vening Meinesz put forward the idea of regional compensation, in which the root has a wider lateral extent than the surface feature it is compensating.

*Testing isostasy by gravity measurements*

The old method of testing isostasy was to compare the observed deflections of the vertical with the theoretically computed values according to a specified hypothesis of isostasy. During the present century, gravity measurements have been used instead because they can be made much more rapidly and they give the same basic information on the subsurface mass distribution.

There are two main problems. The first is to test to what extent isostatic equilibrium occurs, irrespective of the hypothesis. The second is to attempt to distinguish between the different hypotheses and their versions. Gravity measurements are an effective method of tackling the first problem but have only very limited success in dealing with the second.

Before gravity observations can be interpreted, they need to be corrected for latitude and elevation. The three main types of anomaly used for interpretation are as follows:

Bouguer anomaly $= g_{obs} - g_\phi + \text{FAC} - \text{BC} + \text{TC},$

Free air anomaly $= g_{obs} - g_\phi + \text{FAC}$

Isostatic anomaly $=$ Bouguer anomaly $-$ computed anomaly of root.

where $g_{obs}$ = observed value of gravity at a point on the Earth's surface;

$g_\phi$ = theoretical gravity on the spheroid at latitude $\phi$ of the point, as given by the International Gravity Formula;

FAC = the free air correction, allowing for the variation in gravity with height above the spheroid;

BC = the Bouguer correction, which is the attraction of the rock between sea-level and the height of the point, treating it as a slab of uniform thickness;

and   TC = the correction for deviations of the topography from a flat plateau.

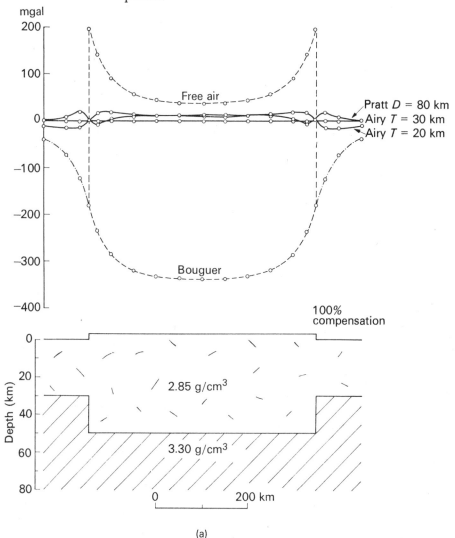

(a)

**Fig. 2.14**  The computed Bouguer, free air and isostatic anomalies over an ideal mountain range with isostatic compensation as follows:

(a) In exact equilibrium according to the Airy hypothesis, with crustal thickness $T$= 30 km;

The Bouguer anomaly shows up the gravitational effect of the lateral variations in density below sea-level. It is the most convenient basis for interpreting local and regional gravity anomalies of the continents and shelf seas in terms of subsurface mass distributions. The free air anomaly is useful for interpreting gravity anomalies of the oceans and across continental margins, and also provides a useful rough test of isostasy.

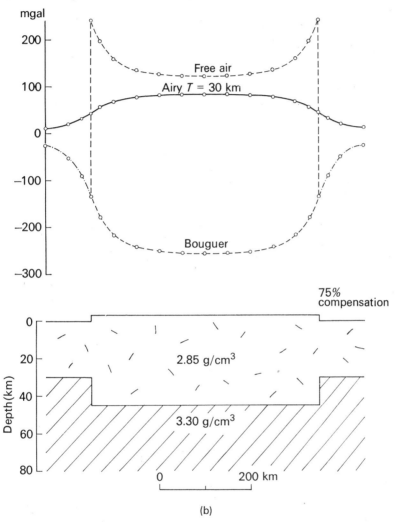

**Fig. 2.14** (continued) (b)  75% compensation according to the Airy hypothesis, the depth extent of the root being 25% less than in (a) ;

Figure 2.14 shows how gravity anomalies are used to test isostasy. Figure 2.14(a) shows all three types of anomaly over a mountain range in perfect isostatic equilibrium according to the Airy hypothesis, with a crustal thickness of 30 km. In Figure 2.14(b) the surface topography is only 75% compensated by the root, and in Figure 2.14(c) there is no compensation at all.

The most thorough method of testing isostasy over a given surface feature, such as a mountain range, is to compare the observed Bouguer anomaly with the predicted gravitational effect of the compensating mass deficiency according to both Airy and Pratt hypotheses, allowing for different depths of compensation. This is equivalent to computing the isostatic anomaly. Thus in Figure 2.14(a) the gravity effect of the predicted root for T = 30 km is exactly equal to the observed Bouguer

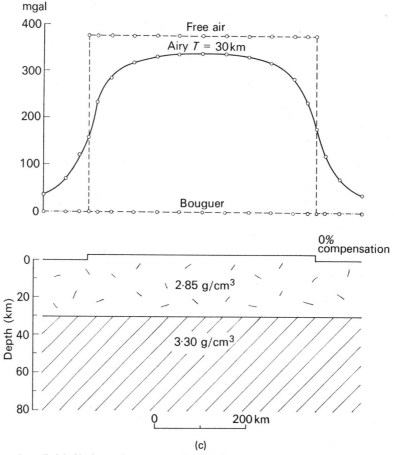

Fig. 2.14 (continued) (c) No isostatic compensation at all.

anomaly, and the corresponding isostatic anomaly is zero everywhere. Ideally this approach should enable us to determine (1) whether or not the mountain range is in isostatic equilibrium, and (2) which hypothesis gives the best agreement with the gravity observations. In practice, the isostatic anomalies provide an accurate test of the extent to which isostatic equilibrium occurs; but they do not clearly distinguish between the different hypotheses and depths of compensation. This is because the gravity anomalies lack sensitivity to the exact geometry of the compensating mass deficiency and because of the ubiquitous presence of disturbing anomalies of shallow origin which are typically of larger amplitude than the differences we would be looking for.

A rough, but effective, method of testing isostasy over a topographic feature which is wide in comparison with the depth of compensation is to use the free air anomaly. This would be approximately zero near the centre of the feature if it is in equilibrium. This is because the gravitational effect of a wide root is approximately equal to the Bouguer correction, and thus the free air anomaly will be approximately zero. This method breaks down near the margin of the feature and also in regions of rugged topography, but is reasonably effective provided that the feature is about ten or more times wider than the depth of compensation.

### Isostasy and crustal structure

Gravity observations have shown that most of the Earth's major surface features are approximately in isostatic equilibrium, but they cannot unambiguously reveal the form the compensation takes. On the other hand, seismic refraction studies give the crustal structure and thickness but cannot give information on isostatic equilibrium. Put together, these two sets of geophysical observations enable us to infer the form of the compensating masses, thereby giving much more information than the two methods treated separately could do. Because of this, the intense seismic investigations of crustal structure since about 1950 enable a new assessment of the hypotheses of isostasy to be made.

Looked at in the most general way, isostatic compensation occurring near the Earth's surface may occur in one or more of three basic ways. These are: (1) by lateral variation in the mean density of the crust; (2) by variation in the thickness of the low density crust; and (3) by lateral variation of density within the upper mantle. (1) and (3) could be regarded as modified forms of the Pratt hypothesis, and (2) is the classical form of the Airy hypothesis.

The oceans and continents are in broad isostatic equilibrium with each other. Seismic crustal studies show that this occurs principally according to variations in crustal thickness, although there may also be some lateral variation in the mean density of the crust and upper mantle. Similarly, young fold mountains have been shown to have thickened crust beneath, approximately as predicted by the Airy hypothesis. In contrast, the ocean ridges and some continental uplifted areas such as East Africa and western U.S.A. are compensated by low density rocks within the upper mantle.

The concentrated crustal structure investigations in the western U.S.A. (STEINHART and MEYER, 1961; PAKISER, 1963) are particularly significant in showing the different forms that isostatic compensation can take. They show that crustal thickness is not necessarily related to topographical elevation as the Airy hypothesis predicts. For example the Great Plains are elevated 1 km above sea-level and are underlain by a crust of 45–50 km thickness; but the Basin and Range province with an average elevation of about 2 km has a crust only 25–30 km thick, quite contrary to the predictions of the Airy hypothesis. Pakiser concluded that changes in crustal thickness across the boundaries of the major geological provinces of western U.S.A. bear little direct relation to the change in altitude unless the velocity of $P_n$ remains constant across the boundary. The whole region is in approximate isostatic equilibrium, the compensation between major provinces occurring at least partly in the upper mantle. On the other hand, within the Basin and Range province $P_n$ remains approximately constant and changes in topographical elevation do appear to be related to changes in crustal thickness. Similarly the Sierra Nevada and Rocky Mountains appear to be underlain by local thickened crust suggesting Airy type compensation.

No single universal hypothesis of isostasy can explain all the Earth's major surface features. The form of compensation beneath any given feature needs to be investigated by combined gravity and seismic studies rather than to be assumed. It is quite invalid, for instance, to assume the Airy hypothesis and then use gravity studies alone to determine variations in crustal thickness as has sometimes been done in the past. Knowledge of the form of the compensation is especially important in discussing the cause of vertical movements at the Earth's surface, and the fact that compensation

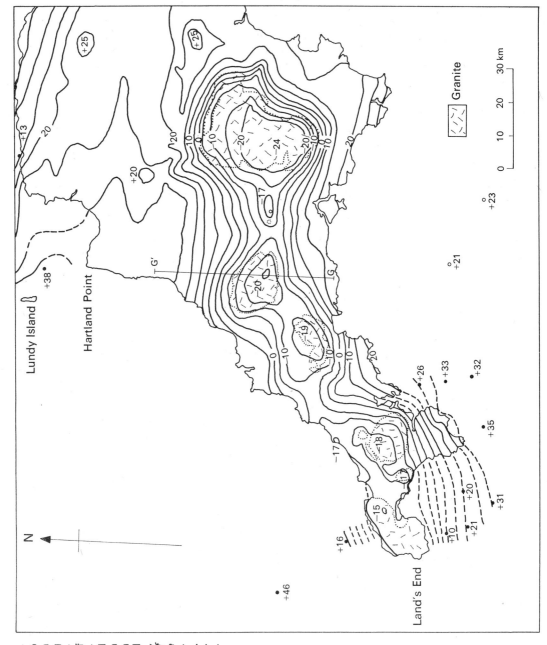

**Fig. 2.15** The Bouguer anomaly map of south-western England, showing the belt of negative anomalies of about –50 mgal amplitude associated with the Armorican granites. Redrawn from BOTT and SCOTT (1964), *Present views of some aspects of the geology of Cornwall and Devon.* p. 28, Royal Geological Society of Cornwall.

can occur in the mantle as well as the crust adds a further dimension to our understanding of the origin of the Earth's surface features.

*Gravity anomalies and the density of the crust*

Quite large gravity anomalies of local areal extent are a characteristic feature of the continental crust. Negative anomalies are caused by thick accumulations of low density sediments in basins, and both positive and negative anomalies occur over large igneous intrusions in the basement. The negative anomalies which occur over granite and granodiorite intrusions are important for crustal structure in that they show that the mean density of upper crustal basement rocks is considerably higher than that of granite, thereby throwing doubt on the old idea of a granitic layer.

The decrease in Bouguer anomaly over a typical post-tectonic granite batholith is between 15 and 60 mgal (1 mgal $= 10^{-3}$ cm/s²). As an example, the Bouguer anomaly map over the granites of south-west England is shown in Figure 2.15. To explain the steep marginal gradients of this $-50$ mgal anomaly, the granite itself must be at least $0.10$ g/cm³ lower in density than the intruded basement rocks—in fact the contrast is probably about $-0.16$ g/cm³. An interpretation of the profile across the Bodmin Moor granite is shown in Figure 2.16, which shows that the granite batholith has outward

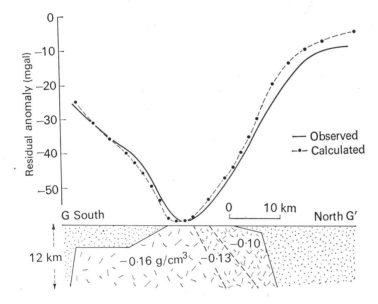

**Fig. 2.16** Interpretation of the gravity anomaly profile across the Bodmin Moor granite (south-western England) in terms of subsurface shape of the low density granite batholith, along line GG' (Figure 2.15). Redrawn from BOTT and SCOTT (1964), *Present views of some aspects of the geology of Cornwall and Devon*, p. 31, Royal Geological Society of Cornwall.

sloping contacts and must extend to a depth of at least 10 km. The granite itself is known to have a density of about $2.60$ g/cm³, which means that the basement rocks have an average density of about $2.75$ g/cm³ or thereabouts, and that this density must extend down to 10 km depth which is a third to half of the crustal thickness in this region (p. 36).

The density differential of $0.10-0.15$ g/cm³ between granite and basement rocks is confirmed by surface measurements of rock density. For instance, WOOLLARD (1966) found the average density of 1158 samples of basement rock from North America to be 2·742 g/cm³. This is substantially higher than the density of 2·67 g/cm³ which used to be assigned to the 'granitic layer'. Clearly, the term 'granitic layer' is a misnomer for the basement rocks of the upper crust, which have a density intermediate between that of acid and basic igneous rocks (BOTT, 1961).

As an example, the structure of the upper third of the crust beneath the northern Pennines, England, is shown in Figure 2.17. This has been deduced from gravity anomalies, and shows two of the common structural features, namely sedimentary basins and granites.

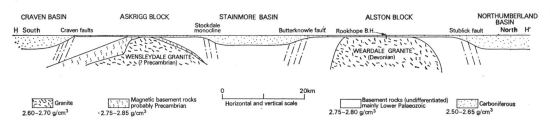

**Fig. 2.17**   The structure of the uppermost third of the continental crust beneath the northern Pennines, England, along line HH′ (Fig. 2.12) as interpreted from gravity and magnetic anomalies. Redrawn from BOTT (1967a), *Proc. Yorks. geol. Soc.,* **36**, 165.

The knowledge of crustal thickness determined by seismic surveys makes it possible to estimate the density difference between crust and topmost mantle. For a long time it was traditionally assumed by isostasists that the contrast is 0·6 g/cm³ (crust = 2·67, topmost mantle = 3·27). However, WOOLLARD (e.g. 1966) has made a careful study of the variations in crustal thickness of regions in isostatic equilibrium, and he finds that on average a topographical elevation of 1 km corresponds to a root at the base of the crust of 7·5 km. This indicates a density differential of 0·39 g/cm³ or thereabouts between crust and mantle. Woollard also found that the density differential varies from region to region, partly depending on the density of the underlying mantle and partly on the density of the crust. Allowing for an increase in density with depth, Woollard estimated the mean density of the continental crust to be in the range 2·87–3·00 g/cm³.

## 2.5  Special regions of the continental crust

### Mountain ranges

There are two present-day belts of young fold mountains, namely the circum-Pacific belt and the Alpine–Himalayan belt. They show geological evidence for strong horizontal and vertical movements affecting the rocks involved during the Tertiary. Other belts of the continental crust have been active as young fold mountains during the geological past, but these have been partly or completely obliterated as surface features by erosion and repeated uplift. In some respects, the crustal structure beneath the modern mountain ranges differs from the normal continental crust, as has been revealed by gravity and seismic observations briefly summarized here.

Gravity anomalies show that the present-day mountain belts are in general in approximate isostatic equilibrium. To give an example, Figure 2.18 shows the gravity anomaly profile across the eastern Alps. Although there are quite large local anomalies disturbing the profile, it is clear that the Airy isostatic anomalies depicted are of much smaller amplitude than the negative Bouguer anomaly. This shows that isostatic equilibrium has been reached to an extent of 90–95% at least. The Airy anomaly for a crustal thickness of 20 km gives the best fit despite later seismic work showing that 30 km would be the better estimate. The reason for this discrepancy is that the sharp negative anomaly near the crest of the Alps is caused by a mass deficiency within the crust, probably a granite (BOTT, 1954), and if this is allowed for it is found that the Airy $T = 30$ km gives a better fit; this serves as a warning against placing too much confidence in use of isostatic anomalies to estimate crustal thickness. Pratt type isostatic anomalies would equally well account for the observed profile.

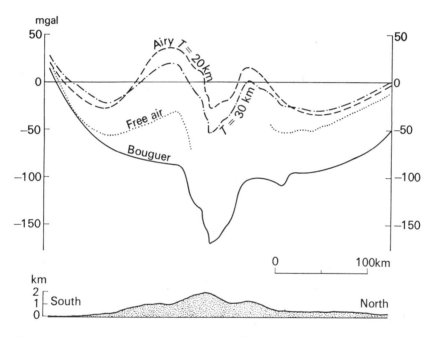

**Fig. 2.18** The Bouguer, free air and Airy isostatic anomalies across the eastern Alps. The central negative anomaly is caused by a shallow source in the upper crust, probably a granite, and its presence shows how local anomalies make it difficult to use gravity to distinguish between isostatic hypotheses. The topography is too rugged for sensible use of the free air anomaly to test isostasy. Comparison of the isostatic and Bouguer anomalies suggests that the mass deficiency beneath is slightly larger than needed to compensate the mountains (i.e. they are overcompensated), but that equilibrium is at least 90% attained. Redrawn from HOLOPAINEN (1947), *Publs isostatic Inst. int. Ass. Geod.,* No. 16.

The gravity anomalies of the western Alps provide a less clear-cut example (Figures 2.19 and 2.21). These anomalies show that the main Alpine mountains are in isostatic equilibrium, but that the belt of strong positive anomalies to the east (known as the Ivrea zone) and the strong negative anomaly over the Po basin further east represent large deviations from equilibrium.

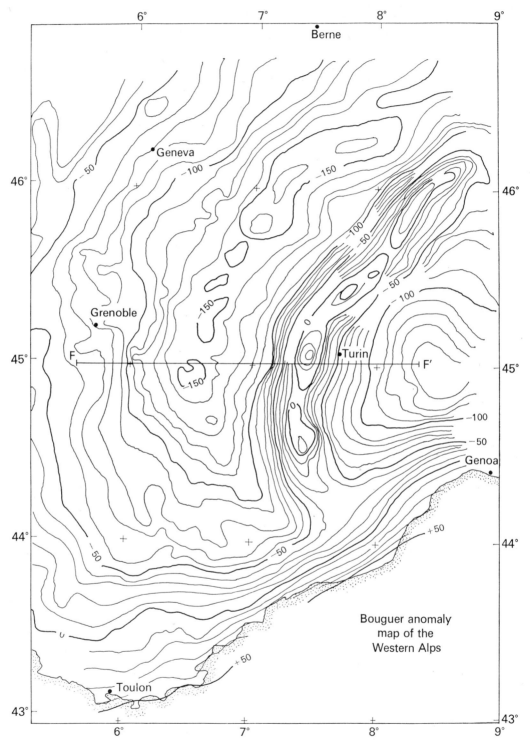

**Fig. 2.19** Bouguer anomaly map of the western Alps. Redrawn from CORON (1963), *Seismologie*, ser. XXII, **2**, 33.

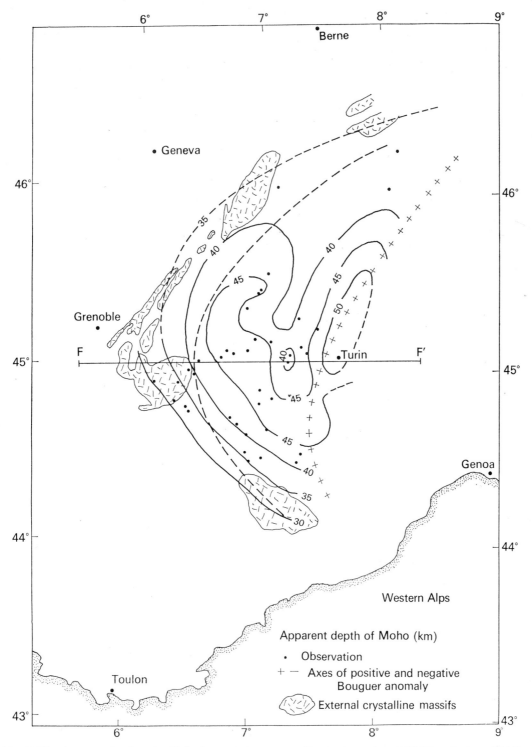

**Fig. 2.20** Apparent depth to the Moho beneath the western Alps. Redrawn from FUCHS and others (1963), *Seismologie*, ser. XXII, **2**, 162.

It has been widely assumed for many years that mountain ranges are compensated by roots of thickened low density crust beneath, according to the Airy hypothesis. But before about 1950 there was scarcely any evidence to confirm this. The isostatic equilibrium could be explained equally well by the Pratt hypothesis. Opinion had been influenced by the surface geological evidence of strong

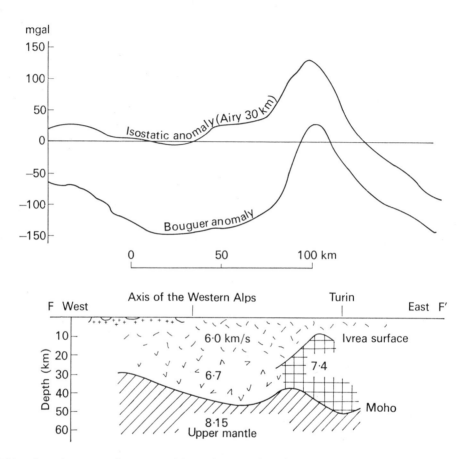

**Fig. 2.21** Crustal structure, Bouguer and isostatic anomalies along FF′ (Figures 2.19 and 2.20) across the western Alps. Adapted from FUCHS and others (1963) and CORON (1963), *Seismologie*, ser. XXII, **2**, 168.

overfolding and thrusting suggesting crustal shortening by compression, but the geological evidence can also be interpreted in terms of vertical movement and gravity tectonics. Seismic evidence on crustal thickness is needed to distinguish between the isostatic hypotheses.

Seismic refraction crustal structure investigations have now been done in some mountain ranges. They do confirm that there are roots of thickened crust beneath. The western Alps have been particularly well studied (e.g. CLOSS and LABROUSTE, 1963). The resulting map of the interpreted depth to the Moho is shown in Figure 2.20, and the section along FF′ showing gravity anomalies and crustal structure is shown in Figure 2.21. The crust beneath the foreland to the north and west is about 30 km thick, and it thickens to about 50 km beneath the Alps. To the east, an anomalous zone of

7·4 km/s material corresponding to the Ivrea zone reaches within 10 km of the surface, and the crust beneath this zone is also unusually thick. Beneath the Alps themselves, the velocity increases with depth in the crust but the Moho appears to be well defined. In the eastern Alps, there is a 7·4 km/s layer near the base of the crust. The thickening of the crust beneath the Alps is about what would be expected according to the Airy hypothesis. There is therefore no need for a large mass deficiency in the upper mantle beneath of the sort which occurs beneath ocean ridges (Chapter 3).

The crust beneath the central Andes has been observed to be 45–56 km thick. The investigations in western U.S.A. (PAKISER, 1963) show thickened crust beneath the Sierra Nevada and beneath the Rocky Mountains. Thus where seismic surveys have been done, the existence of roots of thickened crust have been established. However, much more seismic work on the deep structure of mountain ranges is needed.

The conventional interpretation of the root beneath a mountain range is that it is caused by crustal thickening. The old hypothesis that crustal shortening resulted from a contracting Earth is no longer tenable, because the amount of shortening which could be produced in this way would be quite inadequate. However, crustal shortening can now be readily explained as a complementary pheno-menon to ocean-floor spreading and continental drift, both being the product of convection in the mantle (p. 273). Other interpretations of a root are also possible; it could be produced by differen-tiation of crustal material from the mantle, or if the Moho is a phase change beneath mountain ranges it might migrate downwards as a result of raised temperature. In the writer's view, crustal shortening associated with ocean-floor spreading is the most satisfactory explanation.

*Rift valley systems*

The East African system of interrelated rift valleys stretches from Rhodesia to the Gulf of Aden with eastern and western branches encircling Lake Victoria. It extends further northwards along the Red

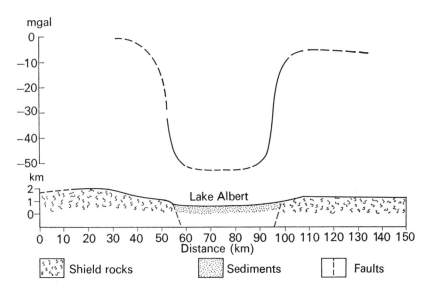

**Fig. 2.22** Observed Bouguer anomaly profile over the Lake Albert rift valley. Redrawn from GIRDLER (1964), *Physics Chem. Earth*, **5**, 123, Pergamon Press.

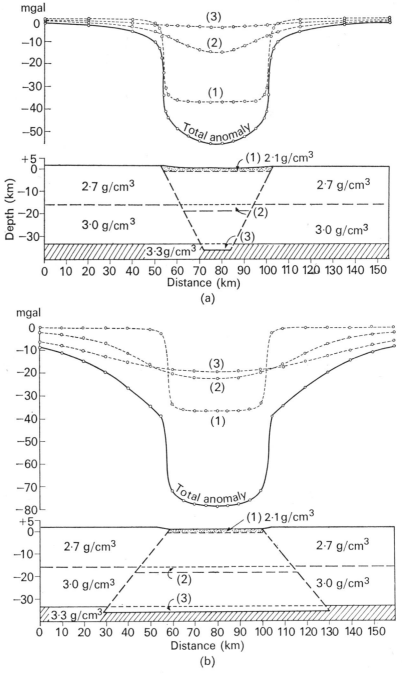

**Fig. 2.23** Computed gravity anomalies for a theoretical sediment-filled rift valley formed by (a) normal faulting, and (b) reverse faulting. The total anomaly is made up from (1) the low density sediments, (2) the downwarp of the Conrad discontinuity (assumed to be present) and (3) the downwarp of the Moho. Observed profiles resemble (a) rather than (b). Redrawn from GIRDLER (1964), *Physics Chem. Earth*, **5**, 125, Pergamon Press.

Sea into the Dead Sea rift system, attaining a total length of over 6000 km. Another linked series of rift depressions extends north-eastwards and also westwards from Lake Baikal in south-central Siberia and is over 2000 km in length. The two rift systems form major linear tectonic features of the continental crust. The East African rift system appears to form a continental extension of the ocean ridge system. These rift systems have been active during the Tertiary and are associated with broad uparching of the crust which is unrelated to folding and is probably caused by development of low density rocks in the upper mantle (p. 148).

The main problem of origin has been to determine whether the faults bounding the individual rift valleys are normal or reverse. It has now been shown beyond reasonable doubt that the faults are normal. This is confirmed by both geological observations on the faults themselves and gravity profiles across them. For instance, the observed Bouguer anomaly profile across the Albert rift (Figure 2.22) agrees well with the computed model for normal faulting but is quite different from that for reverse faulting (Figure 2.23). This therefore shows that rift systems are caused by horizontal tension applied to the crust. This is confirmed by earthquake mechanism studies (p. 263).

Gravity surveys show that negative isostatic anomalies occur locally over the sediment filled rift valleys. For instance, BULLARD (1936) found that the East African plateau is in approximate isostatic equilibrium, but that some individual rift valleys show negative isostatic anomalies reaching down to about $-80$ mgal. Similarly, there is a large negative isostatic anomaly across the Lake Baikal rift depression. The negative anomalies are substantially caused by the low density sedimentary infill, which may reach up to 5–6 km in thickness.

The isostatic problem is to explain the mechanism for such a large local deviation from equilibrium. On the older compression hypothesis which has now been abandoned, it was assumed that horizontal

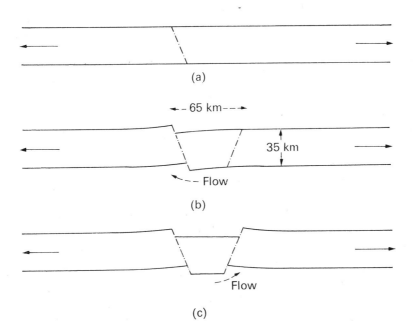

**Fig. 2.24**   Vening Meinesz' hypothesis for the formation of a rift valley by tension and normal faulting. Adapted from HEISKANEN and VENING MEINESZ (1958), *The Earth and its gravity field*, p. 390, McGraw-Hill.

compression would hold down the rift block bounded by reverse faults against the isostatic buoyancy. An explanation in terms of the tension hypothesis has been developed by Vening Meinesz (HEISKANEN and VENING MEINESZ, 1958). Referring to Figure 2.24, the stages in development are as follows:

(1) Under conditions of crustal tension, a normal fault is produced with its outcrop approximately perpendicular to the maximum tension (Figure 2.24(a));

(2) As a result of movement on the fault, the crust is warped and the bending sets up further stresses. Maximum tension is produced at the surface on the downthrown side where the curvature is greatest;

(3) A second normal fault develops at this position (Figure 2.24(b)), which according to the theory of bending of thin sheets would occur about 65 km from the first fault if the crust is 35 km thick. If the two normal faults dip towards each other, a rift valley is formed;

(4) Once the two fault planes have been established, persistent tension will cause repeated subsidence of the rift wedge.

The above mechanism does not violate the principle of isostasy, because a wedge of crust narrowing downwards floats in a fluid at a lower level than a rectangular block would do. Vening Meinesz showed that a subsidence of several kilometres can occur by this mechanism, provided the subsidence is aided by a load of sediments.

Suppose the relatively small crustal extension associated with rift valleys becomes more intense. Then the continental blocks on either side will separate from each other and an incipient ocean may form between them. This is the mechanism by which the Red Sea may have been formed as shown in Figure 2.25 (GIRDLER, 1964). This is probably the mechanism by which continental drift is initiated. It may also explain why we do not see any signs of continent-wide rift systems of pre-Tertiary age; perhaps they have all been swallowed by new oceans, and this may be the ultimate fate of the present-day rift systems.

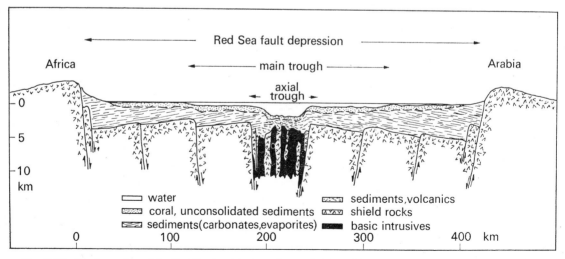

**Fig. 2.25**  The formation of the Red Sea by dyke intrusion, as indicated by geophysical evidence. Redrawn from DRAKE and GIRDLER (1964), *Geophys. J. R. astr. Soc.*, **8**, 489.

## 2.6 Interpreting the continental crust

### Principles of interpretation

The observed pattern of elastic wave velocities, incomplete as it is, provides the main basis for interpreting the chemistry and mineralogy of the crust beneath depths penetrable by boreholes. The link between the observed velocity structure and its interpretation depends on experimental measurements of the physical properties of rock types over the appropriate range of temperature and pressure. A unique interpretation is not possible because of overlap of velocities. At best, we can only hope to obtain a broad picture of the variation of composition with depth.

Both pressure and temperature increase with depth through the crust. The confining pressure is mainly caused by the weight of the overburden, which at depth $d$ is $\rho g d$ where $\rho$ is the mean density of the overlying rocks and $g$ is gravity. In practise the confining pressure at a point may be increased or decreased slightly by tectonic overpressures. Assuming the crustal density to be $2 \cdot 9$ g/cm$^3$, the pressure increases by $0 \cdot 3$ kbar/km* and at the typical Moho depth of 35 km it is 10 kbar. Anticipating the discussion in Chapter 6, the average temperature gradient in the basement rocks is about 25°C/km, but this falls off with depth to about half its surface value at the Moho because of radioactive heat sources within the crust. A realistic estimate of the temperature-depth profile typical of the continental crust is shown in Figure 2.26 (after Birch), the Moho temperature being between 500 and 600°C. Higher than average temperatures in the lower crust would be expected in regions of relatively high heat flow such as western U.S.A. and regions where the crust is abnormally thick such as the Alps; but the anomalously low heat flow of Precambrian shields suggests that the crust beneath is relatively cool, the Moho being within the range 250–450°C.

It is convenient to relate the composition of rocks at different depths in the crust (and upper mantle) to igneous rocks and their high pressure forms. The average compositions of the main groups of igneous rocks and of eclogite, which is the high pressure form of gabbro and basalt, are shown in Table 2.1. The densities of coarse-grained igneous rocks are shown in Tables 2.2 and 2.3. It should

**Table 2.1**  Average composition of igneous rock groups and eclogites. The average compositions of the igneous rock groups was taken from NOCKOLDS (1954), and of eclogites from LAPADU-HARGUES (1953). The number of samples in each group is shown at the head of the column.

|  | Alkali granite (48) | Granodiorite (137) | Intermediate igneous rock† (635) | Gabbro (160) | Peridotite (23) | Eclogite (water free) (34) |
|---|---|---|---|---|---|---|
| $SiO_2$ | 73·9 | 66·9 | 54·6 | 48·4 | 43·5 | 49·0 |
| $TiO_2$ | 0·2 | 0·6 | 1·5 | 1·3 | 0·8 | — |
| $Al_2O_3$ | 13·8 | 15·7 | 16·4 | 16·8 | 4·0 | 14·5 |
| $Fe_2O_3$ | 0·8 | 1·3 | 3·3 | 2·6 | 2·5 | 3·8 |
| $FeO$ | 1·1 | 2·6 | 5·2 | 7·9 | 9·8 | 9·1 |
| $MnO$ | 0·1 | 0·1 | 0·2 | 0·2 | 0·2 | — |
| $MgO$ | 0·3 | 1·6 | 3·8 | 8·1 | 34·0 | 8·9 |
| $CaO$ | 0·7 | 3·6 | 6·5 | 11·1 | 3·5 | 11·5 |
| $Na_2O$ | 3·5 | 3·8 | 4·2 | 2·3 | 0·6 | 2·5 |
| $K_2O$ | 5·1 | 3·1 | 3·2 | 0·6 | 0·3 | 0·7 |
| $H_2O$ | 0·5 | 0·7 | 0·7 | 0·6 | 0·8 | — |
| $P_2O_5$ | 0·1 | 0·2 | 0·4 | 0·2 | 0·1 | — |

† excluding nepheline types.

* 1 kilobar (kbar) = 1000 bar (bar) = $10^9$ dynes/cm$^2$ = 986.9 atmospheres.

be pointed out that the problem of adequate random sampling introduces systematic errors into the quoted average compositions and densities, but the general trends are reliable. In particular, later work has shown that Daly's values for the densities of granite and granodiorite are probably too high.

**Table 2.2** Average densities of coarse-grained igneous rocks and eclogites. These average density estimates have been taken from tables in the *Handbook of physical constants* (CLARK, 1966) and are based on compilations made by R. A. Daly, Francis Birch and S. P. Clark.

| | Number of samples | Mean density (g/cm³) | Range (g/cm³) |
|---|---|---|---|
| granite | 155 | 2·67 | 2·52–2·81 |
| granodiorite | 11 | 2·72 | 2·67–2·79 |
| syenite | 24 | 2·76 | 2·63–2·90 |
| quartz diorite | 21 | 2·81 | 2·68–2·96 |
| diorite | 13 | 2·84 | 2·72–2·96 |
| gabbro and norite | 38 | 2·98 | 2·72–3·12 |
| peridotite | 3 | 3·23 | 3·15–3·28 |
| dunite | 15 | 3·28 | 3·20–3·31 |
| eclogite | 10 | 3·39 | 3·34–3·45 |

**Table 2.3** Compressional wave velocities in rocks. Taken from a table compiled by Frank Press in the *Handbook of physical constants* (CLARK, 1966) and mainly based on measurements made by Francis Birch and Gene Simmons.

| | Number of samples | Mean density (g/cm³) | Mean velocity (km/s) | |
|---|---|---|---|---|
| | | | 1 kbar | 10 kbar |
| granite | 10 | 2·643 | 6·13 | 6·45 |
| granodiorite | mean of locality | 2·705 | 6·27 | 6·56 |
| quartz diorite | 2 | 2·852 | 6·44 | 6·71 |
| gabbro and norite | 3 | 2·988 | 7·02 | 7·24 |
| dunite | 5 | 3·277 | 7·87 | 8·15 |
| eclogite | 4 | 3·383 | 7·52 | 7·87 |
| greywacke | 2 | 2·692 | 5·84 | 6·20 |
| slate | 1 | 2·734 | 5·79 | 6·22 |
| amphibolite | 1 | 3·120 | 7·17 | 7·35 |

The $P$ wave velocities in a wide range of rocks up to pressures of 10 kbar have been determined by BIRCH (1960, 1961a) and his co-workers (Table 2.3 and Figure 2.26). The velocity increases strongly with pressure up to about 0·5–1·0 kbar as the pores collapse, but HUGHES and MAURETTE (1956, 1957) found that increase in temperature causes the velocity to decrease. BIRCH (1958) used the available results to compute the variation of velocity of granite and gabbro with depth in the crust, assuming a realistic geothermal gradient (Figure 2.26). Beneath about 5 km depth, the effects of temperature and pressure tend to cancel each other out. Thus the $P$ velocity in each rock type tends to remain approximately constant with depth, and it may even decrease if the geothermal gradient is locally high.

NAFE and DRAKE (1963) investigated the relationship between $P$ velocity and density by plotting these against each other for water-saturated sediments and sedimentary rocks (Figure 2.27). The observations cluster near a curve and measurements on igneous and metamorphic rocks extend this curve without offset. The Nafe-Drake curve enables us to estimate density of shallow rocks from the $P$ velocity subject to an error of about $0\cdot1$ g/cm³, and it has been widely used in the joint interpretation of gravity and seismic refraction surveys.

BIRCH (1961a) went one stage further by determining an empirical relationship between density, $P$ velocity and chemical composition as represented by mean atomic weight. He used laboratory

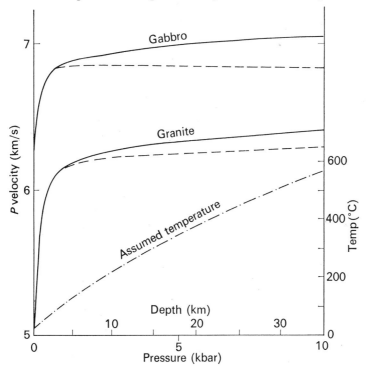

**Fig. 2.26**   The compressional wave velocity ($P$) in granite and gabbro as a function of (1) pressure alone (solid line), and (2) pressure and temperature (dashed line), assuming geothermal gradient as shown. The graphs are based on the experimental results of Birch and his co-workers. Redrawn from PRESS in CLARK (1966), *Handbook of physical constants*, revised edition, p. 208, Geological Society of America.

observations of $P$ velocity at 10 kbar to ensure that the pores had collapsed so that the relationship would be valid beneath a few kilometres depth. He found an approximately linear relationship between velocity and density for a given mean atomic weight m as follows (Figure 2.27):

$$\rho = a(m) + bV_P$$

where   $\rho$ = density,
$V_P$ = $P$ velocity,
$a(m)$ is a constant depending on the mean atomic weight,
and $b$ is another constant.

This relationship appears to remain valid across phase transitions such as the gabbro-eclogite transition. KNOPOFF (1967) has shown that Birch's experimental results at 10 kbar can be fitted to an empirical relationship of the form

$$V_P Z^{-\frac{2}{3}} = A + B\rho Z^{-2}$$

where $A$ and $B$ are constants and $Z$ is the representative atomic number given by the expression

BIRCH

$$Z^{\frac{3}{2}} = \frac{\Sigma n_i Z_i^{\frac{5}{2}}}{\Sigma n_i Z_i},$$

$n_i$ being the percentage by number of the element of atomic number $Z_i$. Knopoff has shown how the relationship he gives is modified by increasing pressure. Birch's and Knopoff's relationships differ somewhat, but both of them and other similar ones are useful in attempting to interpret velocities in the crust and mantle.

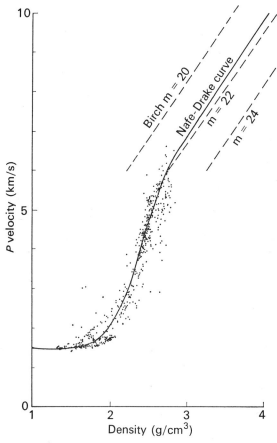

**Fig. 2.27**  Plot of $P$ velocity against density observed for a wide selection of water-saturated sediments and sedimentary rocks, extended for hard rocks. BIRCH's (1961a) empirical relationship between density and $P$ velocity as a function of mean atomic weight m is also shown on the diagram. Partly redrawn from NAFE and DRAKE (1963), *The sea*, vol. 3, p. 807, Interscience Publishers.

In interpreting variations of elastic wave velocity within the Earth, the main problem is to sort out whether they are caused by (1) the normal effect of temperature and pressure gradients on rock of uniform chemical and mineralogical composition, or (2) change in chemical composition, or (3) phase changes affecting the mineralogy but not the chemical composition. Certain guidelines can help the interpretation. (1) First order discontinuities, which should be recognizable on ability to give near-normal incidence reflections, are almost certainly caused by change in chemical composition. This is because phase changes affecting multi-component systems involving solid solution, as most rocks are, would be expected to be spread over a discrete interval of depth. (2) The results of experimental petrology can give the mineral assemblages stable at given temperature, pressure and water vapour pressure. (3) The empirical relations above make possible the prediction of density change accompanying a given change in seismic velocity for the different possible interpretations. Considerations such as isostasy, or the mean density or moment of inertia of the Earth, may restrict the allowable change in density, thereby possibly ruling out some of the suggested explanations.

## Composition of the upper crust

Table 2.4 shows estimates of the mean composition of (1) the crystalline basement rocks of a continental shield, (2) the rocks forming young mountain ranges, and (3) acidic and basic igneous rocks. The table shows that the mean composition of the topmost part of the continental crust is between that of acidic and basic igneous rocks, and more specifically it is between that of granodiorite and quartz diorite.

**Table 2.4**   Estimates of the composition of parts of the topmost crust compared with acid and basic igneous rocks. (1) and (3)–(4) are taken from POLDERVAART (1955); (1) gives an average for the Canadian Shield estimated by Grout and (3) is an estimate of the mean composition of the upper crust in Norway based on the sampling of glacial clays made by Goldschmidt; (5) and (6) are taken from NOCKOLDS (1954). (2) is taken from EADE and others (1966) and is based on direct sampling of 200 000 sq. miles of the New Quebec area. $H_2O$ and $CO_2$ have been excluded from the figures given.

| | Canadian Shield | | Glacial clays from Norway | Average sediment from young fold mountains | Average silicic igneous rock | Average mafic igneous rock |
|---|---|---|---|---|---|---|
| | (1) | (2) | (3) | (4) | (5) | (6) |
| $SiO_2$ | 63·9 | 65·8 | 62·1 | 58·6 | 69·2 | 48·6 |
| $TiO_2$ | 0·8 | 0·5 | 0·8 | 0·6 | 0·5 | 1·8 |
| $Al_2O_3$ | 17·0 | 16·4 | 16·6 | 12·9 | 14·7 | 15·7 |
| $Fe_2O_3$ | 2·4 | 1·5 | 7·3 | 2·9 | 1·7 | 2·8 |
| FeO | 3·0 | 3·0 | — | 2·2 | 2·2 | 8·1 |
| MnO | — | — | — | — | 0·1 | 0·2 |
| MgO | 1·8 | 2·3 | 3·5 | 4·3 | 1·1 | 8·7 |
| CaO | 4·1 | 3·4 | 3·2 | 14·2 | 2·6 | 10·8 |
| $Na_2O$ | 3·7 | 4·1 | 2·2 | 1·4 | 3·9 | 2·3 |
| $K_2O$ | 3·1 | 2·9 | 4·1 | 2·7 | 3·8 | 0·7 |
| $P_2O_5$ | — | — | 0·2 | 0·1 | 0·2 | 0·3 |

It was shown above (p. 53) that the crystalline basement rocks are on average about 0·10–0·15 $g/cm^3$ denser than granite, and that their mean density is about 2·75–2·80 $g/cm^3$, with variation from region to region. This density is what would be expected for a rock with average composition lying between granodiorite and diorite.

The observed $P$ velocity of the upper crust (5·9–6·3 km/s) agrees with the experimental determinations on granites at 1 kbar confining pressure, and taken at its face value this suggests a granitic composition for the upper crust. This is in marked disagreement with the above-mentioned estimates of chemical composition and density which suggest a more basic composition than granite. The reason for the lower than expected $P$ velocity is probably partly that slates and greywackes are abundant in the basement; these rock types possess densities which are typically higher than $2 \cdot 70$ g/cm³ but they have distinctly low $P$ velocities.

A further problem relating to the upper crust is the interpretation of the low velocity zone between 8 and 11 km depth or thereabouts, if it exists. Mueller and Landisman consider that the decrease in velocity with depth is caused by the effect of the steep near-surface temperature gradient outweighing the effect of pressure below a depth of 5 km. The discontinuous increase in velocity at the base of the low velocity channel is interpreted as the boundary between 'granitic' rocks above and dioritic rocks below, which is known to German seismologists as the Fortsch discontinuity. Other possible interpretations of the alleged low velocity zone could be suggested, such as mineral phase changes caused primarily by the higher temperatures at depth, or a compositional change to more granitic rocks at depth (BOTT, 1961).

*Composition of the lower crust*

The lower crustal layer, where it can be distinguished from the upper crust, is recognized by a $P$ velocity of greater than about $6 \cdot 5$ km/s but less than $7 \cdot 6$ km/s. We seek possible mineralogical and chemical explanations of this highly variable and little-known part of the continental crust.

Simple increase in velocity with confining pressure cannot explain velocities as high as $6 \cdot 7$ km/s in the lower crust. *Either* the chemical composition is more basic than that of the upper crust, *or* the stable mineral assemblage differs from that of the upper crust.

It used to be thought that the lower crust is formed of basalt or gabbro. The observed $P*$ velocity of $6 \cdot 7$ km/s matched the experimentally determined velocity of gabbro. It was also supposed that basalt magma was formed by local fusion of the layer. It is now known that the upper mantle is the main source of basalt magma, and therefore the petrological justification for a crustal basaltic layer no longer exists.

Important new evidence on the mineralogical and chemical composition of the lower crust has been provided by an investigation of the stable mineral assemblages of rocks of basaltic composition without water at pressures up to 30 kbar and temperatures between 1000 and 1250°C (GREEN and RINGWOOD, 1967; RINGWOOD and GREEN, 1966). These are:

|  | Rock type | Stable mineral assemblage |
| --- | --- | --- |
| low pressure | basalt<br>gabbro<br>pyroxene granulite | plagioclase<br>pyroxene(s)<br>± olivine<br>± spinel |
| intermediate | garnet granulite | garnet<br>pyroxene(s)<br>plagioclase |
| high pressure | eclogite | garnet<br>pyroxene<br>± quartz |

The effect of increasing the confining pressure at a fixed temperature is to convert gabbro or basalt into eclogite through a transitional stage of garnet granulite. The stability fields for rocks having the composition of the quartz-tholeiite variety of basalt are shown in Figure 2.28. The field boundaries have been extrapolated to lower pressures and are compared with possible geothermal gradients. Green and Ringwood find that the transition to eclogite would occur at slightly lower pressure for rocks of olivine basalt composition. These experimental results imply that under dry conditions it is eclogite, not gabbro, which is the stable form of rocks of basic composition throughout the *whole* of the normal continental crust. Eclogite has a *P* velocity of over 8·0 km/s, which would place it in the upper mantle by definition. Garnet granulite has a velocity of 7·5–8·0 km/s. Thus the interpretation of the lower crust as a layer of basic rock (gabbro or basalt) or its high pressure form under dry conditions appears to be decisively ruled out, provided the equilibrium assemblage is attained and the extrapolation of the field boundaries is correct.

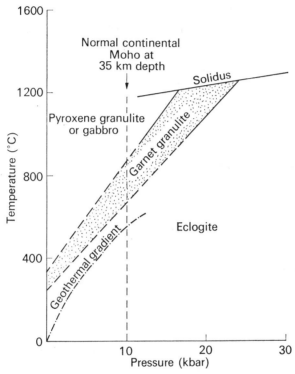

**Fig. 2.28**  The stability fields for rocks having the composition of the quartz-tholeiite variety of basalt, showing also the typical continental geothermal gradient of Birch (from Figure 2.26). Taken partly from RINGWOOD and GREEN (1966), *The Earth beneath the continents*, p. 614, American Geophysical Union.

These results raise the further problem of why gabbro and basalt do occur at and near the Earth's surface, where they are apparently metastable. The reason is that they crystallize from magma at high temperature and low pressure in the stability field of gabbro. Cooling occurs relatively rapidly to low temperature (i.e. 50–200°C) and the metastable assemblage is 'frozen in'. At low temperature, the rate of phase reaction from gabbro to eclogite is probably too slow for significant change to occur during the geological time scale. On the other hand, at temperatures of 300–600°C appropriate to the lower crust, one would expect the mineral reactions to occur fast enough for equilibrium to be established in a matter of years or even days.

If the old idea of a basaltic layer is demonstrably wrong, what are the rocks of the lower crust which exhibit *P* velocities in the range 6·6–7·4 km/s? One possibility is that the lower crust is relatively dry and that high pressure forms of acid and intermediate rocks occur, similar to garnet granulite and eclogite but more silica-rich in composition. Such high pressure modifications of granite and diorite are to be expected, although the presence of sodium-rich feldspar in these rocks would tend to increase the pressure needed to cause the break-down of feldspar to alkali pyroxene (jadeite). RINGWOOD and GREEN (1966) have discussed the possible mineral assemblages which might occur in

**Table 2.5**   Estimated mineral assemblages, densities and compressional wave velocities for typical acid to intermediate rocks at low and high pressures. Taken from RINGWOOD and GREEN (1966); the 'diorite' has 60·4% $SiO_2$ and the 'granodiorite' has 65·6% $SiO_2$. The mineral assemblages are given by weight and the high-pressure assemblages represent probable mineralogies at 30 kbar and 1100°C.

| | 'Diorite' | | 'Granodiorite' | |
|---|---|---|---|---|
| | Low pressure | High pressure | Low pressure | High pressure |
| quartz | 14·8 | 20·2 | 21·7 | 27·8 |
| orthoclase | 7·7 | 7·8 | 13·0 | 12·8 |
| plagioclase | 56·5 | — | 52·0 | — |
| clinopyroxene | 5·8 | 47·2 | 2·3 | 41·6 |
| hypersthene | 11·0 | — | 7·7 | — |
| garnet | — | 8·4 | — | 4·6 |
| kyanite | — | 15·6 | — | 12·5 |
| ore minerals | 4·2 | 0·7 | 3·3 | 0·7 |
| density (g/cm³) | 2·83 | 3·20 | 2·78 | 3·07 |
| P velocity (km/s) | 6·6 | 7·6 | 6·4 | 7·3 |

such rocks (Table 2.5), and have calculated the density and P velocity in them. Gradational velocities would occur in the transition region between the low and high pressure forms of acidic and intermediate rocks. In principle, there is no serious difficulty in understanding velocities in the range 6·7–7·4 km/s in terms of high pressure forms of granodiorite and diorite. The exact velocity would depend on both composition and temperature, leaving plenty of scope for wide variations.

Another possible interpretation of the lower crust is that it is 'wet' and that rocks of basaltic composition occur as amphibolite, which is the stable form in the presence of substantial water vapour pressure below about 500°C. The typical mineral assemblage would be: amphibole, plagioclase, epidote and iron-rich garnet. The P velocity would be about 7·0–7·6 km/s and could be lower if amphibolites were admixed with more silica-rich rocks.

# 3 The oceanic crust

## 3.1 Introduction

One of the important advances in geophysics during the early 1950's was the clear demonstration that fundamentally different types of crust occur beneath oceans and continents. Not only is the oceanic crust much thinner, but the layering is quite distinct. The difference has been highlighted by the recent discovery that most of the oceanic crust has been formed during the last 200 my, in contrast to the continental crust which goes back 3000 my.

The oceans, excepting the continental shelves, cover an area of $332 \times 10^6$ km² which is about 65% of the Earth's surface. The average water depth is 3·8 km. Oceanic regions may conveniently be subdivided into ocean basins, ocean ridges, trenches and continental marings (Figure 3.1). Some continental margins may be further subdivided into the continental shelf, the continental slope and the continental rise.

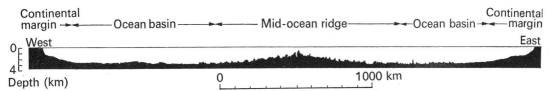

**Fig. 3.1**   The three major morphological subdivisions of the oceans, illustrated by a profile across the North Atlantic from New England to the Spanish Sahara. Redrawn from HEEZEN (1962), *Continental drift*, p. 237. Academic Press.

The main types of bottom topography of the ocean basins are:

(1) *abyssal plains*, which are smooth plains formed of flat-lying sediments with gradients of less than 1/1000;
(2) *abyssal hills*, with a relief of 50–1000 m and widths of 1–10 km, which are particularly common on the floor of the Pacific and cover more than 80% of its total area (MENARD, 1964);
(3) *seamounts*, which are circular or oval-shaped underwater mountains rising abruptly from the deeps and having the shape of volcanoes; dredging and magnetic surveys confirm that nearly all seamounts are underwater basaltic volcanoes;
(4) *archipelagic aprons*, which form very smooth areas of the deep seabed near groups of volcanic islands and are made of submarine lava flows smoothed by an overlying veneer of sediment.

The earliest geophysical measurements relevant to the structure of the oceanic crust were the pendulum measurements of gravity made in submarines since 1923 (VENING MEINESZ, 1948; WORZEL, 1965a). These showed that the oceans, apart from trenches and island arcs, are in approximate isostatic equilibrium with the continents. This was usually interpreted according to the Airy hypothesis, that the crust is much thinner beneath oceans than continents. Suppose a typical continental crust is 35 km thick and that the densities of crust, mantle and sea-water are 2·90, 3·30 and 1·03 g/cm³ respectively. If an ocean basin 5 km deep is in equilibrium with it according to the Airy hypothesis, then the basic condition of isostatic equilibrium (p. 47) shows that the oceanic crust would be 6·6 km thick below the seabed. The seismic refraction work described below has shown that the oceanic crust is on average about 6–7 km thick, in good agreement with the predictions of the Airy hypothesis.

Another early method used to investigate the oceanic crust was to study the dispersion of earthquake surface waves which had traversed oceanic paths. Such a study suggested to GUTENBERG (1924) that the oceanic crust is much thinner than the continental crust. The use of Love and Rayleigh dispersion is still of some importance in assessing crustal structure of inaccessible regions, but in the main it has been superseded in oceanic crustal studies by seismic prospecting methods.

## 3.2 Oceanic crustal structure

Our modern knowledge of the thickness and layering of the oceanic crust has been almost entirely obtained by the refraction and reflection methods of explosion seismology. The methods have been pioneered by Dr. Maurice Ewing and his co-workers at Lamont Geological Observatory and by Dr. M. N. Hill at Cambridge.

### The refraction method

Two main methods of refraction survey have been used at sea. Either separate ships are used for shooting and recording (SHOR, 1963) or the seismic waves are received at buoys and recorded on magnetic tape or telemetered back to the ship (HILL, 1963). As the oceanic Moho is shallower than beneath continents, the refraction lines need only to be about 50–70 km long.

The typical pattern of results obtained is shown in the theoretical model of Figure 3.2 (TALWANI, 1964). The oceanic crust is found to consist of three layers. Layer 1 underlies the seabed and the depth to the top of it is obtained from the ship's precision depth recorder (PDR) or from the reflection $R_1$. Because the $P$ velocity in layer 1 is only slightly greater than in seawater, the refracted head wave $G_1$ only rarely occurs as a first arrival. Consequently the velocity of layer 1 cannot be found in the usual way, and it must be *either* assumed *or* estimated from the critical distance as determined by the maximum amplitude of $R_1$. Similarly, the head wave $G_2$ refracted at the layer 1/layer 2 interface only occurs as a first arrival for a short distance and if layer 2 is thinner than in Figure 3.2 it may not occur as a first arrival at all. It may then be necessary to use the reflection $R'_1$ to determine the depth to the top of layer 2, and to assume a velocity for layer 2 for purposes of interpreting the layers beneath. The two underlying interfaces give rise to easily identifiable first arrivals $G_3$ and $G_4$. The water wave D may be used to determine distance between shot point and receiver.

The results of crustal refraction lines at sea show an unexpected uniformity of oceanic crustal structure wherever measurements have been made, with the exception of the crestal regions of some oceanic ridges. The crust beneath the floor of the ocean consists of three layers above the Moho as follows:

|  | $P$ velocity (km/s) | Average thickness (km) |
|---|---|---|
| water | 1·5 | 4·5 |
| layer 1 | 1·6–2·5 | 0·4 |
| layer 2 | 4·0–6·0 | 1·5 |
| layer 3 | 6·4–7·0 | 5·0 |
| ------------------------------Moho------------------------------ | | |
| upper mantle | 7·4–8·6 | |

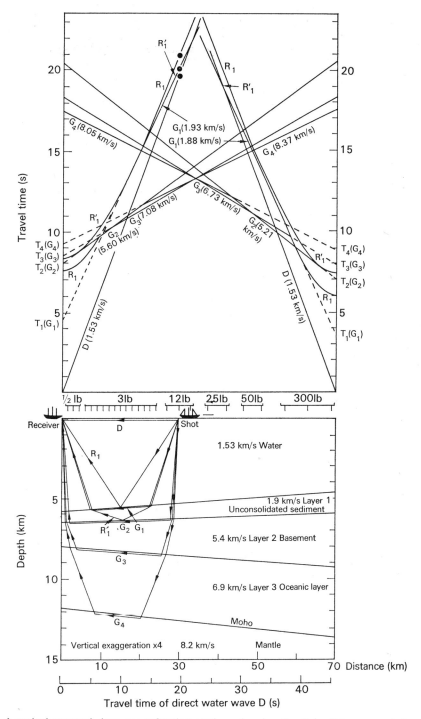

**Fig. 3.2** A typical reversed deep-sea refraction station, showing the time-distance graph for refractions, reflections and the direct water wave. The ray paths are shown for a 30 km separation of shot and receiver. For further explanation see text. Redrawn from TALWANI (1964), *Marine Geol.*, **2**, 37.

Beneath normal ocean basins, the Moho is about 11 km below sea-level and the oceanic crust is about 6–7 km thick. Layer 1 which consists of sediments is locally absent on the flanks and crests of ridges and at the other extreme it reaches over 3 km in parts of the Argentine basin (EWING, 1965). As the velocity within layer 1 increases with depth (see below), the thickness estimates based solely on refraction data must somewhat underestimate the true thickness. Layer 2 is the most variable in thickness and velocity, although accurate estimates of its velocity are difficult to obtain. In the normal parts of the Pacific it averages 1·3 km but it is about 2·4 km thick beneath archipelagic aprons (MENARD, 1964). Layer 3, which is the main oceanic crustal layer, is the most uniform in thickness and velocity.

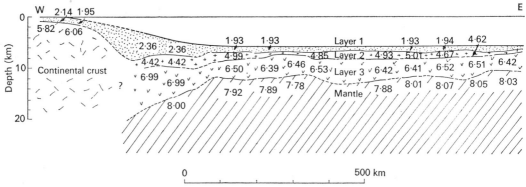

**Fig. 3.3**  Oceanic crustal structure as determined by refraction lines in the Atlantic Ocean east of Argentine. The line runs from (46·0° S, 60·3 °W) on the shelf to (42·7° S, 50·2 °W) in the ocean. Redrawn from EWING (1965), *Q. Jl. R. astr. Soc.*, **6**, 19.

An example of the results of refraction surveys is shown in Figure 3.3. This is a profile over part of the Argentine basin.

*Seismic reflection profiling*

A much more detailed picture of the structure within layer 1 and of the shape of the layer 1/layer 2 interface is obtained by seismic reflection profiling as described by EWING and ZAUNERE (1964). The method is akin to echo-sounding except that the sound source is more powerful and has a lower frequency content, enabling the waves to penetrate below the ocean-floor and to be reflected from interfaces below before attenuation has reduced the amplitude below the detection limit. The early method was to set off small explosive charges at regular time intervals while the ship was under way, but nowadays other types of acoustic sources such as air-guns are mainly used. The return waves are received by a towed hydrophone array and are recorded visually or on magnetic tape. A record is shown in Figure 3.4.

The reflection records yield the two-way travel-time to the reflecting horizons. This can only be converted to depth if the velocity distribution below the seabed is assumed or determined by other methods (see below). The prominence of a given reflection on the record depends on the amplitude of the reflected wave. The relative amplitudes of reflected and incident waves at an interface is called the reflection coefficient $r$, and it is given by

$$r = (\rho_1 V_1 - \rho_2 V_2)/(\rho_1 V_1 + \rho_2 V_2)$$

where $\rho_i$ and $V_i$ are the densities and $P$ velocities of the rocks above and below the interface. A large contrast in $\rho V$ at an interface means a strong reflection.

Seismic profiling records show that the most prominent reflection below the seabed is the interface between layers 1 and 2. This means that the product $\rho V$ changes substantially at this interface and that it is a sharp boundary. It is found that the interface is consistently rougher than the seabed (Figure 3.5). The surface topography of layer 2 may be up to 1 km in amplitude and the wavelength of the undulations is typically 10–20 km. The rough character of the interface extends beneath the abyssal plains. Beneath the abyssal hills of the Pacific Ocean, the interface follows the surface pattern of the hills but the relief is greater (MENARD, 1964).

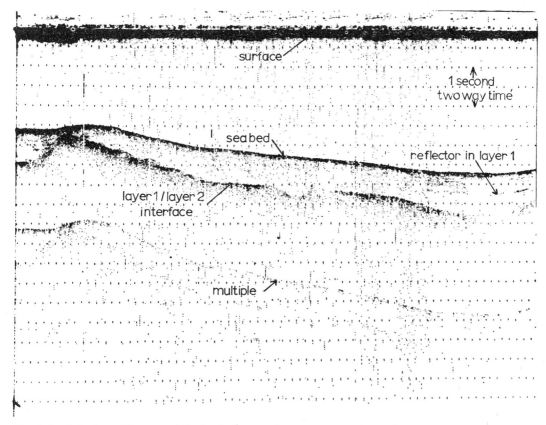

**Fig. 3.4**   Seismic profiling record obtained using an air-gun source on the north-eastern flank of the Iceland-Faroes rise. The record shows (1) the strong reflection from the layer 1/layer 2 interface, (2) the rough topography of the interface in comparison with the ocean-floor, and (3) a weak reflecting horizon within layer 1. Record obtained on a Durham University cruise on R.R.V. *John Murray* during 1967, and reproduced by kind permission of Mr. J. H. Peacock.

Within layer 1 in the American basin of the North Atlantic, three reflecting horizons have been described (EWING, WORZEL, EWING and WINDISCH, 1966). Related horizons have been found in other parts of the Atlantic and in other oceans. *Horizon A* is a prominent and persistent interface occurring 300–500 metres below the ocean bed which cuts out against the layer 1/layer 2 interface at the flank of the mid-Atlantic ridge. It can be underlain by a stratified zone up to 500 metres thick but usually less. It is somewhat smoother than the seabed and much smoother than the layer 1/layer 2 interface.

The horizon is recognized in the South Atlantic (Figure 3.5) and a similar horizon occurs in the Pacific Ocean. *Horizon β* marks the upper surface of a deeper set of stratification which closely overlies the layer 1/layer 2 interface. It is of more restricted occurrence than horizon A and is not seen in Figure 3.5. *Horizon B* is a smooth reflector in the Atlantic which locally replaces the layer 1/layer 2 interface (Figure 3.5). It must mark the upper surface of acoustically opaque rocks which mask underlying reflectors including the layer 1/layer 2 interface. Horizon B occurs only in regions far

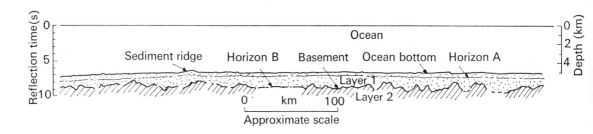

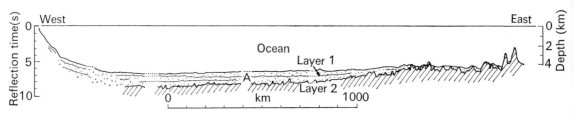

**Fig. 3.5** Seismic profiling results from the Argentine Basin, showing the structure of layer 1 in the southwestern Atlantic. Redrawn from EWING (1965), *Q. Jl. R. astr. Soc.*, **6**, 14 and 17.

removed from the mid-Atlantic ridge. A similar reflector below the Pacific Ocean is known as horizon B'. A remarkable feature of all these horizons, contrasting strongly with the typical continental crust, is the almost complete lack of folding, faulting or tilting shown by them.

### Velocity-depth distribution in layer 1

Neither the refraction method nor the seismic profiling method described above can give the velocity-depth distribution in layer 1. However, in land reflection surveying one method of estimating average velocity down to a reflector is to determine the travel-time $t$ for varying horizontal distance between shot and receiver. If $z$ is the depth to the reflector and $V$ is the average velocity down to it, then it can be shown by simple geometry that $V^2t^2 = 4z^2 + x^2$ provided the angle of incidence is small. The gradient of the straight line obtained by plotting $t^2$ against $x^2$ is $1/V^2$. By applying this method to a series of reflecting horizons, the interval velocity between each pair of adjacent reflectors can be estimated.

A modified version of the $t^2/x^2$ method which allows for large angles of incidence has been used to obtain the velocity-depth structure of layer 1 (LE PICHON, EWING and HOUTZ, 1968; HOUTZ, EWING and LE PICHON, 1968). A disposable free-floating sono-buoy telemeters the reflected signals back to the ship, which steams away from it while firing an air-gun. Over a hundred such profiles have been shot.

The structure of layer 1, showing the interval velocities between reflecting horizons and their inter-pretation, is as follows:

| Horizon | Sub-layer | P velocity (km/s) |
|---------|-----------|-------------------|
| seabed | | |
| | | 1·6–2·2 (unconsolidated sediments) |
| horizon A | | |
| | layer A | 1·7–2·9 (semi-consolidated sediments) |
| horizon β | | |
| | layer β | 2·7–3·7 (consolidated sediments) |
| horizon B or | | |
| layer 1/2 interface | | |

layer 1 {

The velocity just below the seabed is 1·47–1·55 km/s and therefore the velocity probably increases fairly steadily with depth down to horizon A. Both horizons A and β appear to mark discontinuous increases of velocity with depth. Houtz, Ewing and Le Pichon report reception of critical refractions from horizon β.

## 3.3 Coring and drilling into layer 1

Further insight into the composition and age of the sediments of layer 1 has been obtained by widespread coring, and more recently by drilling into the ocean-floor.

Coring has shown that the rocks forming the deep ocean-floor are unconsolidated sediments including pelagic deposits, red clay and turbidites. All core samples which have been obtained are relatively young geologically, going back at most to the late Jurassic or Cretaceous. Core samples near horizon A have been taken at the south-western corner of the Hatteras abyssal plain where it crops out on the ocean bed, and the microfossils obtained are of Upper Cretaceous (Maestrichtian) age (EWING, WORZEL, EWING and WINDISCH, 1966).

The rate of sedimentation during the Pleistocene can be determined from cores using the decay of ionium ($Th^{230}$) which has a half-life of 80 000 years. Measurement of the abundance of $Th^{232}$ in the sample makes it possible to estimate the initial amount of ionium present at the time of deposition. Using this method, the normal rate of sedimentation in the Pacific is found to be 0·3–1·6 mm per thousand years (MENARD, 1964), although rates greater than 10 mm per thousand years occur in the sediment traps near the continental margins. The average rate of sedimentation is higher in the Atlantic, and exceptionally high values of about 50 mm per thousand years occur in the Argentine basin (EWING, 1965). Other important scientific information, such as Pleistocene climatic variations, has been obtained from cores but this is outside the scope of this book.

A few years ago a plan was launched in U.S.A. by Hess and others to drill through the oceanic crust into the topmost mantle beneath. It became known as the Mohole project. As a preliminary, a boring was put down through layer 1 in water 3366 m deep to the east of Guadalupe Island, which is about 250 km west of the Baja California coast. It penetrated 185 m of greenish-grey clays with siliceous and calcareous microfossils indicating a Miocene to Pliocene age before passing into basalt (RIEDEL and others, 1961). Unfortunately the Mohole project was abandoned for lack of funds.

The disappointment resulting from cancellation of the Mohole project has to some extent been alleviated by the initiation of a new programme aimed to drill a large number of holes through part or all of the oceanic layer 1, following successful drilling on the subsided Blake Plateau, off Florida, during 1965. The initial programme (called JOIDES, short for Joint Oceanographical Institute Deep Earth Sampling programme) was planned to drill about 30 holes in the Atlantic Ocean and about 36 in the Pacific Ocean during 1968 and 1969. The programme has now been extended as DSDP

(Deep Sea Drilling Project) and further holes are being drilled in the Atlantic Ocean and Mediterranean Sea during 1970. The specially designed drilling vessel *Glomar Challenger* is being used as the drilling platform. One of the present problems is that the holes have to be abandoned when the first drill bit is worn out, thus limiting penetration through hard beds such as chert. It is hoped that this 're-entry' problem will be solved during 1970.

Initial reports on legs 1 and 2 in the Atlantic Ocean have now been published, and the results have already yielded important new information on the geological history of the ocean basins and on the structure of layer 1. Leg 3 in the South Atlantic Ocean has been exceptionally important in the confirmation it has given of predictions of the ocean-floor spreading hypothesis, as will be described in Chapter 7. Here we discuss the results bearing on the structure of layer 1.

The floor of the Gulf of Mexico and of the North American basin were drilled during leg 1 during the period August to September 1968 (EWING and others, 1969). One highlight of this leg was the discovery of salt dome cap rock and petroleum in a hole drilled in the deep Gulf of Mexico, where thick sediments are apparently underlain by oceanic crust; Upper Jurassic fossils were discovered in the cap rock. Another highlight of leg 1 was the penetration of horizon A at two nearby sites in the North American basin. The horizon was found to contain beds of mid-Eocene chert formed by silicification of turbidites and radiolarian muds. Horizon A was also reached in the North American basin during leg 2, which was run from New Jersey to Senegal during October to November 1968 (PETERSON and others, 1970); it was found to consist of interbedded cherts and clays of the same mid-Eocene age. One hole drilled east of the Bahamas during leg 1 passed through a succession from Cenomanian to Tithonian (Upper Cretaceous to uppermost Jurassic) but failed to reach horizon $\beta$ which was below.

Thus horizon A acts as a seismic reflector because of the chert layers, and not because of a local concentration of turbidites as had been previously suspected (turbidites were found to be abundant throughout the successions). The dating of horizon A as mid-Eocene raises a problem, because coring near the outcrop area (see above) suggested an Upper Cretaceous age. PETERSON and others (1970) put forward three possible explanations of the discrepancy: (1) the mid-Eocene cherts may not be horizon A (highly unlikely in view of reflection surveys at site positions); (2) horizon A may be diachronous (i.e. of varying age) away from the region where it has been proved by drilling; (3) the coring in the outcrop area may have sampled beds older than horizon A.

We can be confident that the later legs of the Deep Sea Drilling Project will continue to produce important evidence bearing on the fundamental problems of the history of the ocean basins.

## 3.4 Oceanic magnetic anomalies and the composition of layer 2

The observed range of $P$ velocities of layer 2 could be caused either by basaltic lavas or by consolidated sediments. Seismic profiling has shown that the layer comes to the seabed on the flanks of the mid-Atlantic ridge where dredging has brought up basalt from the bottom. The trial Mohole reached basalt below the clays of layer 1, and 14 m of it was penetrated before drilling ceased. This basalt may form the top of layer 2, but the evidence is not conclusive because Guadalupe Island is volcanic and the lava may be local. MENARD (1964) has argued that the shape of the layer 1/layer 2 interface below abyssal hills supports an igneous origin for layer 2. All these observations suggest that layer 2 is formed of basaltic rocks rather than of consolidated sediments, but the evidence from oceanic magnetic anomalies outlined below conclusively shows that highly magnetic rocks form a substantial part of the layer.

Magnetic anomalies are measured at sea by means of a proton or fluxgate magnetometer with the sensing element towed two or more ship-lengths behind to be removed from the ship's magnetic field. Both types of instrument measure the total-field magnetic anomalies to an accuracy of about

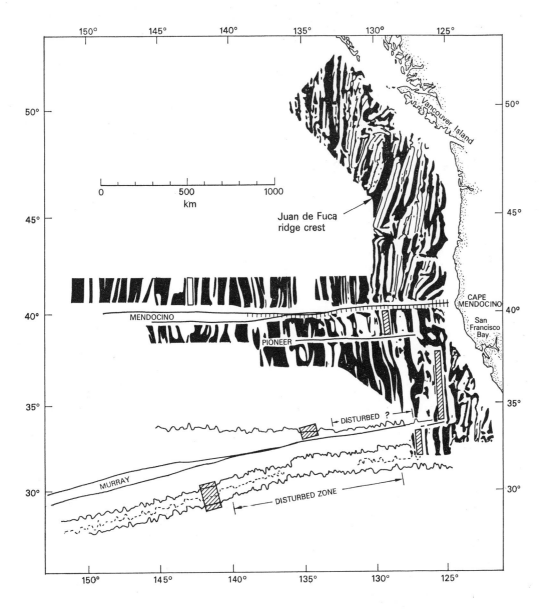

**Fig. 3.6**   Magnetic anomaly lineations off the west coast of North America. Positive anomalies are shown in black. Note the offset of the magnetic anomaly pattern at the fracture zones and the change in the amount of offset along the Murray fracture zone. MENARD (1964), *Marine geology of the Pacific*, p. 46, McGraw-Hill.

$1\gamma\ (=10^{-5}$ oersted). The measurements are easy to make, and since the mid 1950's magnetic profiles and surveys have become numerous.

A most important discovery was made when the magnetic anomalies off the west coast of North America were first mapped in some detail (RAFF and MASON, 1961; MASON and RAFF, 1961). A quite unexpected pattern of north-south orientated strip-like positive and negative magnetic anomalies was found to dominate the whole region surveyed. The pattern of anomalies may be depicted by showing the positive strips in black, as in Figure 3.6. The continuity of the strips is interrupted by the east-west fracture zones, where they are apparently displaced laterally by up to more than 1100 km. The individual strips are about 10–20 km wide and the peak-to-peak amplitudes reach up to $1000\gamma$ (Figure 3.7).

Later work has shown that strip-like magnetic anomalies are typical of oceanic regions, although the pattern is not everywhere as regular as off the west coast of North America. Furthermore, it has been found that the strips run parallel to the crests of the ocean ridges and that the pattern of anomalies is symmetrical about the crests (Figures 3.7 and 3.8), allowing for lack of symmetry in the shapes of anomalies caused by the local magnetic field direction not being vertical in general.

Oceanic magnetic anomalies have been a key to understanding the mechanism and history of ocean-floor spreading (Chapter 7). They reveal the past history of reversal of the Earth's main

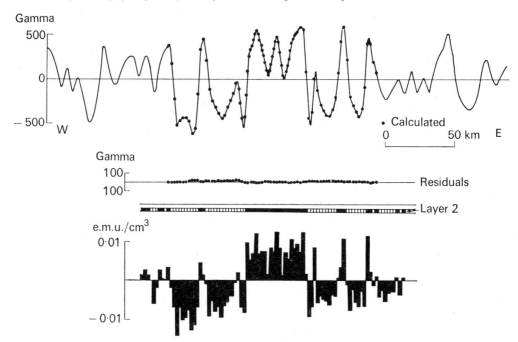

**Fig. 3.7**  Interpretation of a magnetic anomaly profile across the Juan de Fuca ridge, north-eastern Pacific, in terms of a distribution of magnetization within layer 2, obtained using a matrix interpretation method. Layer 2 has been subdivided into two-dimensional rectangular blocks, and the magnetization of each block has been computed and is shown graphically below. The residuals are the observed minus calculated values of magnetic anomaly. Note the symmetry of the anomalies and of the interpreted magnetization distribution about the centre of the ridge (which is at the centre of the profile). Redrawn from BOTT (1967b), *Geophys. J. R. astr. Soc.*, **13**, 320.

magnetic field (Chapter 5). They have revealed large apparent lateral displacements associated with fracture zones, later to be interpreted as transform faults (Chapter 7). They also show that the rocks forming layer 2 of the oceanic crust must be highly magnetic. The last aspect only will be treated here. The others are discussed in later chapters as indicated above and in **3.5** below.

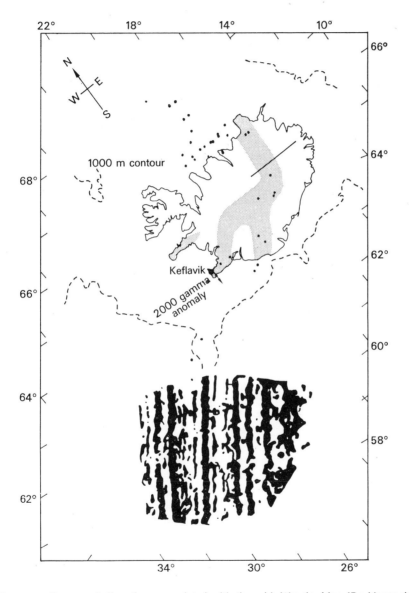

**Fig. 3.8** The magnetic anomaly lineations associated with the mid-Atlantic ridge (Reykjanes ridge) south-west of Iceland. Positive anomalies are shown in black, and earthquake epicentres north of the survey region are shown as solid dots. This map results from a carefully controlled aeromagnetic survey described by HEIRTZLER, LE PICHON and BARON (1966), *Deep-Sea Res.*, **13**, 428.

These large amplitude magnetic anomalies with steep gradients must be caused by highly magnetic rocks at relatively shallow depth. The sediments of layer 1 are effectively non-magnetic and cannot contribute significantly to the anomalies. An interpretation of the profile across the Juan de Fuca ridge in the north-eastern Pacific in terms of lateral variations of magnetization within layer 2 is shown in Figure 3.7. In the model, the positive anomalies are underlain by magnetization in the direction of the Earth's field, and the negative anomalies are underlain by reverse magnetization; however, a layer of uniform magnetization could be added without affecting the agreement so that the anomalies could be equally well explained by alternate strips of more and less strong magnetization. A contrast in magnetization of the order of 0·02 emu/cm³ is needed to explain the observed amplitudes. If one attempts to interpret such large amplitude anomalies in terms of magnetic rocks within layer 3, it is found that excessively strong and highly irregular patterns of magnetization are needed (BOTT, 1967b). It is therefore clear that the anomalies are caused mainly by strong lateral variations in magnetization within layer 2. Basalt lava or intrusive rocks of similar composition are the only common rocks which could cause such large and consistent anomalies.

## 3.5 Ocean-floor spreading and formation of oceanic crust

*The idea of ocean-floor spreading*

The modern concept of the formation of oceanic crust dates from the suggestion of HESS (1962) and DIETZ (1961) that new crust is progressively formed by magmatic processes in the narrow crestal zone of ocean ridges. The ocean-floor spreads laterally in both directions to accommodate the newly formed crust (Figure 3.9). It was suggested that this is the process by which continental drift occurs (p. 208). North America and Europe, for instance, gradually drifted apart as new oceanic crust was

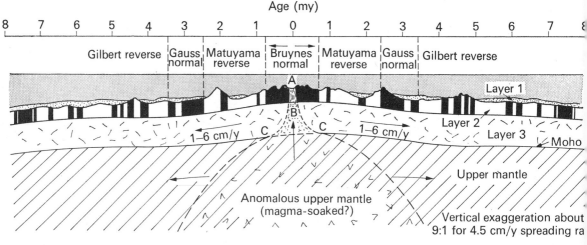

**Fig. 3.9**   The ocean-floor spreading hypothesis and the related mechanism by which the oceanic crust may be formed as the lithosphere on either side spreads laterally at about 1–6 cm/y:
   At A layer 2 is formed by solidification of basic magma;
   At B there is a mush of ultrabasic upper mantle rock, basic magma and water;
   At C the Moho is frozen in at the 500°C isotherm as water and olivine react above to produce serpentine.
In layer 2, normal magnetic polarity is shown black and reverse polarity is shown white, using the geomagnetic time-scale described in Chapter 7.

formed between them producing the Atlantic Ocean. On the other hand, ocean-floor spreading in the Pacific must be related to sinking of oceanic crust into the mantle at or near the continental margins around, which form the circum-Pacific belt. The ocean-floor spreading hypothesis is usually linked to the mantle convection hypothesis, the convection currents rising beneath the ocean ridges and either dragging or forcing the crust apart laterally (Chapter 9).

Spectacular support for the ocean-floor spreading hypothesis has come from the study of oceanic magnetic anomalies. The first important step was taken by VINE and MATTHEWS (1963). They suggested that the alternating strips of positive and negative magnetic anomaly are caused by underlying blocks of layer 2 alternatively magnetized in the normal and reverse directions of the Earth's present-day field (Figure 3.7). They pointed out that this interpretation follows almost as a corollary from the combination of the ideas of ocean-floor spreading and of periodic reversal of the Earth's magnetic field as discovered by palaeomagnetic observations (p. 166). As new crust is formed in the crestal zone of an ocean ridge by igneous processes, shortly after solidification it cools through the Curie point and picks up a strong component of permanent (remanent) magnetization in the direction of the ambient field. This swamps the induced magnetization, so that blocks magnetized when the field was reversed produce magnetic anomalies of opposite sign. The newly magnetized block is then split and forced apart to make room for fresh injections of new crust (Figure 3.9). The hypothesis suggests that the past history of the Earth's magnetic field reversals may be fossilized as oceanic magnetic anomalies caused by layer 2. That this is so has been strikingly confirmed by palaeomagnetic measurements on thick cores from the ocean floor (p. 213).

As well as confirming the validity of the ocean-floor spreading theory, the Vine-Matthews hypothesis has produced a simple explanation of the strip-like pattern of oceanic magnetic anomalies and of their symmetry in relation to the crests of ocean ridges. Furthermore, oceanic magnetic anomalies can now be used to trace the history of ocean-floor spreading and continental drift over the last 200 my (Chapter 7).

*Formation and composition of layer 3*

Layer 3 forms the main thickness of the oceanic crust. It is distinguished by a remarkable uniformity of *P* velocity and thickness. This layer used to be interpreted as a consolidated basic igneous rock such as gabbro. However, the ocean-floor spreading hypothesis has opened up a new approach to the interpretation of the layer.

HESS (e.g. 1965) suggested that layer 3 is formed of partially serpentinized peridotite and that it has been produced by hydration of olivine in the topmost mantle during the process of crust formation beneath the crests of ridges. The reaction

$$\text{olivine} + \text{water} \rightarrow \text{serpentine}$$

can only occur below about 500°C and it requires an adequate supply of water. Hess suggested that the 500°C isotherm beneath the crests of ocean ridges occurs at a depth of about 6 km below the seabed, and that olivine is converted to serpentine above this level. As the newly formed crust is carried laterally by the spreading process, the temperature at a given depth falls and the water supply becomes inadequate for further reaction. The oceanic Moho according to this hypothesis represents the boundary between the normal rock of the uppermost mantle below and its serpentinized alteration product above. Once it has been removed from the crestal region of the ridge, the Moho is thus frozen in.

An alternative explanation of layer 3 has been suggested by CANN (1968). He considered that it has formed by the metamorphism of basaltic crust to amphibolite in the crestal region of the ridge where temperature is high. The ocean-floor spreading carries the crust laterally and the metamorphic boundary between the basalts of layer 2 and the amphibolites of layer 3 is frozen in. In contrast to Hess's hypothesis, the Moho is a compositional boundary according to Cann.

## 3.6 Ocean ridges

It was shown above that new oceanic crust is probably formed beneath the crest zones of oceanic ridges. If true, the ocean ridges are of great importance in understanding the origin of the oceans and the process of continental drift. We now briefly describe their bathymetry and geology and the structure of the crust and uppermost mantle beneath them.

*Bathymetry and geology*

The ocean ridge system forms the largest uplifted linear surface feature of the Earth (Figure 3.10). Intensive study of the ridge system dates from about 1956, when Ewing and Heezen found that the belt of shallow focus earthquakes following the crest is continuous and that the individual ridges form an interconnected world-wide system. It has a total length of 80 000 km and individual ridges are typically about 500–1000 km wide. The crests rise about 2–3 km above the average depth of the adjacent ocean basins. On a broad scale the ridges are gently arched uplifts but locally the topography can be extremely rugged. The seamounts forming the sub-bottom terrain tend to be elongated parallel to the crest.

In the Indian and Atlantic Oceans, where the areal extent is being increased by ocean-floor spreading, the crests of the ridges lie close to the median line of the ocean; the ridges are relatively narrow and rugged, and a prominent deep trench known as the median rift is typically present along the crest. In contrast, the East Pacific rise is located on the east side of the almost certainly shrinking Pacific Ocean (p. 223); it is much wider and less rugged than the other ridges and it has no prominent trench along the crest of it. These contrasts between the Atlantic and Pacific type of ridge are probably fundamental and are also reflected in the more rapid spreading rate associated with the East Pacific rise (p. 213).

Transverse fracture zones form prominent topographical features crossing certain parts of all the ridges. These are typically depressions with associated uplifted blocks which cut across the ridges, displacing the crest laterally and in some places vertically also. A series of parallel fracture zones crosses the mid-Atlantic ridge in equatorial latitudes, displacing the crest by about 4000 km in a left-lateral sense, and enabling it to maintain its median position in both the North and South Atlantic. As mentioned above, a series of fracture zones which interrupt the magnetic lineation pattern cross the East Pacific rise.

The East Pacific rise passes into North America at the Gulf of California and the extension of the Carlsberg ridge of the Indian Ocean passes into the Gulf of Aden, where it divides, one branch continuing up the Red Sea and the other passing into the East African rift system (p. 61). Both western North America and East Africa are regions where plateau uplift of about 1·5–2 km has occurred during the Tertiary.

Rock samples dredged from the ridges show that basalt is by far the commonest rock. Gabbro, serpentinite and other igneous rock types, including some metamorphosed varieties, are also found. The islands along the crests are almost all basaltic volcanoes, although St. Paul Rocks on the equatorial mid-Atlantic ridge is formed of coarse-grained ultrabasic rocks including serpentinite. Iceland is the largest 'outcrop' of an ocean ridge and it is formed of basalt and other igneous types

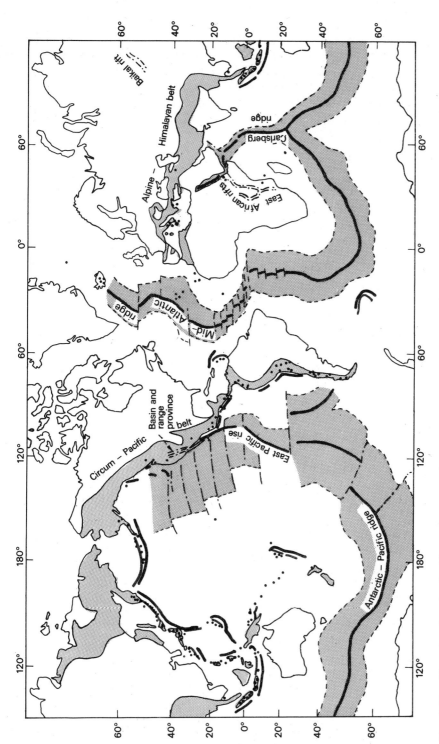

**Fig. 3.10** Map showing the Earth's major surface features. Tertiary and recent mountain ranges and island arcs are shown in red. Ocean trenches are shown as thick black lines. Ocean ridges and their crests are also shown. Major active or recently active volcanoes are marked as black dots. Compiled from various sources.

of late Tertiary to Recent age. Recent volcanic activity occurs in the central graben, and the rocks become progressively older towards the east and west from it. Recent volcanic activity is typical of the whole crestal region of ocean ridges, and must account for nearly all of the volcanism which occurs; in contrast the continental volcanism is relatively insignificant. However, continental volcanism on the edges of the oceans has been important (and still is in island arcs, etc.) and may give us an indication of when continental drift episodes started.

## Structure of crust and upper mantle beneath ridges

The pioneer submarine gravity measurements made by VENING MEINESZ (1948) showed that the mid-Atlantic ridge at 40° N is in approximate isostatic equilibrium. Subsequently, surface ship gravimeter traverses across the ridges have confirmed that they are in approximate isostatic equilibrium, although the local topographic features are uncompensated and cause the gravity profiles to be irregular. Over the East Pacific rise (Figure 3.11) and the mid-Atlantic ridge (Figure 3.12) there

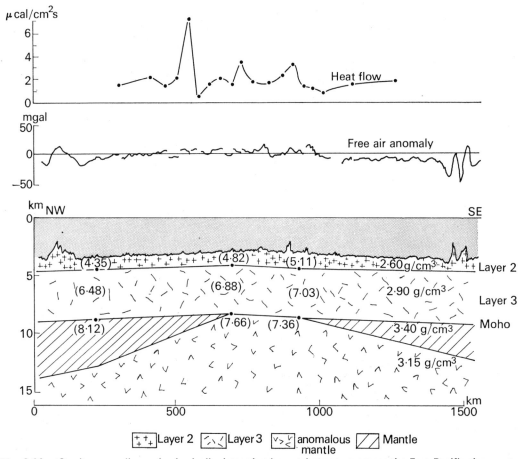

**Fig. 3.11** Gravity anomalies and seismically determined crustal structure across the East Pacific rise west of Peru, about at latitude 15°–17°S. *P* velocities in km/s are shown in parentheses. Heat flow measurements are also shown. The boundaries of the anomalous region in the upper mantle beneath the crest need to slope outwards as shown to satisfy the gravity anomalies. Redrawn from TALWANI, LE PICHON and EWING (1965), *J. geophys. Res.*, **70**, 350.

are small positive free air anomalies over the axial region and slightly negative anomalies on and beyond the flanks. This would be expected for relatively deep isostatic compensation.

Seismic refraction measurements of crustal structure across the East Pacific rise show that the layers 2 and 3 are apparently continuous across the crest (Figure 3.11). The crust beneath the crest tends to be slightly thinner than beneath the adjacent ocean basins, and the topmost mantle velocity below the crest is anomalously low. Across the mid-Atlantic ridge (Figure 3.12), layer 1 rocks are present only in intermontane basins. Layer 2 continues uninterrupted across the crest, but layer 3 becomes confused with the topmost mantle beneath the crestal region, where the $P$ velocities are

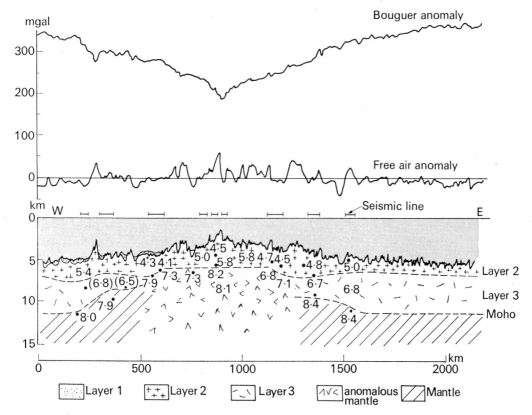

**Fig. 3.12** Gravity anomalies and seismically determined crustal structure across the mid-Atlantic ridge, from (36·6° N, 48·3° W) to (25·5° N, 29·8° W) approximately. The Bouguer anomalies were obtained by filling up the sediment basins and the ocean with rock of density 2·60 g/cm³. $P$ velocities are shown in km/s. Redrawn from TALWANI, LE PICHON and EWING (1965), *J. geophys. Res.*, **70**, 343.

highly variable. The crust certainly does not thicken—if anything it is slightly thinner beneath the ridge than beneath the adjacent ocean basins. The different types of crustal structure beneath the East Pacific rise and the mid-Atlantic ridge may result from different spreading rates (p. 213) and may possibly indirectly reflect the different types of ridge associated with shrinking and expanding oceans.

The seismic evidence conclusively rules out the idea that the isostatic compensation beneath ridges is caused by crustal thickening. It must be caused by an anomalously low density region in the under-lying upper mantle. It is probable that these low density rocks of the topmost mantle can be equated with the region of anomalously low $P$ velocities underlying the Moho at the crest. On this assumption, TALWANI, LE PICHON and EWING (1965) have constructed models of the deep structure beneath the mid-Atlantic ridge satisfying both the gravity and seismic observations, using the empirical Nafe-Drake curve (Figure 2.27) to relate density and velocity. One of their models is shown in Figure 3.13. Although the models cannot be a unique interpretation, they do indicate two indisputable features: (1) the underlying upper mantle must be anomalous and a substantial thickness of the anomalous low density rocks are needed to explain the gravity anomalies; and (2) the boundary between the anomalous rocks and the more normal mantle slopes outwards from the crest.

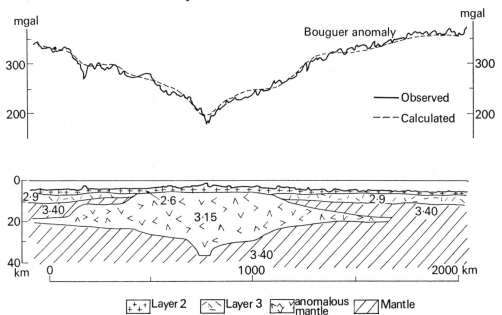

**Fig. 3.13** A possible density model of the crustal and upper mantle structure beneath the mid-Atlantic ridge along the profile of Figure 3.12. The densities (in g/cm³) assigned to the various layers in the model are shown. Redrawn from TALWANI, LE PICHON and EWING (1965), *J. geophys. Res.*, **70**, 348.

It will be shown in the next chapter (p. 148) that the anomalously low density rocks are probably caused by phase changes, including partial fusion, and to a lesser extent by thermal expansion associated with a hotter than average upper mantle. This is exactly what would be expected if the ridges are underlain by upwelling convection currents in the upper mantle. The very active basaltic volcanism can also be understood in this way (BOTT, 1965a).

## 3.7 Continental margins

Continental margins form the junction between the fundamentally different types of crust beneath oceans and continents. The two main types are (1) seismically inactive margins, and (2) seismically active margins. Less well defined continental margins occur in the anomalous regions where an ocean ridge impinges on a continent, such as the Gulfs of Aden and California.

*Inactive margins*

Inactive margins are not associated with any unusual concentration of earthquake activity. They are typical of the oceans which are increasing in area by ocean-floor spreading, including the Atlantic and Indian Oceans. The inactive margins may be subdivided into continental shelf, continental slope and continental rise.

The continental shelf forms the seaward extension of the adjacent continent out as far as the 200 m (100 fathom) depth contour. Seismic refraction surveys show that the crust beneath the shelf is continental in character but somewhat thinner than average. The crust starts to thin abruptly towards the ocean near the 200 m depth contour. Deep sedimentary troughs running parallel to the margin are characteristic of the shelf (DRAKE, EWING and SUTTON, 1959).

The continental slope is the steep boundary between the shelf and the deep ocean. Most continental slopes are less than 50 km across, possessing a gradient steeper than 1 in 10. The slope may be cut by submarine canyons through which turbidity currents carry sediments from the shelf down to the abyssal plains. Most of the change in crustal structure occurs beneath the slope.

The continental rise forms the gentle slope of the ocean floor towards the abyssal plains. It may be several hundred kilometres wide or even absent. The crust beneath the rise is typically oceanic but layer 1 is thicker than usual because of sediments dumped over the margin (Figure 3.3).

Seismic methods are not yet sufficiently refined to show exactly how the oceanic and continental types of crust merge together. The normal procedure for investigating the structure of a margin is to use the seismic refraction stations on either side of the slope to fix the crustal thickness at two or more points; then the gravity anomalies are used to deduce the shape of the Moho across the margin. It is found that the complete transition from continental to oceanic crust occurs over a horizontal distance of 200 km or less (Figure 3.14).

One particularly interesting feature of the Atlantic type of margin was pointed out by WORZEL (1965b). The isostatic anomalies tend to show a negative anomaly of about − 50 mgal over the slope

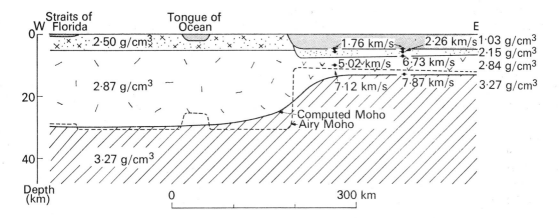

**Fig. 3.14**  Crustal structure section across the Bahamas continental margin as determined from limited seismic refraction data and gravity observations. The shape of the Moho computed from the gravity anomalies is compared with the shape it would have if in perfect isostatic equilibrium according to the Airy hypothesis, showing that there may have been a substantial transfer of low density crust towards the ocean, akin to a landslip at the base of the crust. Adapted from WORZEL (1965b), *Submarine geology and geophysics*, p. 342, Butterworths.

and the adjacent part of the rise, even when thick sedimentary accumulations are allowed for. There is a deficiency of mass beneath the slope and adjacent part of the rise and a corresponding surplus of mass beneath the adjacent part of the shelf, as shown in Figure 3.14. Thus the typical inactive margin is not strictly in isostatic balance according to the Airy hypothesis. This could be readily explained as a steady creep of the low density continental crust towards the ocean, rather like a landslide at the base of the crust, without full isostatic adjustment. This could occur as a result of hot creep near the base of the crust. Other factors, such as a change in underlying mantle densities across the margin, may possibly contribute to this gravity anomaly.

*Active margins and ocean trenches*

A seismically active margin is typical of the shrinking Pacific Ocean, where according to ocean-floor spreading *either* the surrounding continents are being thrust over the Pacific floor, *or* the ocean-floor is underthrusting the continent, *or* both are occurring together. It is characterized by the circum-Pacific belt of shallow, intermediate and deep focus earthquakes which dips at about 45° beneath the continents, reaching the surface at the foot of the slope; about 85% of all earthquake energy is released in this belt (p. 264). The typical active margin has a deep trench on the oceanward side and either an island arc or a mountain range on the landward side. The continental shelf is narrow or absent although sea areas on the landward side of island arcs may be large. Our main interest here centres on the structure of the trenches.

Trenches (Figure 3.10) are the largest linear subsidence features affecting the surface of the solid Earth. Nearly all of them occur near the margin of the Pacific; the exceptions are the South Sandwich and Puerto Rico trenches of the Atlantic Ocean, and the East Indian trenches bordering the Indian Ocean. Trenches and associated island arcs are mostly convex towards the ocean.

Trenches are remarkable for their length and continuity. The Peru–Chile trench is 4500 km long without serious interruption and the Tonga trench is continuous and straight at a depth of 9 km for a length of 700 km. They reach depths about 2–4 km below the adjacent ocean floor. The ocean's greatest depths of 10–11 km are all found in the trenches of the deep west Pacific. The average width is less than 100 km and in cross-section they have an asymmetrical V-shape with the steeper slope (about 8–20°) towards the land. The apex of the V may be truncated by a flat bottom a few kilometres across formed by infilling sediments.

The submarine gravity measurements of Vening Meinesz showed that the trenches and tectonic arcs of the East and West Indies are associated with the largest known negative isostatic anomalies. The negative anomalies were attributed by Vening Meinesz to a downbuckle of the underlying crust, known as a tectogene, which was held down against gravity by compression in the crust or by the drag of converging convection currents.

More recently, combined gravity and seismic refraction studies of trenches have been made (HAYES, 1966; TALWANI, SUTTON and WORZEL, 1959). These give a much more reliable interpretation than is possible using gravity anomalies by themselves. Seismic refraction lines are used to give control points of crustal structure and the gravity anomalies are used to interpolate the structure between them, using the empirical Nafe-Drake relationship (p. 64).

The results of applying combined gravity and seismic interpretation to the structure beneath the Puerto Rico trench and the Peru–Chile trench, and their associated margins, are shown in Figures 3.15 and 3.16. The main conclusions from these and other similar studies are as follows:

(1) The trenches are strongly out of isostatic equilibrium and they represent a deficient mass.

(2) The crust beneath the trench is oceanic and is about as thick as beneath the adjacent ocean, or in some places slightly thicker. The negative gravity anomaly is caused simply by depression of the crust, resulting in the dense underlying mantle being replaced by the volume of low density water in the trench itself. There is no need for or indication of a tectogene beneath.

(3) A local thinning of the crust on the oceanward side of the trench is indicated by some gravity profiles across trenches, including the Peru–Chile trench.

(4) The sediment infill varies greatly in thickness, but is typically thin. The main part of the Peru–Chile trench contains less than 500 m of sediment despite the proximity to the Andes, but in contrast the southern end of this trench is swamped by 2 km or more of sediments

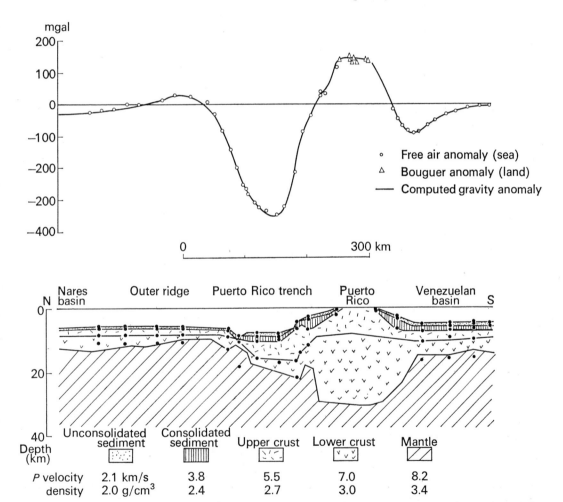

Fig. 3.15   Crustal structure profile across the Puerto Rico trench, based on seismic refraction results (shown by solid circles) and gravity observations. The interpretation is based on the empirical Nafe-Drake relationship between P velocity and density (Figure 2.27) and approximately satisfies both gravity and seismic observations. Adapted from TALWANI, SUTTON and WORZEL (1959), J. geophys. Res., 64, 1550.

(HAYES, 1966). It is apparent that trenches subside quite independently of their sediment infill and that sediment thickness in them depends on availability.

There have been two main theories of origin, (1) that trenches and arcs are formed by compression involving thrust and strike-slip faulting, and (2) that they are formed by tension and normal faulting. The old form of the compression hypothesis involving downbuckling of the crust is untenable on mechanical grounds (p. 238) and is inconsistent with the crustal structure beneath trenches. The tension hypothesis attributes the trench subsidence to downsinking of a wedge bounded by normal faults;

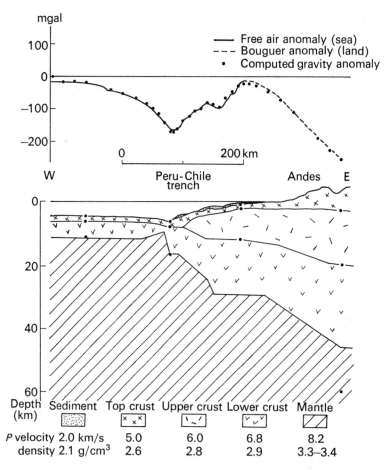

**Fig. 3.16** Crustal structure profile across the Peru-Chile trench near 13°S, interpreted from seismic refraction and gravity observations, as in Figure 3.15. Redrawn from HAYES (1966), *Marine Geol.*, **4**, 332.

this idea has been successful in explaining rift valleys, but it can only produce large subsidence if the wedge of crust is thick. Simple calculations show that it cannot account for subsidence of more than a few hundred metres to oceanic crust, and hence it is ineffective as a mechanism for trench formation.

The most satisfactory explanation of trenches and island arcs is that they form where continental crust is overriding oceanic crust, or oceanic crust is underriding a continent or another piece of oceanic crust. They form as a result of thrust faulting and the depressed block forming the trench is held down against isostatic buoyancy forces by compression in a relatively strong lithosphere. GUNN (1947) showed that the crust would need a strength of about 3 kbar to hold down the deepest trenches by compression acting across thrust faults. This hypothesis fits best with current ideas on ocean-floor spreading. It is further elaborated in Chapter 8.

# 4  The mantle

## 4.1 Introduction

The mantle is the largest of the three major subdivisions of the Earth, forming 84% by volume and 69% by mass. Its upper boundary is the Moho and it is separated from the core by the Gutenberg discontinuity at a depth of about 2900 km.

As mentioned in Chapter 1, Bullen has subdivided the mantle into three zones on the basis of the elastic wave velocity distribution of Jeffreys. These are:

|  |  |  |
|---|---|---|
| zone B | 33–400 km | upper mantle |
| zone C | 400–1000 km | transition zone |
| zone D | 1000–2900 km | lower mantle |

Despite our greatly increased knowledge of the seismic velocities in the mantle, Bullen's subdivisions still provide a good framework for discussing the mantle. The boundaries between the zones are marked by a change in the velocity-depth gradient and their exact depths are not known with precision. In particular, the depth of 33 km for the top of the upper mantle is a formal figure based on continents and is not to be regarded as very meaningful.

Our knowledge of the mantle has been much increased as a result of the 'Upper Mantle Project'. This was initiated at the Helsinki meeting of the International Union of Geodesy and Geophysics in 1960, where it was resolved that particular attention of Earth scientists should be devoted to the outer 1000 km of the Earth. Since then, many important discoveries have been made. They have an important bearing on the role of the mantle in Earth processes, and may provide the key to understanding how the Earth's major surface features have been formed.

This chapter aims to describe the seismological and electrical properties of the mantle and the interpretation of them in terms of the composition. The discussion of the thermal and non-elastic processes which occur within the mantle is deferred to later chapters.

## 4.2 Seismological methods of investigating mantle structure

*Body waves*

The use of body waves to investigate the structure of the mantle leans heavily on the classical Herglotz-Wiechert method of determining the velocity-depth distribution (Chapter 1). This method assumes radial symmetry. Then the travel-time $T$ of a body wave travelling between two points on the surface separated by an angular distance of $\Delta$ (Figure 4.1) is related to the velocity-depth function by the following equation, which is derived by applying the methods of differential geometry to the ray paths (BULLEN, 1963):

$$\Delta = 2p \int_{r_p}^{r_a} r^{-1}(\eta^2 - p^2)^{-\frac{1}{2}} dr,$$

where $v(r)$ = velocity as a function of radius $r$

$r_a$ = radius of Earth's surface,
$r_p$ = radius at depth of greatest penetration of ray,
$\eta = r/v(r)$

and   $p$ = a ray parameter relating to the angle of launching of the ray
(it can be shown that $p = dT/d\Delta$ provided there is radial symmetry).

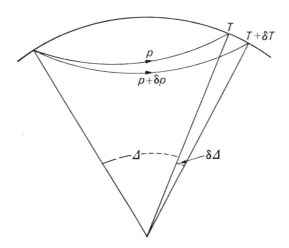

**Fig. 4.1**   Properties of a seismic ray.

This equation determines $T$ as a function of $\Delta$ if the velocity-depth distribution is known. It also enables the variation of amplitude with $\Delta$ resulting from geometrical spreading to be estimated for a given velocity distribution, because it tells us how much $\Delta$ changes for the cone of rays from $p$ to $p + \delta p$—a relatively small change would correspond to a high amplitude and vice versa. In practice, $T$ is known as a function of $\Delta$ and we wish to use the equation to determine the velocity-depth function. Because the unknown function $v(r)$ occurs under the integral sign, this involves solution of an integral equation. It can be shown to reduce to Abel's integral equation and a unique solution can be obtained by numerical methods provided that $dv/dr < v/r$ over the appropriate range of depth. BULLEN (1963) gives the theory of the solution.

This method was used by Jeffreys and Gutenberg to derive the velocity-depth distribution for $P$ and $S$ waves through the mantle. When applied in greater detail, however, there are certain characteristics of the velocity-depth distribution in the mantle which may preclude unambiguous interpretation of the time-distance observations for body waves. The more important difficulties are described below, and methods by which some of them can be overcome are given.

*Firstly*, the Herglotz-Wiechert method fails when the velocity decreases with depth more rapidly than $v/r$. Other methods of investigation (see below) have revealed just such a velocity decrease with depth in the upper mantle for $S$ waves and locally for $P$ waves. Hence this classical method of seismology is of limited value for investigating upper mantle structure. Under this condition the solution of the integral equation is indeterminate for the velocity-depth function in the low velocity zone and below it. A typical time-distance graph showing this is given in Figure 4.2(a). The characteristic feature caused by the low velocity zone is the shadow zone where no rays emerge at the surface. The recognition of a shadow zone points to the existence of a low velocity zone but it does not remove the ambiguity in interpreting the velocity-depth distribution.

   If both the source of the waves and the receiving station are situated above a low velocity layer, all that can be done is to assume a velocity distribution across the layer possibly based on other evidence such as surface waves; the underlying velocity distribution can then be determined uniquely.

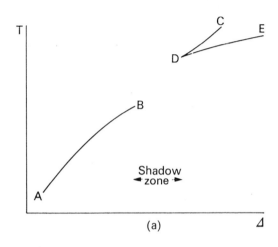

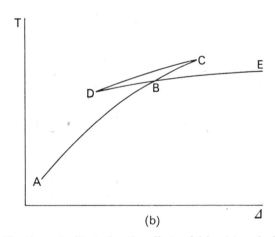

**Fig. 4.2**   Time-distance ($T$—$\Delta$) graphs illustrating the effects of (a) a low velocity zone, and (b) a sudden increase of velocity with depth causing triplication.

   A *second* type of difficulty inherent in the Herglotz-Wiechert method is caused by sudden increases in the velocity gradient with increasing depth. These are known as second order discontinuities. They cause a triplication of the time-distance graph as shown in Figure 4.2(b). If the complete curve including the segments BC, CD and DB is known, then the velocity-depth distribution can be deduced without ambiguity using the Herglotz-Wiechert method; but if the segment ABE alone is known the uniqueness is lost. The complete curve is difficult to observe using conventional seismological stations, because late arrivals closely following the first arrival tend to be masked. The problem of fully defining the time-distance curve in the vicinity of a triplication can be overcome to some extent by making use of seismological array stations (see below).

A *third* difficulty arises from lack of spherical symmetry within the Earth. Allowance can be made for the spheroidal shape of the Earth and for variations in crustal structure. But it is now known that there are considerable lateral variations in $P$ and $S$ velocities within the upper mantle and possibly also at greater depths. It is usually assumed that these lateral variations are averaged out in broad velocity-depth distributions such as those of Jeffreys and Gutenberg.

This difficulty can be turned to advantage. Use can be made of systematic deviations of travel-time from the values predicted by the Jeffreys-Bullen tables to show up regional velocity variations within the mantle. 'Rays' from distant earthquakes emerge steeply through the upper mantle and crust. If the travel-times for a series of distant earthquakes covering a range of azimuths are combined, then any systematic deviation from the predicted arrival times should reflect abnormality in the velocity structure of the underlying crust or upper mantle. Correction can usually be applied for the crust, leaving the contribution from the upper mantle.

*Fourthly* and lastly, observational errors cause uncertainty in the velocity-depth structure deduced from body waves. Important improvement can be gained through use of artificial explosions including nuclear shots, for which the time and exact position of the source are accurately known. Modern instrumentation, including array stations, and the much improved world-wide network has also substantially improved accuracy in the recognition and timing of phases.

The problem of observational errors is much more serious for $S$ than for $P$. It still remains more difficult to recognize the onset of $S$, despite modern improvements. $S$ is also affected by a pronounced low velocity channel in the upper mantle. Consequently investigation of the $S$ velocity-depth distribution in the upper mantle and transition zone rests heavily on the use of surface wave dispersion.

## Phased seismological array stations

An important recent innovation is seismology has been the introduction of phased array stations. These consist of arrays of individual seismometers recording one (or more) components of ground motion, spread over the ground in an appropriate pattern. The output from each individual seismometer is recorded on a separate track of magnetic tape at a central installation, making it possible to apply versatile processing methods using analogue or digital computers, either at the time of recording using on-line computers or at a later date. Phased array stations were originally established to aid the detection of underground nuclear explosions, but they also provide us with an important new seismological tool for investigating the Earth's interior.

The first phased array station, now dismantled, was built by the United Kingdom Atomic Energy Authority (UKAEA) in 1961 at Pole Mountain in Wyoming. The UKAEA seismology group under the leadership of Dr. H. I. S. Thirlaway has now built four permanent array stations at Eskdalemuir in south Scotland, Yellowknife in Canada, Warramunga in Australia and Gauribidanur in India. These stations essentially consist of two lines of short-period vertical seismometers which cross at right angles, as shown in Figure 4.3(a). A particularly large and sophisticated array station incorporating seismometers of different type has been built in Montana, U.S.A. This is known as LASA, standing for Large Aperture Seismic Array, and it incorporates a series of sub-arrays which are deployed as shown in Figure 4.3(b).

The versatility of a phased array station depends on the ability to apply time delays to the recordings

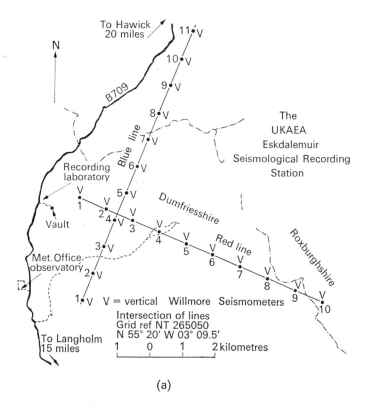

(a)

**Fig. 4.3**(a)   A map showing the geometry of the United Kingdom Atomic Energy Authority phased seismo-logical array station at Eskdalemuir, south Scotland. Redrawn from TRUSCOTT (1964), *Geophys. J. R. astr. Soc.*, **9**, 61.

from individual seismometers before combining their outputs in various ways. This makes it possible to steer the array to search for seismic signals coming from a specified direction. The delays are applied to individual seismometers so that the wave arriving from the specified azimuth and dip direction reinforce each other when their signals are combined, while arrivals from other directions are as far as possible suppressed. By repeating this operation for a series of different directions, it is possible to determine the delays which give the strongest reinforcement, and thus to determine the direction of approach of the signal both in azimuth and in dip. If a fast enough computer is available, this can be done in real-time, i.e. as fast as the event is recorded.

The simplest use of an array station is to improve the clarity of seismic signals by increasing the signal-to-noise ratio. However, the opportunity to steer the array enables the direction of approach of a wave to be determined. As the broad velocity-depth distribution in the Earth is well known, this makes it possible to locate the position of a given event. It also enables us to measure $dT/d\Delta$ directly, because this quantity is given by the velocity of the wavefront across the array (Figure 4.1). Array stations can also be used to separate interfering signals coming from different directions. An example is shown in Figure 4.4, in which LASA has been used to extract the long period record of an Argentine earthquake which was received at the same time as an event from the Kuriles. As will be seen below,

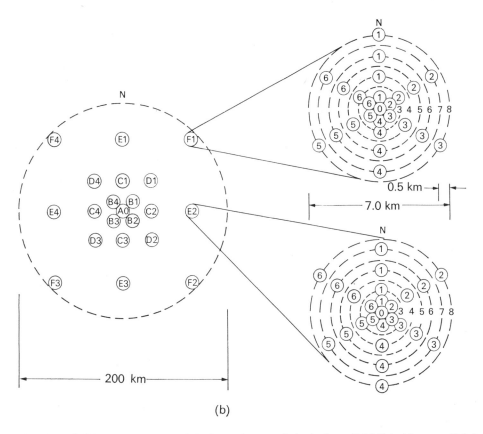

(b)

**Fig. 4.3** (continued) (b)   The geometry of the Large Aperture Seismic Array (LASA) in Montana, U.S.A. This consists of 21 sub-arrays each consisting of 25 short-period vertical seismometers, and a three-component long-period set of seismometers at the centre of each sub-array. Redrawn from CAPON, GREENFIELD and LACOSS (1969), *Geophysics*, **34**, 306. Used with the permission of the Society of Exploration Geophysicists.

results of processing array records have been of considerable importance in unravelling the velocity structure of the mantle.

*Surface waves*

Two types of surface elastic wave can propagate in the presence of a free surface such as the Earth's surface. These are Rayleigh and Love waves, named after the scientists who predicted their existence. Early in the history of instrumental seismology, it was observed that both Rayleigh and Love waves are generated by earthquakes and that the resulting wave-trains are dispersed. The study of the dispersion of long-period surface waves is of fundamental importance in assessing the $S$ velocity structure of the upper mantle.

Surface waves which are sensitive to the structure of the upper mantle have periods ranging from 30 s to 600 s. Modern improvements in the design of long-period seismographs have made it possible to record such long-period surface waves.

Rayleigh waves are the only type of surface wave which can occur in a uniform elastic half-space

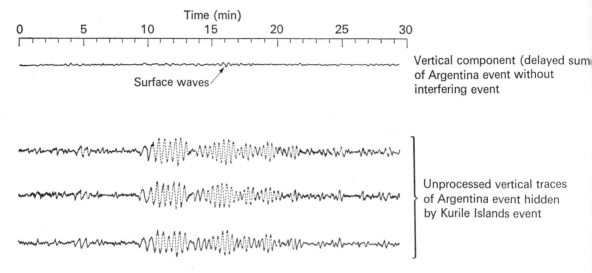

**Fig. 4.4**   Use of LASA to suppress long-period interfering teleseism. Redrawn from CAPON, GREENFIELD and LACOSS (1969), *Geophysics*, **34**, 317. Used with the permission of the Society of Exploration Geophysicists.

(Figure 4.5(a)). Particle displacement is confined to the vertical plane containing the direction of propagation. The amplitude of the displacement decreases with increasing distance from the free surface. For Poisson's ratio of 0·25 the velocity of propagation is $0·92\beta$ for all wavelengths, where $\beta$ is the $S$ wave velocity. The motion of a particle at the free surface is a retrograde ellipse with its major axis vertical, the ratio of the axes being about 1·47.

The Rayleigh wave train becomes 'dispersed' if the elastic moduli and density vary with distance from the free surface. Waves with wavelength $\lambda$ are sensitive to the elastic properties and density over a depth of about $\lambda/5$. Consequently the velocity of propagation depends on wavelength, the longer wavelengths sampling the properties over a greater depth range.

The simplest type of structure which can propagate Love waves is a uniform layer with one free surface and the other surface in contact with a uniform half-space, such that the $S$ wave velocity in the layer $(\beta_1)$ is less than in the half-space $(\beta_2)$ (Figure 4.5(b)). Particle motion is perpendicular to the direction of propagation (as for $S$ waves). Within the half-space the amplitude decreases exponentially with distance from the boundary. Such Love waves are dispersed and the phase velocity varies from $\beta_1$ for the very short wavelengths to $\beta_2$ for the very long wavelengths. They are closely similar to waveguide waves in radar, quantum theory, etc. Unlike Rayleigh waves, Love waves are not affected by the sea. Love waves can occur in more complicated structures provided velocity increases initially with depth; the character of the dispersion curve reflects the layering.

Suppose we have a dispersed train of waves as shown in Figure 4.6, containing a packet of waves having a spread of wavelengths close to $\lambda$ (for simplicity we could consider just two wavelengths, $\lambda + d\lambda$ and $\lambda - d\lambda$). If the velocity depends on wavelength, then the waves travel in a packet as shown, but the individual peaks and troughs travel with a different velocity from the packet itself.

The peaks and troughs travel with the *phase velocity*, which is the velocity with which an unmixed wave would travel. The packet travels with the *group velocity* which represents the velocity with which

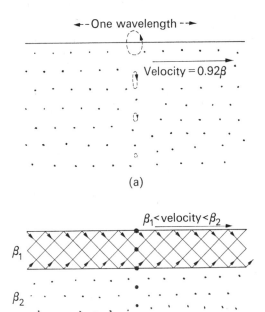

(a)

(b)

**Fig. 4.5**    Sketches to illustrate the propagation of surface waves:
  (a) *Rayleigh waves,* showing how the particle motion changes with depth from the free surface; particle motion is within the plane of the diagram. $S$ velocity is represented by $\beta$.
  (b) *Love waves*, which can be represented by the constructive interference of rays in the upper layer which are repeatedly reflected between the surface and the interface (at supercritical incidence); particle motion is perpendicular to the plane of the diagram.

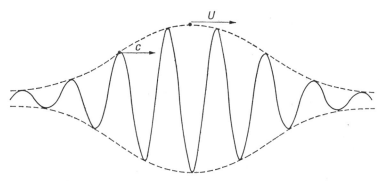

**Fig. 4.6**    A gaussian wave packet (which is dispersed), illustrating the meaning of *phase velocity c* which is the velocity of the individual wave peaks, and *group velocity U* which is the velocity of the envelope of the wave packet.

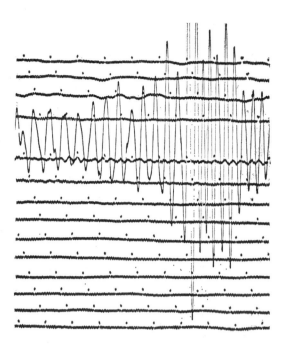

**Fig. 4.7**   East-west component of the long-period record observed at College, Alaska, of an earthquake in the Himalayan region on October 21, 1964. Timing marks on the record are at one-minute intervals. The part of the record shown here gives a good example of a dispersed train of Rayleigh waves. The record is not confused by Love waves because the seismograph is orientated along the direction of propagation of the waves. Note the large amplitude arrivals of about 20 s period towards the end of the train, which are known as the Airy phase. By courtesy of U.S. Coast and Geodetic Survey.

the energy is transmitted. The group velocity $U$ is related to the phase velocity $c$ by the equation
$$U = c - \lambda dc/d\lambda,$$ $\lambda$ being the wavelength.
If the phase velocity dispersion curve is known, the group velocity dispersion curve can be obtained by differentiation. The reverse process requires an integration and introduces an arbitrary constant. An example of a dispersed train of Rayleigh waves is shown in Figure 4.7.

A group velocity dispersion curve can be obtained for an earthquake of known epicentre and time by observing the passage of a dispersed wave-train passing a seismograph station. The group velocity for a given wavelength is $d/t$, where $d$ is the distance along the great circle between the epicentre and the station and $t$ is the time between the event and the passage of the wave. Phase velocity can be determined by using an array of three long-period seismographs and observing the velocity of a peak or trough as it crosses the array. The phase velocity curve can be deduced from the group velocity curve provided assumptions about the motion at source are made.

The method of interpretation is to compare observed phase or group velocity dispersion curves with theoretically computed curves for assumed models of the Earth. Before the computer era, theoretical curves could only be constructed for relatively simple models such as one or two plane layers overlying a uniform half-space. These were adequate for early investigations of crustal structure such as distinguishing oceanic and continental crusts, but not for application to the mantle. A method for computing dispersion curves for multi-layered models using matrix methods to apply the boundary conditions between layers was given by HASKELL (1953). This method is readily applicable to computers. It can be modified to take into account the Earth's curvature. The computer adaptation of Haskell's method enables theoretical dispersion curves to be calculated for a realistic model of the mantle. Most modern studies of surface wave dispersion are based on this or similar methods and their developments.

The Love wave dispersion curve depends on the rigidity modulus and density of each layer, or more conveniently on the $S$ velocity and density. The curve is several times more sensitive to $S$ velocity than to density. Unique interpretation of observed results cannot be obtained unless either density or $S$ velocity is assumed. In dealing with the upper mantle, it is usual to assume the density and make use of the dispersion to derive an $S$ velocity distribution. An acceptable $S$ distribution must satisfy the observed body wave travel-times for $S$, and if agreement is not reached the density assumptions would need to be changed. Rayleigh wave dispersion is dependent on $P$ and $S$ velocities and on density, although the waves which penetrate the upper mantle are most sensitive to $S$ velocity. The very long-period Rayleigh waves and their counterpart, the spheroidal free oscillations, are more strongly dependent on $P$ velocity and density; as the $P$ velocity distribution in the lower mantle is known fairly well, potentially they can give information on the density of the lower mantle and even of the core.

The Rayleigh and Love waves discussed above belong to the fundamental mode, which means that there are no nodal planes at which the displacements are zero. Higher mode surface waves, for which one or more nodal planes occur, are recognizably excited by some earthquakes. They are potentially an important tool for investigating the $S$ velocity structure of the upper mantle.

## Free oscillations

It is known that free natural vibrations of the Earth are excited by earthquakes. They were first convincingly recognized after the Chile earthquake of May 22, 1960. There are two types, the *torsional vibrations* in which the periodic displacement is everywhere perpendicular to the radius vector, and the *spheroidal vibrations* which involve radial and tangential displacement. Each type can be subdivided into an indefinite number of modes, which depend on the disposition of the nodal surfaces. Three types of nodal surface can occur (Figure 4.8): (1) concentric spherical surfaces; (2) systems of concentric cones with apices at the centre; and (3) equally spaced diameters which intersect the surface at two poles. The fundamental modes have a node of type (1) only at the centre.

The observation of free vibrations depends on use of instruments sensitive to ultra-long period oscillations. Strain seismometers, which measure strain in place of ground displacement or velocity, respond to both torsional and spheroidal vibrations. Earth tide gravimeters can detect the spheroidal vibrations. The various modes are recognized on the trace by carrying out a power spectrum analysis. If a given mode is sufficiently strongly excited, it appears as a peak on the power spectrum, the position of the peak giving an estimate of the period. The power spectral densities for strain seismometer records of the Alaskan and Chilean earthquakes are shown in Figure 4.9. The fundamental modes and a few higher overtones are most strongly excited in these records.

The spheroidal vibrations are equivalent to the standing wave set up by equal trains of Rayleigh waves travelling in opposite direction round the Earth. The torsional vibrations are equivalent to interfering Love wave trains. The period of a given mode of free vibration can be re-expressed as the phase velocity of the equivalent surface wave. Thus the observation of the periods of the free oscillations enable the Love and Rayleigh wave phase velocity dispersion curves to be extended from about

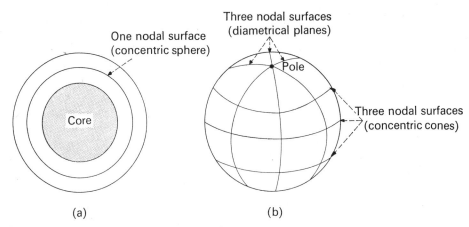

**Fig. 4.8** Diagram to illustrate the three possible types of nodal surface for free vibration of the Earth:

(a) shows a single spherical surface (first overtone) for torsional vibrations, the core-mantle boundary and the free surface being antinodes. Torsional vibrations are restricted to the crust and mantle, but the core participates in spheroidal vibrations.

(b) shows three conical and three diametrical nodal planes appropriate to torsional or spheroidal vibrations. Note that the position of the pole is arbitrary.

The torsional mode incorporating all seven nodal planes shown would be referred to as $_1T_6^3$ and the corresponding spheroidal mode would be $_1S_6^3$. In general, the $_lT_n^m$ and $_lS_n^m$ modes possess l spherical, m diametrical and (n−m) conical nodal surfaces.

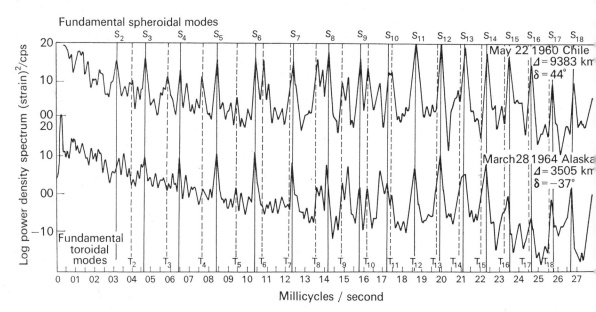

**Fig. 4.9** Power spectral density of the Alaskan and Chilean earthquakes recorded on a strain seismometer at Isabella, California. The angle δ is the deviation of the great circle path from the axis of the strain seismometer. Redrawn from SMITH (1966), *J. Geophys. Res.,* **71**, 1187.

300 s to 2576 s (torsional) and to 3229 s (spheroidal). The longest periods correspond to the second order fundamental modes, and the higher order modes correspond to progressively shorter periods.

## 4.3 The *P* and *S* velocity structure of the mantle

The *P* and *S* velocities are the most accurately known physical properties of the mantle. Over most of the mantle, they are probably known to better than 2%. They provide a firm basis for discussion of the physical properties of the mantle. Let us take as a starting point the velocity-depth curves of Gutenberg and Jeffreys (Chapter 1, Figure 1.5). The main features of these curves have stood the test of time, but modern work has shown up further important detail in the radial distribution. It has also revealed regional variations in the velocity-depth structure of the upper mantle.

The determination of the *P* velocity distribution still depends mainly on body wave studies. In contrast, investigation of the *S* velocity structure now leans heavily on evidence from surface waves and free oscillations. The outcome is: *for P*, it is becoming clear that there is no standard *P* structure for the upper mantle but local detail has been revealed and some new features in the lower mantle have been pinpointed; *for S*, it has also been found that there are broad regional variations in the *S* structure of the upper mantle but some of the fine detail *P* waves have shown is not revealed by surface waves.

### *Radial distribution of P*

Our knowledge of the radial distribution of *P* through the mantle has been substantially improved in recent years by making use of array stations and by studying amplitudes as well as arrival times. A modern assessment of the *P* velocity distribution through the mantle is shown in Fig. 4.10. The salient features are discussed below, starting with the upper mantle and working downwards.

The Jeffreys and Gutenberg distributions for *P* and *S* differ most strikingly in the upper mantle. Jeffreys shows a relatively gentle increase in velocity with depth down to 400 km, where the velocity

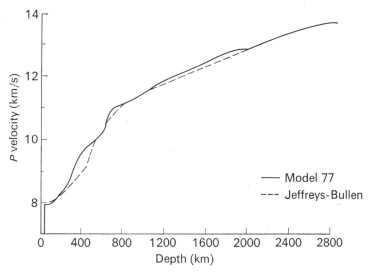

**Fig. 4.10**  The model 77 compressional velocity profile in the mantle, showing the Jeffreys-Bullen profile for comparison. Redrawn from TOKSÖZ, CHINNERY and ANDERSON (1967), *Geophys. J. R. astr. Soc.,* **13**, 44

gradient increases abruptly. In contrast, Gutenberg shows a decrease in velocity with depth beneath the Moho, reaching a minimum at about 100 km depth; beneath the low velocity channel, the velocity increases rapidly with depth starting at about 200 km. Our first problem is to find whether the low velocity channel for *P* waves does really exist.

Gutenberg suggested the existence of a low velocity zone in the upper mantle by making use of amplitudes of *P* waves from shallow focus earthquakes. He found that the amplitude fell, with increasing distance, more rapidly than would be expected up to about 15° (1600 km) from the epicentre. Beyond this distance, the amplitude increases sharply by a factor greater than ten. Gutenberg attributed this shadow zone effect to a low velocity channel for *P* in the upper mantle (Figure 4.2(b)). Although the amplitude studies show up the existence of the low velocity channel, the distribution of *P* across the channel still cannot be uniquely determined.

One method of overcoming the low velocity layer problem, applicable only to a restricted number of regions, is to make use of intermediate and deep focus earthquakes to give direct estimates of the velocity at the focus. The method was introduced by GUTENBERG (1953). The principle is to plot the time-distance curve for a range of angular distances from the epicentre and to pick out the minimum value of $dT/d\Delta$ and its distance $x$ from the epicentre from the curve. Gutenberg showed that the velocity at the focus is given by

$$V_f = x(dT/d\Delta)_{min}$$

Satisfactory use of the method requires accurate knowledge of the position of the focus. The method needs to be used with caution because it can lead to erroneous results for foci near the top of the low velocity channel. Gutenberg used this method to show that *P* has a minimum velocity at about 100 km depth and *S* has one between 100 and 200 km depth. More recently, BROOKS (1962) applied the method to the New Guinea/Solomon Islands region, where he found that the *P* velocity falls from 7·9 km/s at 25 km depth to a minimum of 7·5 km/s at 115 km. These conclusions must be treated with reserve, because it is now known that there are large lateral variations in seismic properties beneath island arc-trench regions (OLIVER and ISACKS, 1967; DAVIES and MCKENZIE, 1969).

Body wave studies of upper mantle structure suffer from the snags that only a relatively small part of the mantle is sampled in each study, and that investigations are mostly restricted to the vicinity of earthquake belts. Gutenberg based his argument for a low velocity *P* channel on investigations near active tectonic zones such as the circum-Pacific belt and western Europe. Despite this, in 1959 his opinion was that the *P* low velocity channel is universally present in the upper mantle. Since 1959, new evidence has been obtained which suggests that this may be incorrect.

Study of *P* waves resulting from nuclear explosions in western United States has provided some important new evidence. Most underground nuclear explosions are detonated at the Nevada Test Site, but one large shot called GNOME was set off in a salt dome in New Mexico. MISS LEHMANN (1964) made a study of the travel-time for *P* from GNOME, as recorded in some detail along lines extending up to 2500 km to the north-west and north-east (Figure 4.11); the north-western line is effectively reversed for 700 km since it crosses the Nevada Test Site. In each direction, $P_n$ lies on straight line segments up to beyond 1000 km (beyond 1750 km on the north-eastern line). This shows that the *P* velocity does not increase significantly with depth for some distance beneath the Moho on either line. To the *north-west*, the first *P* arrival which has penetrated deeper into the upper mantle is observed just beyond 1000 km; it is weak and substantially delayed relative to $P_n$. The only reasonable interpretation of this reversed line is that a low velocity channel for *P* must occur in the upper mantle beneath the line. To the *north-east*, *P* is first observed at 1400 km, and both $P_n$ and *P* arrive earlier than at a corresponding distance to the north-west. Miss Lehmann's interpretation is that no low velocity channel occurs at shallow depth, but that a shallow channel may occur between 150 and 215 km depth. Beneath the Canadian Shield, BRUNE and DORMAN (1963) find no evidence for any low velocity *P* zone although a slight decrease in velocity with depth is not ruled out.

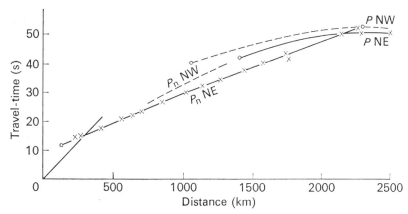

**Fig. 4.11**    Travel-times of *P* waves to the NE and NW of the GNOME nuclear explosion. Redrawn from LEHMANN (1967), *The Earth's mantle*, p. 55, Academic Press.

MISS LEHMANN (1962) has suggested that there is a discontinuous but small increase in velocity at a depth of about 220 km beneath Europe and North America; this would form the lower boundary of the low velocity channel where it exists. Apparent reflections from discontinuities within the upper mantle have also been recorded. For instance HOFFMAN, BERG and COOK (1961) found evidence for reflections at 190, 210, 520, 555 and 910 km depths, although further work is needed if weight is to be placed on the interpretation of such reflections.

Figure 4.10 shows a low velocity channel in the upper mantle. Its existence is undisputed beneath some regions, but it is doubtful whether it is of world-wide extent.

The *transition zone* is marked by a rapid increase in *P* and *S* velocity with depth, contrasting with the relatively small velocity gradients in the upper mantle. According to Jeffreys it starts with a second order discontinuity at 400 km depth known as the 20° discontinuity; according to Gutenberg, a smooth increase in velocity gradient beneath the low velocity channel starts at about 200 km depth. Both these interpretations place the lower limit of the transition zone at about 1000 km depth.

New insight into the fine structure of *P* through the transition zone has come from careful studies of recordings at distances of about 10° to 30° from shallow earthquakes and nuclear events. It had been found from surface wave studies that the steep rise in *S* through the transition zone takes place in two steps (see below) and a corresponding structure has subsequently been found for *P* waves. NIAZI and ANDERSON (1965), for instance, used the Tonto Forest Array Station in Arizona to determine directly the apparent velocity across the array $(dT/d\Delta)$ for 70 earthquakes with epicentres between 10° and 30° from the station. They found that the plot of $dT/d\Delta$ against distance $\Delta$ (Figure 4.12) is best fitted by three segments. By assuming the Gutenberg *P* velocities down to 225 km and by applying the Herglotz-Wiechert method below this depth, they were able to invert the observations to give a velocity depth distribution down to 800 km (known as model Z). They found that the rapid increase in *P* with depth occurs in two steps, starting at about 320 km and 640 km respectively.

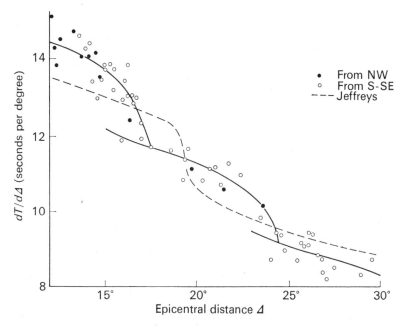

**Fig 4.12**   $dT/d\Delta$ as a function of epicentral distance $\Delta$ determined for earthquakes up to 30° distance at the Tonto Forest Array Station, Arizona. The observations are compared with the theoretical curves for the Jeffreys-Bullen $P$ velocity distribution and for the model Z obtained by direct inversion of the data. Redrawn from NIAZI and ANDERSON (1965), *J. geophys. Res.,* **70,** 4636.

The existence of a double step in the transition zone beneath the United States is also indicated by observed triplications in the time-distance curve for $P$ (as in Figure 4.2(b)). ANDERSON (1967a) has reviewed such investigations and similar work done in Asia, Japan, Australia and the Pacific. He found that there is widespread evidence for the existence of the double 'discontinuity'. Each step causes a strong bending of the seismic rays towards the Earth's surface, as shown by the ray paths computed for the surface wave model CIT11GB of the upper mantle and transition zone (Figure 4.13). In the model, ray concentrations indicating strong amplitudes emerge at 15° and 20°, and there are corresponding shadow zones and triplications of the time-distance graph. This is the sort of evidence one looks out for. Anderson concluded that 'The data is all in substantial support of two discontinuities; one between 300–400 km, and one between 600–700 km'.

In the *lower mantle,* the $P$ and $S$ velocity profiles of Jeffreys and Gutenberg are closely similar to each other. Beneath 1000 km depth the velocities rise gently with depth until the bottom 200 km where the gradient decreases. The profiles show only slight irregularities.

Some recent investigations, however, have shown up some significant irregularities in the $P$ velocity profile of the lower mantle. CHINNERY and TOKSÖZ (1967) have made use of $dT/d\Delta$ observations obtained by the Large Aperture Seismic Array (LASA) in Montana, supplemented by travel-times from the Longshot nuclear explosion, to show that there are two further 'discontinuities' below 700 km. At 1200 km depth they find a local steepening of the gradient, and at 1900 km depth there is a local decrease in gradient possibly associated with a shallow low velocity zone. These features are shown in Figure 4.10. CARPENTER, MARSHALL and DOUGLAS (1967) have constructed an amplitude-distance curve for $P$ between 30° and 102° using nuclear explosions only. Their curve shows a sharp

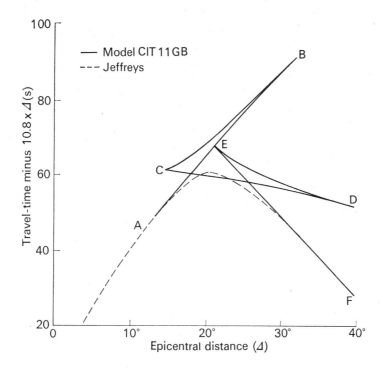

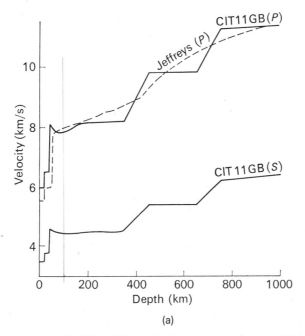

(a)

**Fig. 4.13**(a)   The CIT11GB model of *P* and *S* velocity in the mantle down to 1000 km depth, and the corresponding time-distance curve for *P* shown above. The Jeffreys *P* distribution is shown for comparison. Redrawn from JULIAN and ANDERSON (1968), *Bull. seism. Soc. Am.,* **58**, 341 and 342.

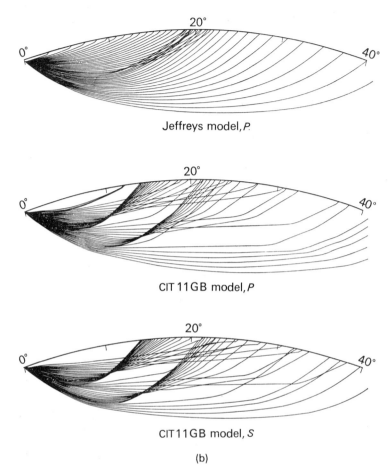

**Fig. 4.13** (continued) (b)    Ray paths for the Jeffreys and CIT11GB models. Redrawn from JULIAN and ANDERSON (1968), *Bull. seism. Soc. Am.,* **58**, 348 and 351.

peak between 33° and 36°, a minimum followed by a peak at 75°, and a sharp minimum between 93° and 96°. These represent local irregularities within the lower mantle. A further step in providing the tools to comb the finer structure of the lower mantle has been taken by DOUGLAS and CORBISHLEY (1968), who have introduced a method of using all four UKAEA array stations (in Scotland, Canada, India and Australia) to remove station corrections and obtain accurate estimates of $dT/d\Delta$. Their preliminary results for $dT/d\Delta$ between 67° and 82° seem to confirm the decrease in gradient at about 1900 km depth which was postulated by Chinnery and Toksöz.

*Radial distribution of S*

Gutenberg's *S* distribution incorporates a low velocity zone in the upper mantle. It starts just beneath the Moho and extends down to about 200 km. He discovered the channel from body wave studies on *S* from shallow earthquakes. He concluded that the *S* channel is more pronounced than the corresponding feature for *P*, because *S* waves penetrating into the upper mantle were more strongly delayed than *P*.

The existence of a universal low velocity channel for *S* in the upper mantle has been amply confirmed by surface wave studies. In contrast to body waves, these can sample the average structure of the upper mantle and transition zone relatively easily. DORMAN, EWING and OLIVER (1960) were first to apply the method to investigate the average global structure of the upper mantle. They used Rayleigh waves up to 250 s period and found that a universal low velocity channel for *S* in the upper mantle was needed to explain the observed group velocity dispersion curve.

Subsequent to this pioneer work, the long period end of the spectrum has been extended by using improved instruments and by incorporating free oscillation data. This has led to the important discovery that the steep rise in *S* velocity through the transition zone takes place in two steps, the upper one between about 350 and 500 km depth and the lower one between about 650 and 750 km depth (ANDERSON and TOKSÖZ, 1963).

Recent models of *S* velocity distribution in the mantle incorporate a low velocity zone in the upper mantle and two regions of steep velocity gradient in the transition zone. Published models do vary somewhat in the search for a perfect fit to Love and Rayleigh wave dispersion curves, free oscillation data and body wave time-distance observations for *S*. A completely unique solution is not possible since assumptions need to be made. But the main features described above are becoming firmly established, and we can expect the models to become progressively nearer to the actual distribution as observations increase and improve.

One recent model is known as CIT-12 (TOKSÖZ, CHINNERY and ANDERSON, 1967). It is based on surface wave dispersion along paths which are 60% beneath oceans and 40% beneath continents, and represents an average structure. The two assumptions which were made are: (1) if the *S* velocity is

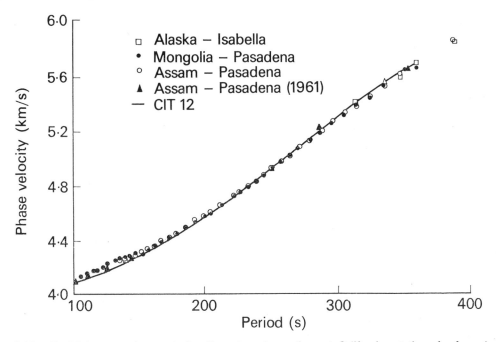

**Fig. 4.14**   Rayleigh wave phase velocity dispersion observations at Californian stations for four closely similar paths, compared with the theoretical dispersion curve for model CIT-12 (Figure 4.16). Redrawn from TOKSÖZ, CHINNERY and ANDERSON (1967), *Geophys. J. R. astr. Soc.*, **13**, 34.

known, *P* can be calculated from it using GUTENBERG'S (1959) estimates of Poisson's ratio; and (2) density can be computed from the *P* velocity using the relation of BIRCH (1964). Thus *S* is the only independent variable. CIT-12 has been obtained by searching for a structure which satisfies Rayleigh and Love wave dispersion and free oscillation data, as shown in Figures 4.14 and 4.15. The model itself is shown in Figure 4.16. At the time of writing, it represents one of the best estimates of *S*

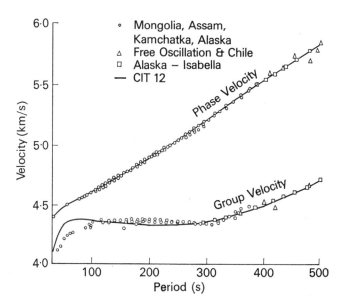

**Fig. 4.15**   Love wave dispersion observations for four close-lying great circle paths, compared with the theoretical curves for model CIT-12. Redrawn from TOKSÖZ, CHINNERY and ANDERSON (1967), *Geophys. J. R. astr. Soc.,* **13,** 34.

structure down to about 1000 km. Beneath this depth, we still have to rely on the Gutenberg and Jeffreys distributions, although improvements are to be expected in the future.

## *Regional variations in P*

About fifteen years ago hardly anyone suspected that the radial velocity structure of the mantle varied from region to region. It is now known that there are substantial lateral variations in the *P* and *S* structure of the upper mantle. It is also suspected that regional variations occur within the lower mantle, although these are necessarily more difficult to detect.

Crustal seismic refraction studies on land and at sea have shown up large variations just beneath the Moho, ranging from about 7·3 to 8·6 km/s. Spectacular variations in $P_n$ even occur from one structural province to another within a continent or an ocean. For instance, $P_n$ is substantially lower beneath the western United States than beneath the central and eastern parts of the country; within western United States it varies from one structural province to another (PAKISER, 1963).

The variation of $P_n$ beneath U.S.A. is sufficiently well known to produce a map of it. This was done by HERRIN and TAGGART (1962) who used the extensive recordings of $P_n$ for the GNOME nuclear

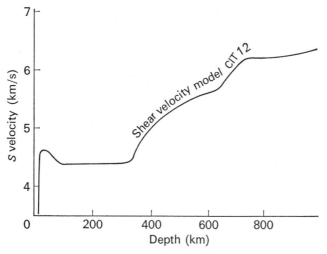

**Fig. 4.16**   The shear velocity distribution for the CIT-12 mixed path model of upper mantle structure. Redrawn from TOKSÖZ, CHINNERY and ANDERSON (1967), *Geophys. J. R. astr. Soc.,* **13**, 35.

explosion in New Mexico to determine the interval velocity from one station to another. Their results agree well with refraction determinations of $P_n$. A revision of their earlier map is shown in Figure 4.17. Similar variations in $P_n$ are known to occur elsewhere but they have not been investigated in such detail.

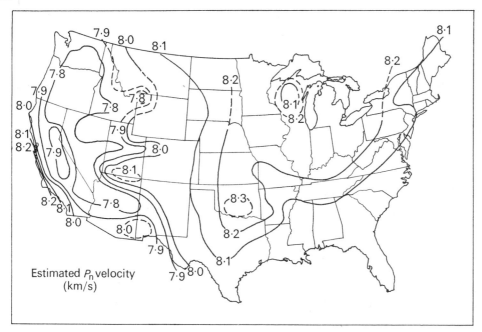

**Fig. 4.17**   Estimated $P_n$ velocity beneath the United States based on a flat earth approximation. After TUCKER, HERRIN and FREEDMAN (1968), *Bull. seism. Soc. Am.,* **58**, 1256.

To give an idea of the sort of variations in $P_n$ which occur, estimates for some regions are given below:

|  | $P$ velocity km/s |  |
|---|---|---|
| western United States | 7·8–8·0 | (HERRIN and TAGGART, 1962) |
| eastern United States | 8·0–8·3 | (HERRIN and TAGGART, 1962) |
| Great Britain | 8·0 | (AGGER and CARPENTER, 1964) |
| western Alps | 8·1–8·2 | (CLOSS and LABROUSTE, 1963) |
| Finland | 8·2 | (PENTTILA, 1969) |
| Japan | 7·5–7·9 | (JAMES and STEINHART, 1966) |
| ocean basins | 8·0–8·5 |  |
| ocean ridges | 7·3–8·2 |  |
| Iceland | 7·4 | (TRYGGVASON, 1964) |

Typical values for the normal continental crust and for ocean basins are 8·0–8·2 km/s; similar values are found beneath fold mountain ranges and ocean trenches. Low values are associated with ocean ridges, regions of plateau uplift such as western U.S.A. and volcanic regions such as Japan.

It also comes within the scope of the seismic refraction method to determine whether $P_n$ depends on direction, i.e. if the topmost mantle is anisotropic. If orientated crystals of olivine occur, then one would expect anisotropy because the compression wave velocity in olivine depends on direction; the lowest value occurs parallel to the $\beta$-crystallographic axis which typically occurs perpendicular to banding in olivine rich rocks. HESS (1964) studied refraction profiles in the Pacific made by Raitt and Shor to see if anistropy could be detected. Raitt's profiles were in the vicinity of the Mendocino fracture zone off the west coast of U.S.A., and here $P_n$ varies from about 8·0 km/s in a north-south direction to 8·6 km/s in an east-west direction. The north-south direction is perpendicular to the fracture zone and parallel to the magnetic strip anomalies (p. 79). More recent measurements by RAITT, SHOR, FRANCIS and MORRIS (1969) confirm this anisotropy but show that the velocity difference at any single station is about 0·3 km/s.

Refraction measurements of $P_n$ are easy to perform, but they give no information on the velocity structure beneath the topmost few kilometres of the mantle. It is much more difficult to determine how the velocity-depth structure through the upper mantle varies from region to region. One method which has proved successful is to determine differences in travel-time and amplitude of $P$ from earthquakes and nuclear explosions up to about 30°. This is exactly the same approach as has been used to determine average radial $P$ velocity distributions through the upper mantle, except that emphasis is placed on regional differences. Various investigations using nuclear explosions in the United States, such as Miss Lehmann's study of $P$ from GNOME described above, have shown that the whole $P$ structure of the upper mantle differs strikingly between western and eastern U.S.A. The $P$ low velocity zone is pronounced in the west where it starts just beneath the Moho; in the central and eastern parts it is much less conspicuous and is deeper where it occurs; it may hardly exist at all beneath the Canadian Shield. The average value of $P$ down to 300 or 400 km is much higher in the east than in the west.

Another method of determining regional variations in the mean value of $P$ through the upper mantle is to make use of anomalies in the arrival time of $P$ from distant earthquakes, using the method introduced on p. 97. To give an example, using the travel-times predicted by the Jeffreys-Bullen tables as a standard, CLEARY and HALES (1966) have shown that anomalies in $P$ arrival time for the United States range from − 1 to + 1 seconds. The arrivals are early in the central and eastern parts and late in the western part, as would be expected. Such large residuals can only arise from large variations in the average *P* velocity in the upper mantle beneath. If the velocity anomalies are spread over a vertical range of 400 km, then the average *P* velocity over this range must differ by about 0·4 km/s between west and east to account for the 2 second differences. This is a substantial difference.

TRYGGVASON (1964) has used this same method to compare the upper mantle beneath Iceland and Sweden. *P* arrivals reaching Reykjavik from distant earthquakes, after applying a crustal correction, are delayed by an average of 1·3 seconds relative to the Jeffreys-Bullen tables. Arrivals at Kiruna are systematically early by 1·4 seconds. To explain the delay of 2·7 seconds at Reykjavik relative to Kiruna, the upper mantle beneath Iceland needs a velocity as low as 7·4 km/s extending from the Moho down to 200 km depth (Figure 4.18). Whatever the details, this must represent a major anomalous structure in the upper mantle.

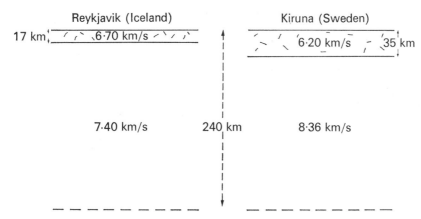

**Fig. 4.18**    Differences in upper mantle *P* velocity structure beneath Reykjavik and Kiruna which would explain the differences in *P* velocity arrival times, relative to the Jeffreys-Bullen values, at these two stations.

Thus important variations in the *P* velocity structure at the top of the mantle and throughout the upper mantle do occur. A large amount of systematic study will be needed to reveal the worldwide pattern of such variations and to extend investigations through the transition zone into the lower mantle.

## Regional variations in S

Regional variations in the *S* velocity structure of the upper mantle were originally discovered by DORMAN, EWING and OLIVER (1960) using Rayleigh wave dispersion. Using wave trains which had taken continental paths, they found the observed dispersion to be in broad agreement with models based on *S* wave studies given by Gutenberg and Lehmann, with a prominent low velocity channel between depths of 100 and 200 km. For oceanic paths, the dispersion curves require the upper surface of the low velocity zone to be shallower. The oceanic dispersion is well simulated by their model 8099, which has an *S* velocity of 4·6 km/s at the top of the mantle dropping to a minimum of 4·3 km/s between depths of 75 and 200 km. The *S* structures beneath oceans and continents appears to converge below about 300 km depth.

AKI and PRESS (1961) came to similar conclusions about the broad differences between sub-continental and sub-oceanic upper mantles using surface wave dispersion. They suggested that the differences in dispersion curves could be explained either by a shallower top to the oceanic low velocity zone or by lower velocities within the channel.

Using Rayleigh and Love wave dispersion in conjunction with *S* wave travel-time information, BRUNE and DORMAN (1963) found abnormally high *S* velocities in the crust and mantle beneath the Canadian Shield. The mantle down to 115 km depth had an *S* velocity of 4·72 km/s. Beneath this, they found a shallow low velocity channel with *S* = 4·5 km/s down to 315 km, below which the velocities converged with the Gutenberg curve.

Making use of improved instrumentation, Anderson and his co-workers at the California Institute of Technology have studied the dispersion of Love and Rayleigh wave-trains which have passed round the Earth more than once. They used periods ranging from 80 to 666 seconds. A wide range of great-circle paths have been used, and this has enabled the dispersion curves to be separated for oceans, shields and continental tectonic regions. The resulting interpretation of *S* velocity structure down to 500 km depth is shown in Figure 4.19. The range of periods used only permits interpretation to 500 km depth, and differences beneath this depth are caused by forcing the dispersion curves

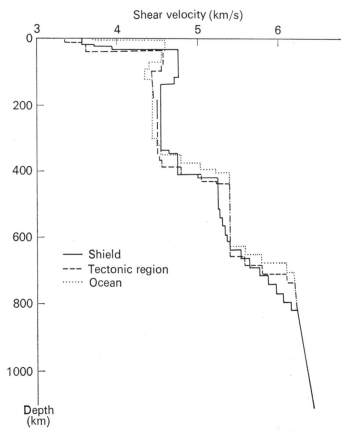

**Fig. 4.19**　Upper mantle shear velocity models for oceanic, continental shield and tectonic regions, based on dispersion of surface waves. The profiles are uncertain below 500 km depth because of insufficient data. Redrawn from TOKSÖZ, CHINNERY and ANDERSON (1967), *Geophys. J. R. astr. Soc.,* **13**, 48.

together at 500 second period. This does not, of course, rule out possible lateral variation in $S$ structure beneath this depth, but such variation cannot be detected by this study.

These results of TOKSÖZ and ANDERSON (1966) confirm the earlier results for the difference between the upper mantle beneath oceans and continents. They also bring out the existence of substantial differences in $S$ velocity structure beneath shields and tectonic regions of continents.

Anomalies in the arrival times of $S$ from distant earthquakes, taking the Jeffreys-Bullen times as standard, have been used by HALES and DOYLE (1967) to investigate regional variation of $S$ structure in the upper mantle beneath the United States. The anomalies in $S$ are much larger than those in $P$, ranging from about $-4$ to $+3$ seconds; early arrivals occur in the centre and east, and late arrivals in the west, as for $P$. The observed ratio of $S$ delay to $P$ delay is $3 \cdot 72 \pm 0 \cdot 43$. To explain the 7 second difference, very large variations in the average $S$ velocity structure of the upper mantle are needed.

## 4.4 Electrical conductivity of the mantle

The electrical conductivity distribution down to about 1000 km depth can be investigated by studying the short-period variations of the Earth's magnetic field ranging from a few seconds to several years. The spectrum of the secular variation (p. 164) gives some indication of the conductivity of the lower mantle.

The variations in the Earth's magnetic field with periods of less than a year originate outside the solid Earth, but they include a secondary internal component caused by induced currents flowing in crust and mantle. The currents are known as telluric currents. The most conspicuous short-period variation is the diurnal variation. A typical record of diurnal (or daily) variation is shown in Figure 4.20(a). The same pattern tends to be repeated with some variation from day to day. If a record covering a period of many days is subjected to spectral analysis, it is found that the dominant period is one day and that there are conspicuous harmonics of period 12 hours, 8 hours and so on, as shown in Figure 4.21.

The diurnal variation is believed to be caused by the interaction of the conducting ionospheric layers with the main magnetic field (p. 161). The ionospheric layers move upwards and downwards in response to solar and lunar tidal forces; this causes a varying pattern of horizontal current loops in the ionosphere as the conducting layers cut the magnetic lines of force. If harmonic analysis of the diurnal variation is done over a long enough time, it is possible to separate the lunar and the solar components of the variation.

Another type of short-period magnetic variation is the *magnetic storm*. It is caused by strong current systems in the ionosphere, produced by streams of particles emitted by solar flares. The record of a magnetic storm is shown in Figure 4.20(b), where it can be seen that there are harmonic components with periods of up to a few days. Other long-period variations are (1) a 27 day period and its harmonics which may be related to tides caused by the Moon's orbital motion, and (2) an annual period. Conspicuous short-period variations include 'bays' of about one hour's period and micropulsations which last a few seconds. The whole spectrum of external magnetic events extending from a few seconds to a year or more is potentially useful for probing the conductivity of the Earth down to about 1000–1500 km depth. There is no need to know the cause of the various types of variation to use them for this purpose (BULLARD, 1967).

The magnetic variations can be used to estimate the conductivity within the Earth because the strength of the induced currents depends on the electrical conductivity distribution. The depth to which the telluric currents penetrate depends on the period of the variation—short-period

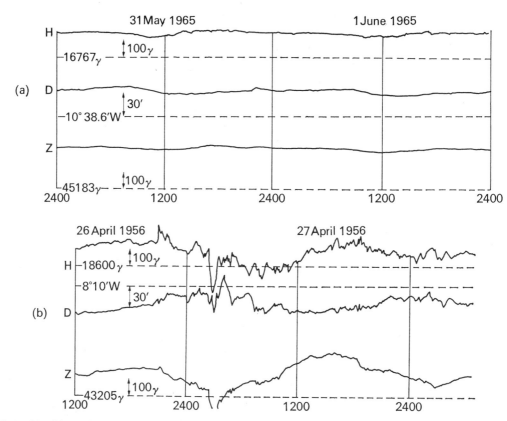

**Fig. 4.20** Magnetic observatory records showing short-period variations in horizontal field (*H*), declination (*D*) and vertical field (*Z*). (a) is typical quiet day variation at Eskdalemuir observatory, south Scotland, and (b) is a magnetic storm at Abinger. (a) redrawn from the original record, and (b) redrawn from BULLARD (1967), *Q. Jl R. astr. Soc.*, **8**, 149.

variations only penetrate to shallow depths and longer periods penetrate deeper. The method depends critically on the possibility of separating the external and internal parts of the field by spherical harmonic analysis. By determining the relative amplitudes and phase difference of the external and internal components of the variations, it is possible to estimate the conductivity down to the depth of penetration by the currents. By carrying out such an analysis for a range of different period variations, a crude conductivity-depth distribution can be built up down to the depth penetrated by the longest period variation available for analysis.

*Radial distribution of electrical conductivity*

The electrical conductivity of sea-water is about 4 ohm$^{-1}$ m$^{-1}$* and for water saturated sedimentary rocks it ranges between $10^{-3}$ and 1 ohm$^{-1}$ m$^{-1}$. The conductivity of the dry crustal rocks beneath the sediments is about $10^{-6} - 10^{-3}$ ohm$^{-1}$ m$^{-1}$. The outermost layers of the Earth are poorly conducting apart from the more highly conducting skin formed by the oceans and sediments.

* Conductivity is the reciprocal of specific resistance (resistivity); 1 ohm$^{-1}$ m$^{-1}$ = $10^{-2}$ ohm$^{-1}$ cm$^{-1}$ = $10^{-11}$ e.m.u.

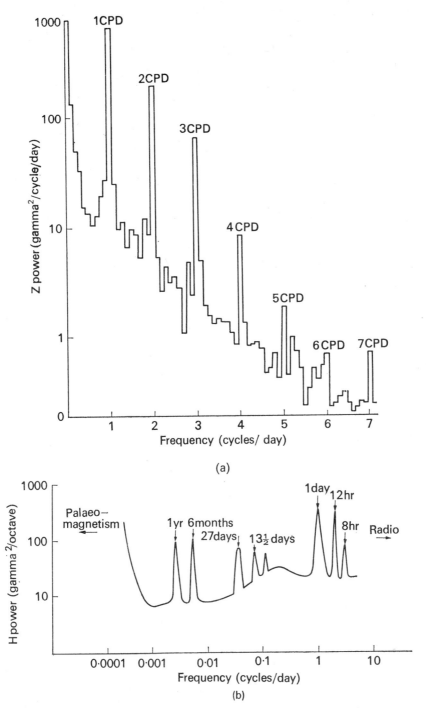

**Fig. 4.21**   The spectrum of the short-period variations of the geomagnetic field.
(a) Power spectrum of the vertical field at Tucson (September to December 1957) showing the daily
variation and its harmonics;
(b) Power spectrum of the horizontal component at British Observatories (to be regarded as a generalized
diagram).
Redrawn from BULLARD (1967), *Q. Jl R. astr. Soc.*, **8**, 144 and 146.

The broad radial distribution of electrical conductivity down to about 1000 km is investigated by the following procedure. It is described for the daily variation but equally well applies to other short period variations. The average daily variation for quiet days is obtained for the vertical and a horizontal component of magnetic field from magnetic observatories over the world. The records from each observatory are Fourier analysed into simple harmonic variations of 24, 12, 8 and 6 hour periods. A spherical harmonic analysis is then carried out on the worldwide spread of data for each time period. This makes it possible to separate the internal and external parts of the variation, giving their amplitude ratio and the phase difference between them. CHAPMAN (1919) was the first person to make this analysis. For the 24 hour period variation he found the external to internal amplitude ratio to be 2·8 and the phase difference to be $-13°$. Similar results were obtained by Chapman for the higher harmonics of the daily variation, and by later workers (LAHIRI and PRICE, 1939; BANKS and BULLARD, 1966; BANKS, 1969) for the longer period variations such as those associated with magnetic storms. This is the raw material for interpreting the conductivity distribution.

Chapman interpreted his results for the daily variation in terms of a uniformly conducting 'core' surrounded by a non-conducting shell. He estimated that the shell is 250 km thick and that the conductivity of the core is $3·6 \times 10^{-2}$ ohm$^{-1}$ m$^{-1}$. This model fails to satisfy the analysis of the longer period variations associated with magnetic storms. A series of models constructed to explain both the daily and storm time variations were proposed by Lahiri and Price. Within the 'core' the conductivity was taken to vary as a power of the radius, and a thin conducting surface layer representing the oceans and sediments was also incorporated. These models show that the conductivity increases at about 600 km depth to at least 1 ohm$^{-1}$ m$^{-1}$. More recently, Banks and Bullard found that the annual variation yields an estimate of 2 ohm$^{-1}$ m$^{-1}$ at 1300 km depth.

The relatively short-period magnetic variations of external origin do not penetrate much below 1000 km. A rough indication of the order of magnitude of the conductivity in the lower mantle can be obtained by using the spectrum of the secular variation. The secular variation is thought to involve variations in the magnetic field at the core-mantle interface. The shortest period variation which is observed at the Earth's surface is four years. If it is assumed that shorter period variations do occur but are prevented from penetrating to the surface by the relatively highly conducting lower mantle then the conductivity can be very roughly estimated. A further estimate can be obtained from the small changes in length of day if these are assumed to result from electromagnetic coupling between core and mantle; such coupling places a lower limit on the conductivity of the lowermost mantle. These and similar methods suggest that the electrical conductivity of the lower mantle lies between about 10 and 1000 ohm$^{-1}$ m$^{-1}$.

A somewhat different approach has been used successfully to estimate the electrical conductivity between depths of about 10 and 150 km. This is the magneto-telluric method of CAGNIARD (1953). The variation of one (or both) components of the horizontal magnetic field is recorded continuously at a station. At the same station, the electric field component perpendicular to this is measured by placing probes into the ground and recording variations in the e.m.f. between them. The variation in electric field is caused by the induced currents. Direct measurement of the electric field makes it possible to separate the internal and external parts of the horizontal magnetic component, provided it is assumed that the field variations are uniform over a much wider extent than the depth penetrated by the induced currents, and that the underlying conductivity distribution is a function of depth only. From the records, the ratios of the amplitudes of the electric to magnetic fields are determined for variations ranging from about 20 second to 1000 second periods. These ratios can then be interpreted in terms of a conductivity-depth profile between depths of about 10 and 100 km.

To give an example, magneto-telluric measurements at Meanook, Alberta, show that the conductivity falls from about $10^{-2}$ ohm$^{-1}$ m$^{-1}$ at 15 km depth to about $10^{-3}$ between 50 and 75 km depths (NIBLETT and SAYN-WITTGENSTEIN, 1960). There is indication that the conductivity increases sharply to $10^{-2}$ at 80 km depth. Magneto-telluric measurements made in Massachusetts also indicate an increase in conductivity to about $10^{-2}$ ohm$^{-1}$ m$^{-1}$ at about 70 km depth (CANTWELL and MADDEN, 1960). PRICE (1967) has remarked that if there is a widespread layer of high conductivity starting at about 75 km depth, then other observations suggest that the conductivity must decrease again before the major rise starts at about 400 km depth.

An interpretation of the radial conductivity profile down to the base of the mantle is shown in Figure 4.22. This is mainly based on the curve of McDONALD (1957), who used the secular variation

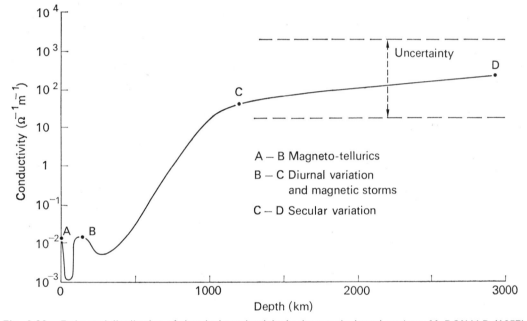

**Fig. 4.22**   Estimated distribution of electrical conductivity in the mantle, based partly on McDONALD (1957).

for the lower mantle and adopted a compromise between two of Lahiri's and Price's models for the upper mantle. The rise in conductivity at about 75 km depth suggested by magneto-telluric measure- ments and the decrease beneath this layer suggested by Price are also shown in Figure 4.22.

*Regional variations in conductivity of the upper mantle*

It has been known for some time that there are some local anomalies in the character of short-period magnetic variations which must be caused by lateral variation in electrical conductivity within the top 100 km of the Earth. Some of these anomalies have been detected through nearby magnetic observatories. They can be investigated by using mobile three-component magnetometer stations or by using magneto-telluric measurements.

One well known type of magnetic variation anomaly occurs near the margins of oceans. Here there are large lateral variations in conductivity down to 5 km depth because sea-water is more than 10 000 times better a conductor than crustal rocks. Because of this, strong fluctuating telluric currents flow in the oceans in response to the short-period magnetic variations. Near the edge of an ocean, the strongest currents tend to flow parallel to the coast. The magnetic field vector associated with an electric current is perpendicular to the plane containing the current and the point of observation. Because the depth of oceans is relatively small compared with the width of most continental shelves, the magnetic vector on the adjacent continent associated with the oceanic telluric currents would be expected to be nearly vertical. Such strong variations in the vertical component are indeed observed within 50–100 km of the coasts.

The high conductivity of sea-water does not completely explain the anomalous magnetic variations near some continental margins. For instance, PARKINSON (1962) has shown that the horizontal field variations near the Australian coast are larger than could be caused simply by the ocean; the observations also require a deep-seated change in conductivity associated with the Australian continental margin, presumably in the underlying upper mantle. Another well known geomagnetic anomaly is associated with the Pacific coast of Japan, noticeably affecting vertical field observations of 'bays' and sudden commencements of magnetic storms (RIKITAKE, 1964, 1966). This is only partly explained by the ocean effect and seems to indicate a highly conducting region in the upper mantle beneath part of Japan.

In western North America, the coast effect appears to be caused both by the ocean and by an underlying lateral change in conductivity in the upper mantle. A further change in the pattern of magnetic variations occurs about 1000 km inland. This was observed by SCHMUCKER (1964) from observations along an east-west traverse in south-western United States. CANER and CANNON (1965) have added observations from Canada and have interpreted all the results using the ratio of the vertical to horizontal variation amplitudes for periods of 10–120 minutes (Figure 4.23). Between the coast and about 1000 km inland the ratio is substantially smaller than further east, indicating that the top 100 km of the mantle beneath western North America is relatively highly conducting. It is interesting to note that the highly conducting mantle may correspond to the region of anomalously low $P$ velocity in the upper mantle and of high heat flow. This all suggests that the high conductivity is related to relatively high temperature in the upper mantle.

Other local anomalies in geomagnetic variation have been discovered in Germany, the Canadian Arctic and south Scotland. The Mould Bay anomaly in the Canadian Arctic (WHITHAM, 1963) indicates a strong increase in conductivity beneath. Whitham estimated that a 20 km thick layer with a conductivity of 1 ohm$^{-1}$ m$^{-1}$ (100 times the normal value) near the crust-mantle boundary is needed to give a preliminary explanation of this anomaly. At Eskdalemuir in south Scotland and at one station to the south-west of it the vertical field variations at short-period are strongly suppressed as can be seen in Figure 4.20 (OSEMEKHIAN and EVERETT, 1968). This indicates a high conductivity region at less than about 100 km beneath Eskdalemuir.

Some anomalies in geomagnetic variation are partly caused by the high conductivity of sea-water. A few, such as one in north Germany, may be caused by highly conducting sedimentary rocks such as salt deposits. Some of the anomalies, however, are probably partly or wholly caused by lateral variation of conductivity within the topmost 100–200 km of the Earth. In this category we have the broad contrast in conductivity between the sub-oceanic and sub-continental upper mantle and the more localized variations beneath continents. Although it is possible that lateral change in the composition of the mantle may cause some anomalies, the more generally favoured explanation is that anomalously high conductivity is caused by raised temperature, or by the presence of magma.

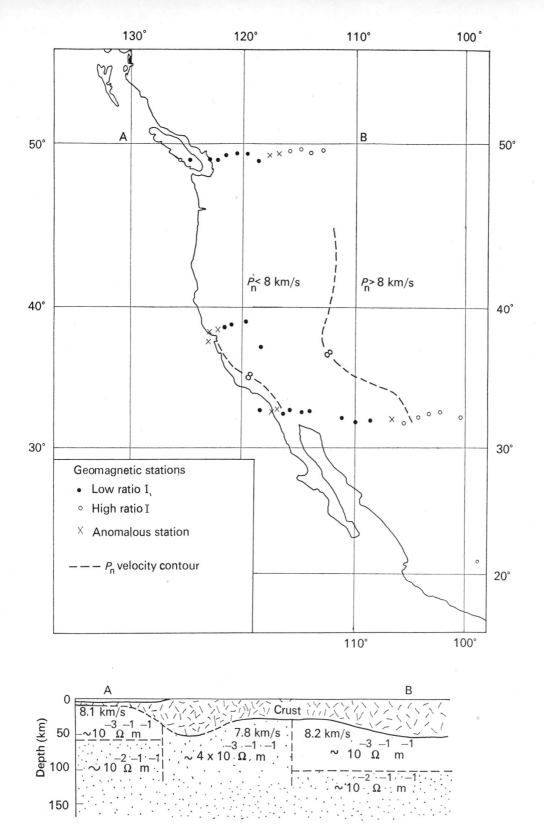

**Fig. 4.23** Geomagnetic depth-sounding in western North America. I is the ratio of vertical to horizontal component amplitudes for variations in the period range 10-120 minutes. Regions of low I are interpreted in terms of relatively high conductivity in the underlying upper mantle, as indicated in the section along line AB shown below. Adapted from CANER and CANNON (1965), *Nature, Lond.*, **207**, 927 and 928.

Our knowledge of regional anomalies in geomagnetic variation is still very scanty, and the interpretation of known anomalies is a puzzle. Much more work is needed, both in obtaining satisfactory detail of the anomalies and in developing methods of interpreting them.

### Interpretation of the electrical conductivity distribution of the mantle

It will be shown below (p. 129) that the mantle down to about 400 km probably consists mainly of silicates amongst which magnesium-rich olivine is predominant. In the lower mantle the stable phases are probably oxides of magnesium and iron and metasilicates.

Silicate minerals such as olivine are practically insulators at room temperature but become semi-conductors as the temperature is raised. Three types of conduction can occur in them. *Impurity semi-conductivity* is caused by the presence of foreign atoms with a misfitting valency in a crystal lattice; these produce either excess electrons or 'holes' (missing electrons) which migrate through the lattice when an e.m.f. is applied. *Electronic semi-conductivity* is caused by movement of free electrons raised into the conduction band through thermal agitation. Semi-conduction can also occur through movement of *ions*, as in an electrolyte. The observed conductivity is the sum of the three types of conductivity, although one type is usually dominant.

Each type of conductivity is thermally activated, so that

$$\sigma = \sigma_0 e^{-E/kT}$$

where $\sigma$ is the conductivity, $T$ is the temperature, $\sigma_0$ and $E$ are constants (which may depend on pressure) and $k$ is Boltzmann's constant. The most rapid increase in conductivity with temperature occurs when $T = \frac{1}{2}E/k$. Much below this temperature the conductivity is small relative to $\sigma_0$, and above $T = E/k$ it asymptotically approaches its maximum possible value $\sigma_0$. The activation energy $E$ controls the temperature range at which the conductivity increases strongly with temperature.

The conductivity of olivine rises from less than $10^{-5}$ ohm$^{-1}$ m$^{-1}$ at room temperature to $10^{-2}$ at 1000°C (BULLARD, 1967). Impurity semi-conduction ($\sigma_0 = 10^{-4} - 10^{-2}$ ohm$^{-1}$ m$^{-1}$) is dominant up to a few hundred degrees. Above about 700°C electronic semi-conduction ($\sigma_0 \doteq 5$ ohm$^{-1}$ m$^{-1}$) increases strongly and swamps the impurity conduction. Above about 1100°C ionic conductivity ($\sigma_0 \doteq 10^6$ ohm$^{-1}$ m$^{-1}$) becomes dominant.

Impurity semi-conduction is believed to be the main conduction process in dry rocks of the crust and topmost mantle. Since it cannot exceed about $10^{-3}$ ohm$^{-1}$ m$^{-1}$ in olivine, it is not adequate to explain the observed increase in conductivity with depth at 70 km or deeper. The other possibilities are ionic and electronic conductivities.

Ionic conduction is strongly inhibited by pressure. TOZER (1959) considered that it is probably unimportant below 500 km depth because temperatures are unlikely to be high enough for it to outweigh electronic conduction at such pressures. However, ionic conductivity may explain the relatively high conductivity layer starting at about 75 km depth. This layer is possibly identical with the low velocity channel, both effects being caused by proximity to the melting point. Ionic conductivity is likely to be predominant in magma and in weakened lattices near the melting point. Ionic conductivity may also explain some of the local anomalies in conductivity within the topmost part of the mantle.

Electronic semi-conduction is probably the predominant conduction process in the transition zone and lower mantle (TOZER, 1959). Tozer interpreted the increase in conductivity through the transition zone as the steep rise in electronic conductivity with temperature. Using experimentally determined constants to estimate $E$ and $\sigma_0$ for olivine, he went so far as to use McDonald's conductivity-depth curve to predict the temperature distribution in the mantle (assuming it to be olivine). However, AKIMOTO and FUJISAWA (1965) have observed a hundredfold increase in the conductivity of the iron olivine fayalite as it undergoes phase transition to spinel at 600°C (Figure 4.24). By analogy, a similar increase in conductivity with depth would be expected as magnesium rich olivine gives way to spinel in the transition zone.

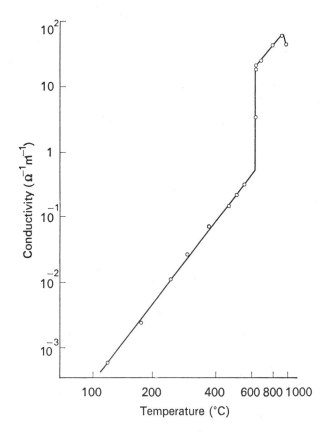

**Fig. 4.24**    Electrical conductivity of the iron-olivine fayalite ($Fe_2SiO_4$) at 43·8 kbar confining pressure as a function of temperature. Adapted from AKIMOTO and FUJISAWA (1965), *J. geophys. Res.,* **70,** 447.

To summarize, the oceans and waterlogged sediments locally form a highly conducting layer near the Earth's surface. At a depth of about 40 km the conductivity is about $10^{-3}$ ohm$^{-1}$ m$^{-1}$ and is probably due to impurities. A layer of higher conductivity (about $10^{-2}$) starts at about 75 km and may be caused by ionic conductivity; locally this layer may start at shallower depth accounting for local anomalies in the geomagnetic variations. We may speculate that the bottom of this layer is equivalent to the base of the low velocity channel at about 200 km depth, and that the conductivity decreases again to about $10^{-3}$ beneath it. Between about 400 and 1000 km depths the conductivity increases by a factor of about a thousand, partly because of the increase in electronic conductivity with temperature rise and partly because of the phase changes in the transition zone. The average conductivity of the lower mantle is 10 ohm$^{-1}$ m$^{-1}$, the uncertainty being a factor of 10 either way.

## 4.5 Temperature-depth distribution in the mantle

Extrapolation of the surface heat flow gives some indication of the temperature distribution of the topmost 50 km of the mantle. Below this depth, the temperature distribution is particularly uncertain, although some limits can be placed on it from the known physical properties of the mantle.

Let us start by anticipating some conclusions of Chapter 6. The mean heat flow of the continental and oceanic regions are both about 1·5 $\mu$cal/cm$^2$ s. A substantial fraction of the continental heat flow (probably over 50%) is caused by radioactive decay within the crust, while the contribution from the thin crust typical of the oceans is negligible in comparison. Consequently the heat flow and the temperature gradient and the temperature must all be significantly lower in the topmost mantle beneath continents than at the same depth beneath oceans. Let us suppose that the heat flow in the uppermost mantle beneath oceans is double that beneath continents. Then at 35 km depth the oceanic mantle would be hotter by about 100°C, and in the absence of heat sources the difference would be over 200°C at 50 km depth. We cannot know the exact temperature regimes in the upper mantle without having full knowledge of the distribution of heat sources and the mechanisms of heat transfer (which we cannot get). Whatever the details, the temperatures at corresponding depths in the upper mantle beneath continents and oceans must differ by the order of one or more hundred degrees. Equally

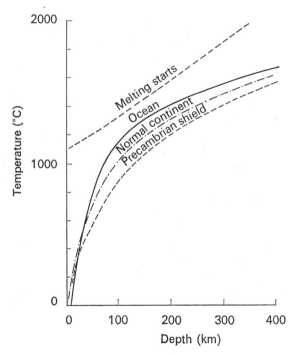

**Fig. 4.25** Estimates of the upper mantle temperature-depth distribution beneath oceans, normal continents and Precambrian shields, based on heat transfer by thermal conduction (including radiative heat transfer) and appropriate distributions of radioactivity. Redrawn from CLARK and RINGWOOD (1964), *Rev. Geophys.*, **2**, 53.

certainly, the sub-continental and sub-oceanic temperatures must converge together at a depth of a few hundred kilometres because otherwise larger gravitational effects than are observed would result from the density differences caused by thermal expansion. The upper mantle temperature distributions of CLARK and RINGWOOD (1964) are shown in Figure 4.25. These do depend on assumed distributions of radioactivity and thermal conductivity and are therefore not unique. But they do bring out the main features discussed above and in the next paragraph.

A second important conclusion stemming from the geothermal gradient is that the temperature gradient must decrease substantially below about 50–80 km, otherwise there would be wholesale melting of the mantle which is known not to occur. The temperature gradient in the crust and topmost mantle below ocean basins is about 20°C/km. If this continued downwards unmodified, then the temperature at 70 km depth would be 1300°C and at 110 km depth it would be 2100°C which exceeds the melting point of olivine at that pressure. Thus there must be a considerable reduction in the temperature-depth gradient below about 60 km depth; this fall-off with depth of the temperature gradient in the upper part of the upper mantle is an important feature of the upper mantle; possible explanations of it are discussed in Chapter 6.

The material of the mantle is known to be in the solid state apart from some local partial fusion in the upper mantle. The melting point therefore places an upper limit on the temperature. Olivine is the most refractory major constituent of the upper mantle. At the surface, the magnesium olivine forsterite melts at 1900°C. The melting point increases with pressure and is 2140°C at 50 kbar (about 160 km depth). This places an absolute upper limit on upper mantle temperature. A more realistic upper limit, which is locally reached, is given by the temperature at which basalt magma forms by partial fusion of an ultrabasic rock. This is about 1100°C at the surface and increases with pressure to about 1300°C at 100 km depth and 2000°C at 350 km depth. The seismic low velocity channel of the upper mantle is commonly identified with the region where the melting point of basalt is most closely reached; this suggests that at a point between about 100 and 200 km depths the temperature-depth gradient is equal to the melting point gradient of basalt which is about 3°C/km, and that below about 250 km it is less than this.

The oceanic Moho is at about 150°C and the continental Moho is unlikely normally to exceed about 700°C. Thus most, if not all, basalt magma must form in the upper mantle. Beneath Hawaii, earthquake foci associated with eruption of basalt occur between 45 and 60 km depth (EATON and MURATA, 1960) suggesting that the magma is formed at a depth of about 60 km. A similar depth has been obtained by GORSHKOV (1958) for Kamchatka. Beneath these two regions the temperature at about 60 km depth is probably about 1250–1300°C. Using earthquakes to detect the source of basalt magma is the most reliable method of determining the actual temperature at a point in the upper mantle, but this method is limited to regions of active volcanism.

Olivine and pyroxenes are believed to undergo drastic high pressure breakdown in the transition zone (p. 141) and because of this the melting point distribution of the lower mantle cannot be obtained by simple extrapolation of the upper mantle melting point distribution. However, UFFEN (1952) used solid state theory to show that the melting temperature of uniform material is related to the seismic velocities by the following equation

$$T_{m} = C(1/V_{P}^{3} + 1/V_{s}^{3})^{-\frac{1}{3}}$$

where $C$ is a constant. Even this equation is not valid across the transition zone because $C$ depends on the molecular weight. Uffen assumed the melting point at 100 km to be 1530°C and thereby computed the melting point at 1000 km depth to be about 3430°C and at 2900 km depth (the core-mantle boundary) to be about 5030°C. The true melting point in the lower mantle may be somewhat greater than Uffen's estimate, but nevertheless Uffen's curve probably does give a reasonably realistic upper limit on lower mantle temperatures.

The temperature at the core-mantle boundary can be estimated by extrapolating the melting point of iron to high pressure, assuming that the outer core is iron in the liquid state and that the inner core consists of solid iron. The method depends on the use of Simon's empirical relationship between melting point and pressure, which is

$$p = a((T/T_0)^c - 1)$$

where $a$ and $c$ are constants and $T_0$ is a reference temperature. STRONG (1959) used experimental measurements of the melting point of iron up to 100 kbar to determine the constants and then used the equation to estimate the melting point at the inner/outer core boundary as 2600°C, later revised up to 2730°C. Assuming an adiabatic gradient in the outer core, the temperature at the core-mantle boundary would be about 2480°C. No great reliance can be placed on the estimate because of the large extrapolation and because the core is unlikely to consist of pure iron (p. 160).

TOZER (1959) has attempted to estimate the temperature-depth distribution of the mantle from the electrical conductivity curve of McDONALD (1957). He assumed the mantle to consist of olivine

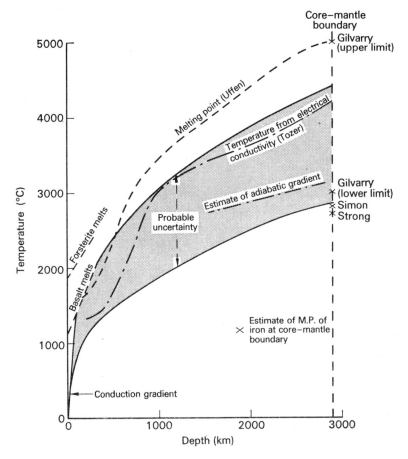

**Fig. 4.26**   Estimates of the temperature-depth distribution in the mantle, showing range of probable uncertainty.

throughout and used experimentally determined properties to estimate the parameters such as excitation energy in the relationship between temperature and ionic conductivity at high pressure. Tozer's temperature distribution shows a rise from 1700°C to 2700°C across the transition zone (corresponding to the steep rise in conductivity) and a temperature of about 4200°C at the core-mantle boundary. It is now known, however, that the rise in conductivity through the transition zone is probably mainly caused by the olivine-spinel transition and other phase changes. Because of this, Tozer's temperature distribution can be treated as an estimate of the upper limit (Figure 4.26).

If a thermally isolated volume of rock is subjected to a change in pressure, the temperature also changes. The slope $dT/dp$ is called the adiabatic gradient. It has often been suggested that the temperature gradient in the lower mantle is near to the adiabatic gradient. The adiabatic gradient is related to the coefficient of volume expansion $\alpha$, the specific heat at constant pressure $c_p$ and the density $\rho$ by

$$dT/dp = \alpha T/c_p\rho.$$

Within the Earth, $dp/dr = -g\rho$, where $r$ is radius and $g$ is gravity. Putting $g = 10^3$ cm/s$^2$ (p. 147), $\alpha = 10^{-5}/°C$ (p. 185), $T = 2500°K$, and $c_p = 0\cdot3$ cal/g°C (p. 185), these equations give $-dT/dr = 0\cdot3°C/km$ in the lower mantle. This is probably accurate to a factor of two. In the upper mantle, the temperature is lower but the coefficient of expansion is probably higher so that the adiabatic gradient would also be about $0\cdot3°C/km$.

Some of the information on the radial temperature distribution of the mantle (and outer core) is summarized in Figure 4.26.

## 4.6 Composition of the mantle

It was mentioned in Chapter 1 that chondritic meteorites are probably the best sample we can get of the non-volatile material from which the inner planets have been built. It was also pointed out that it is unlikely that the Earth has identically the same composition as the chondrites; rather the chondritic model provides an hypothesis for discussion and testing. It is in this vein that the silicate phase in chondrites has been used as a guide to the chemical and mineralogical composition of the mantle.

Chondritic meteorites are ultrabasic in composition (p. 11), the average minerology being 46% olivine, 25% pyroxene, 11% plagioclase and 12% nickel-iron metallic phase. The olivine is a solid solution of about 70–85% forsterite ($Mg_2SiO_4$) and 15–30% fayalite ($Fe_2SiO_4$). The olivine is commonly accompanied by magnesium-rich orthopyroxene, $(Mg, Fe)SiO_3$, although clinopyroxene may occur. The plagioclase feldspar is close to oligoclase. Although the chondrites are remarkably uniform in composition, they have been subdivided into five or more groups depending on mineralogy and chemistry. The hypersthene and bronzite chondrites are by far the most common, and it is these groups which are usually taken as the basis for the chondritic model of the Earth. However, some geochemists think that carbonaceous chondrites are the most primitive form and give the best approximation to the non-volatile fraction of the planetary nebula; these contain abundant serpentine (hydrated olivine) admixed with black carbonaceous matter.

Another well-established approach is to seek common rock types which possess the observed seismic velocity of the topmost mantle. The $P$ velocity just below the Moho is about $8\cdot0$–$8\cdot2$ km/s. Two common rock types have comparable $P$ velocity within the appropriate temperature and pressure range. These are (1) an *ultrabasic rock* such as peridotite, and (2) *eclogite* which is the high pressure form of gabbro. Peridotite is predominantly composed of olivine which may be accompanied by other ferromagnesian minerals, commonly pyroxenes. The olivine of peridotites is characteristically 90%

forsterite and 10% fayalite or therabouts. Dunite is the variety of peridotite consisting almost entirely of olivine. The silicate phase of chondrites is a form of peridotite, although it differs from terrestrial peridotite in being richer in iron and containing sodium-rich feldspar.

Because of the above evidence, it has been conventional to assume that the mantle is ultrabasic in composition similar to peridotite or to the silicate phase of chondritic meteorites. This suggests that at shallow depths magnesium-rich olivine is the prevalent mineral and that it is followed in abundance by magnesium-rich pyroxenes. However, LOVERING (1958) suggested that the achondrites probably give a better model for the mantle and that eclogite, not peridotite, underlies the Moho. We therefore need to look for other sources of information on the composition of the upper mantle, such as a discussion of the nature of the Moho in the light of experimental evidence and a review of igneous rocks which are thought to come direct from the mantle.

*Nature of the Moho*

If the mantle is ultrabasic, then the Moho could be a compositional boundary separating ultrabasic rocks from more silica-rich rocks of the overlying crust; alternatively it could mark the boundary between peridotite and its hydration product serpentinite, as Hess suggested to explain the bottom layer of the crust beneath oceans. The alternative hypothesis that eclogite forms the topmost mantle is usually linked to an interpretation of the Moho as a phase change between gabbro and eclogite. Since Lovering's suggestion, nearly all the new evidence is against the interpretation of the Moho as a phase change.

A cogent argument against the idea that the same phase change can explain both the oceanic and the continental Moho is based on the significantly different temperature-depth profiles beneath oceans and continents (HARRIS and ROWELL, 1960; BULLARD and GRIGGS, 1961). The average temperature gradient at the top of the oceanic and continental crust is about the same, but because the oceanic crust is covered by 5 km of water, the oceanic Moho (11 km) must be substantially cooler than a point at the same depth or at the same pressure beneath the continents. On the other hand, the strong concentration of radioactivity within the continental crust reduces the temperature gradient in the lower part of the crust, causing the continental Moho to be cooler than a point at the same depth beneath an ocean. The temperature-depth profiles must cross each other somewhere between the two Moho depths (Figure 4.27). The unavoidable logical conclusion is that if the Moho beneath continents and oceans is the same phase change, then one of them must represent a downward transition to the dense phase and the other a downward transition to the light phase!

Another argument of Bullard and Griggs against the interpretation of the continental Moho as a phase change depends on lack of correspondence between Moho depth and temperature as inferred from heat flow. This was placed on a firm experimental basis by RINGWOOD and GREEN (1964), who investigated the basalt/eclogite transition at 1100°C and determined its dependence on temperature and pressure. The average heat flow over a large area of eastern Australia is nearly twice the average for the Precambrian shield of western Australia (Figure 6.13). This means that the temperature at 30–40 km depth must be at least 200°C higher beneath eastern Australia. If the Moho beneath Australia is caused by a phase change, then the experimentally determined pressure-temperature relationships of the phase boundary imply that the Moho should be more than 5 km deeper in the east because the higher temperature raises the transition pressure. The crustal thicknesses of the two regions are about the same and certainly do not differ by more than 5 km.

There are two further experimental observations on the stability fields of basalt and eclogite which virtually rule out the interpretation of the Moho in terms of this phase change. Firstly,

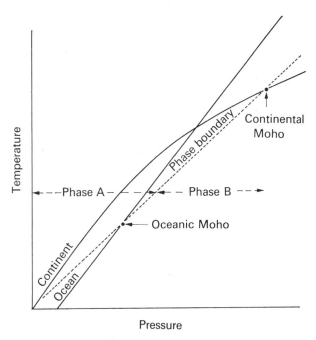

**Fig. 4.27** Diagram showing the intersection of the temperature-depth distributions below oceans and continents with a phase boundary supposedly representing the Moho. The topology of the diagram shows that the phase change must occur in opposite directions going down beneath oceans and continents, thereby suggesting that the Moho cannot be caused by the same phase change beneath both types of region.

Ringwood and Green showed that eclogite is the stable assemblage at conditions of temperature and pressure normally prevailing in the crust; therefore the Moho cannot be interpreted as a downward transition from gabbro or basalt to eclogite. Secondly, the transition from basalt to eclogite takes place through an intermediate granulitic mineral assemblage over a depth interval of the order of 15 km; this is quite incompatible with the seismic evidence which shows that the Moho is a much sharper boundary than this.

The overwhelming weight of evidence is against interpreting the Moho as the basalt/eclogite phase boundary. Beneath the continents, the most reasonable interpretation consistent with our ideas on the lower crust (p. 68) is that the Moho marks the boundary between the acidic to intermediate granulites of the lower crust and the ultrabasic rocks of the topmost mantle. Beneath the oceans, it may mark the boundary between the olivine-rich ultrabasic rock of the mantle and its hydration product serpentinite above.

The United States plan to drill a hole through the oceanic crust into the topmost mantle has been suspended for need of adequate funding. No amount of deduction and speculation about the composition of the topmost mantle can get as far as direct sampling, and the renewal of 'project Mohole' is awaited with the greatest interest.

*Igneous rocks and upper mantle composition*

Some igneous rocks are known to come from the upper mantle. Basalt is the commonest of these but some ultrabasic rocks and some of the ultrabasic nodules in basalts have probably also been derived directly from the mantle. It has been suggested that andesites and even granites have their primary magma sources in the mantle, but this is much more speculative. The petrology of those rocks which have fairly definitely originated below the Moho is one of the important sources of information on the composition of the upper mantle.

Basalt is overwhelmingly the most common igneous rock of oceanic regions and is about equally abundant with granite in the continents. The two main types of basalt are tholeiite (49–51% $SiO_2$) and alkali basalt (45–48% $SiO_2$). The source of most basalt magma is unquestionably within the upper mantle and is probably at about 60 km depth or below (p. 127).

In theory, basalt magma could form *either* by complete or nearly complete fusion of an eclogitic source *or* by partial fusion of a wide variety of ultrabasic rocks other than the pure olivine rock dunite (YODER and TILLEY, 1962; KUSHIRO and KUNO, 1963). Partial fusion of a typical peridotite would yield a 2–9% fraction of basalt magma. However, there is a serious difficulty in attempting to derive the magma from an upper mantle which is merely a high pressure form of basalt. The radioactive content of the upper mantle must be at least five to ten times less than that of basalts, because otherwise the Earth's heat flow would be much greater than is observed. This suggests that the magma has formed through selective fusion of a small fraction of the primary mantle material, which is therefore probably ultrabasic. The radioactive elements have been concentrated into the magma.

How the various forms of basalt magma are produced is one of the main problems of modern petrology. Possible factors are the composition of the source rock, the temperature and pressure conditions at the source, and the history of fractional crystallization as the magma rises directly or in stages towards the Earth's surface. It is usually assumed that the upper mantle has a uniform composition, and that the other two factors cause the main variation in magma composition. KUNO (1967) took the view that the depth of magma formation controls the composition of the magma, alkali basalt magma forming between about 200 and 400 km depth and tholeiite forming at shallower depth. On the other hand, YODER and TILLEY (1962) considered that the primary magma is of picrite basalt composition and that the other types are derived through processes of fractional crystallization. Figure 4.28 shows these two extreme hypotheses applied to the interpretation of the basalt provinces of Japan.

More direct information on the upper mantle comes from the olivine nodules and peridotite nodules which are often found in basalts. Some of these may represent crystal cumulates in basaltic magma chambers, but many of them are believed to be chunks of upper mantle from near the magma source or from above it. KUNO (1967) showed that these peridotite nodules contain a characteristic ortho-pyroxene with a higher content of $Al_2O_3$ than from intrusive peridotite suggesting high pressure at the source. The nodules have a fairly uniform mineralogical and chemical composition (Table 4.1); according to Kuno the average mineralogical composition is olivine (60–70%), Al-rich enstatite (about 30%), Al-rich Na-bearing diopside (less than 10%) and chrome spinel. Eclogite nodules are relatively rare in comparison. Thus the nodules in basalts support the idea of an ultrabasic upper mantle, although they may represent the 'barren mantle' which is left after magma has been removed rather than primary mantle material. Although it is fairly widely thought that the nodules mainly come from the upper mantle, some petrologists consider that they have formed by crystal settling in magma chambers and are not therefore relevant to mantle composition.

It is of great interest to search for larger masses of rock which may have been brought to the Earth's surface from the upper mantle by tectonic or igneous processes. Ultrabasic intrusions, especially peridotites, dunites and the hydrated rock serpentinite are the most promising candidates. Some of these are known to have formed through crystal accumulation in a magma chamber, possibly followed

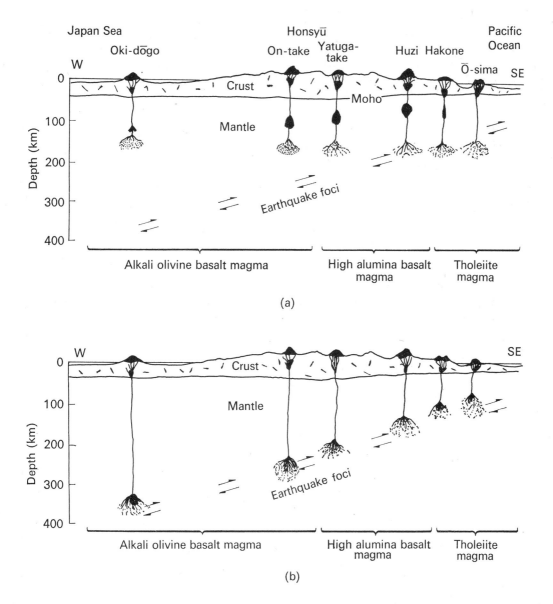

(a)

(b)

**Fig. 4.28** Sections across the circum-Pacific belt at Japan showing the depths of generation of basalt magma according to (a) YODER and TILLEY (1962) and (b) KUNO (1967). Redrawn from KUNO (1967), *The Earth's mantle*, p. 108, Academic Press.

by movement of the crystal mush. But another group, notably the Alpine type peridotites and serpentinites which are intruded into orogenic belts, are believed by Hess and others to be large chunks of mantle which have been forced into the crust. As an example, St. Paul Rocks on the mid-Atlantic

**Table 4.1**   Composition of rock types relevant to the upper mantle.

| | Dunite[1] (anhydrous) | Peridotite[2] | Peridotite nodules[3] | Pyrolite[4] | Kimberlite xenoliths[5] | Eclogite[2] (water free) |
|---|---|---|---|---|---|---|
| $SiO_2$ | 41·3 | 43·5 | 43·9 | 43·1 | 46·4 | 49·0 |
| $TiO_2$ | trace | 0·8 | 0·1 | 0·6 | 0·1 | — |
| $Al_2O_3$ | 0·5 | 4·0 | 2·8 | 4·0 | 2·6 | 14·5 |
| $Fe_2O_3$ | 1·2 | 2·5 | 1·1 | 1·7 | 1·1 | 3·8 |
| FeO | 5·9 | 9·8 | 7·2 | 6·7 | 5·8 | 9·1 |
| MnO | 0·1 | 0·2 | 0·1 | 0·1 | 0·1 | — |
| MgO | 49·8 | 34·0 | 40·9 | 39·3 | 41·2 | 8·9 |
| CaO | trace | 3·5 | 2·4 | 2·7 | 2·0 | 11·5 |
| $Na_2O$ | trace | 0·6 | 0·2 | 0·6 | 0·2 | 2·5 |
| $K_2O$ | trace | 0·3 | trace | 0·2 | 0·2 | 0·7 |
| $H_2O$ | — | 0·8 | 0·1 | 0·2 | — | — |
| $P_2O_5$ | trace | 0·1 | trace | 0·1 | trace | — |

1 From GREEN and RINGWOOD (1963) giving average composition;
2 From Table 2.1;
3 Mean of four analyses quoted by KUNO (1967);
4 1:3 ratio of basalt to dunite, as given by GREEN and RINGWOOD (1963);
5 Mean of two analyses quoted by GREEN and RINGWOOD (1963).

ridge just north of the equator consist of sheared ultrabasic rocks, and WISEMAN (1966) considered that the petrological and geochemical characteristics are not inconsistent with the idea that the island is an outcrop of the upper mantle; but Wiseman is cautious in not ruling out origin through crystal accumulation.

Let us take one more example from igneous petrology. Kimberlite is a type of mica-peridotite which occurs in small pipes and which is much sought after because some of them contain diamonds. Kimberlite is almost certainly intruded as an ultrabasic magma which, according to the evidence of the diamonds, formed between depths of about 100 and 150 km. Kimberlite commonly contains nodules of garnetiferous peridotite and eclogite. The nodules have been carried up from depth; it is usually thought by petrologists that the garnet peridotite nodules represent the unfused parent rock between 100 and 150 km depth and that the eclogite has been formed through crystal settling at this depth.

A specific model for the composition of the upper mantle has been suggested by Ringwood and his co-workers (GREEN and RINGWOOD, 1963). The suggested composition is three parts of average anhydrous dunite to one part of average basalt. The mixture produces a hypothetical type of peridotite which is named 'pyrolite' (Table 4.1). Pyrolite has a closely similar composition to Ringwood's estimate of mantle composition based on a chondritic model of the Earth. To obtain this, Ringwood started with the average composition of the silicate phase of chondrites, he reduced the iron content until the ratio of iron to magnesium was 1:5 in harmony with the evidence from nodules in basalts, and he finally reduced the content of silica by 15% to obtain a realistic balance between olivine and pyroxene; the excess iron and silicon is assumed to have gone to the core.

## Strontium isotopes and the upper mantle

It was explained in Chapter 1 (p. 13) that lead isotope ratios in basalts and some lead deposits of young age are sufficiently uniform to regard them as having been derived from a primary source which has not undergone significant fractionation of uranium and thorium from lead since early in the Earth's history. This uniform and undisturbed source region is probably the upper mantle.

The study of strontium isotope ratios in basalts tells a similar story. $Sr^{86}$ is non-radiogenic and its abundance has remained constant through the Earth's history. But $Sr^{87}$ is the decay product of $Rb^{87}$ and its abundance has progressively increased (at a nearly constant rate since the half-life is 50,000 my). The ratio of $Sr^{87}/Sr^{86}$ in a rock increases at a rate very nearly proportional to the relative abundance of rubidium and strontium. The ratio $Sr^{87}/Sr^{86}$ which an igneous rock had at its time of crystallization can be determined as a by-product of age-dating by the rubidium-strontium method. As strontium isotopes are probably not significantly fractionated during the formation and crystallization of a magma, this gives a direct estimate of the strontium isotope ratio of the region where the magma was formed.

Using the above method, the $Sr^{87}/Sr^{86}$ ratio in meteorites when they last crystallized about 4500 my ago is found to be 0·698. FAURE and HURLEY (1963) found this ratio for 25 young or recent basalts from widely separated continental and oceanic localities to be $0·7078 \pm 0·003$, now revised down to 0·705. This is remarkably uniform and not much above the initial meteorite value. It suggests that the rubidium-strontium ratio at the source of basalt magma, presumably the upper mantle, is low and fairly uniformly distributed and has remained so through most of geological time. If it is assumed that the initial ratio in the mantle 4500 my ago was 0·698 as in meteorites and that the modern ratio in the upper mantle is 0·705, then we can compute that the rubidium-strontium ratio in the upper mantle is about 0·04. On the other hand, the geochemically determined ratio in the upper crust is about 0·25; in harmony with this the igneous rocks of crustal origin show higher $Sr^{87}/Sr^{86}$ ratios than basalt, commonly ranging between about 0·704 and 0·720. Figure 4.29 shows how the strontium isotope ratio has developed in the mantle, and in part of the crust assumed to be differentiated from the mantle 3000 my ago, according to the above estimates.

It might be thought that the strontium isotope ratios give a means of distinguishing igneous rocks of crustal and upper mantle origin. Undoubtedly the mantle derived ones should show ratios below about 0·708, but there is also some suggestion that a low concentration of rubidium is characteristic of the deeper parts of the crust in some regions (such as the Scourian rocks of north-west Scotland).

GAST (1960) has pointed out some important limitations on the upper mantle composition imposed by observed abundances of Sr, Rb, K, Cs, Ba and U in basalts and chondrites. The ratio Rb/Sr in chondrites is about 0·15. In the upper part of the crust it is 0·25 and in the upper mantle about 0·04. Putting the crust and upper mantle together, the upper few hundred kilometres of the Earth must have a much lower overall Rb/Sr ratio than the chondrites. Gast extended his observations to show that the upper mantle abundances of the alkali metals K, Rb and Cs relative to Ba, Sr and U is significantly less than in chondrites. The two possible interpretations are (1) the Earth is not exactly chondritic, or (2) the alkalis are strongly concentrated in the deeper parts of the Earth, which is not what geochemists would expect. This important piece of evidence throws doubt on the chondritic hypothesis and suggests that at best we should treat it with caution. Gast's conclusions have been supported by recent work along the same lines.

## Interpretation of seismic velocity variations in the upper mantle

Our ultimate aim in studying the Earth's interior is to describe the physical and chemical properties

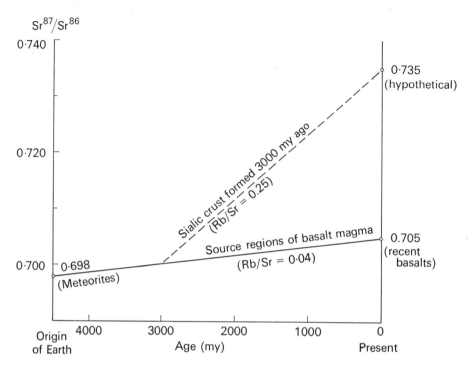

**Fig. 4.29**   Diagram showing the suggested variation with time of the strontium isotope ratio in (a) the mantle since the origin of the Earth (solid line), and (b) a hypothetical sialic crust formed 3000 my ago with concentration of rubidium relative to strontium. Based on FAURE and HURLEY (1963) and redrawn from MOORBATH (1965), *Controls of metamorphism*, p. 261, Oliver and Boyd.

as best we can and to interpret them in terms of past and present processes within the Earth. Modern seismology has revealed some quite unexpected features in the upper mantle such as the low velocity zone and the large regional variations in $P$ and $S$ velocities. Can these features be explained by variations in physical conditions, especially temperature, affecting a chemically homogeneous upper mantle or do they require radial and/or lateral variations in chemistry?

There are two opposing groups of opinion on the processes which occur in the upper mantle. The classical opinion, which is linked to the hypothesis that the continents and oceans are permanent, is that the upper mantle has been static through geological time apart from some differentiation and other relatively minor movements (such as isostatic flow). This view is consistent with quite large radial and regional variations in the chemical composition of the upper mantle; indeed it is often coupled with the hypothesis that the mantle beneath continents has differentiated to form the overlying sialic crust but the mantle beneath oceans remains substantially undifferentiated. The other opinion, with increasing adherence ( > 95% ?), accepts the hypothesis of continental drift and is consequently forced towards the view that large relative horizontal movement occurs between the crust and the mantle below 70 km depth. Some small variations in the chemical composition, such as between primary mantle material and the barren residue after basalt magma has been abstracted, might be expected; but large differences in the chemistry of the mantle beneath continents and oceans would not be expected if the continents have drifted on a large scale. Whichever opinion one takes, it is bound to influence one's ideas about the composition of the upper mantle.

The observed variations in seismic velocity both with depth and with lateral position could be caused by one or more of the following effects: (1) the influence of pressure and temperature on a rock of uniform chemical and mineralogical composition; (2) mineralogical phase changes; (3) partial fusion; (4) anisotropy; and (5) chemical inhomogeneity.

Increase in pressure with depth causes a rise in seismic velocity but the geothermal rise in temperature causes a corresponding reduction. If the temperature gradient is higher than about 8°C/km, the influence of temperature outweighs that of pressure and the velocity decreases slightly with depth (MACDONALD and NESS, 1960). The effect is more strongly marked for shear velocity. The geothermal gradient between the oceanic Moho and a depth of about 50 km is almost certainly much steeper than 8°C/km, but below about 50–100 km it must fall off to less than 8°C/km or otherwise there would be wholesale fusion. This means that a small decrease in velocity with depth down to about 50–100 km and an increase below this depth is to be expected in the sub-oceanic upper mantle simply because of the temperature-pressure regime. Because of the substantial heat sources in the continental crust, both the temperatures and the temperature-depth gradient are probably much less than beneath oceans, giving a possible explanation of the less well-developed low velocity layer beneath continents.

Self-compression and thermal expansion are probably not the complete explanation of the low velocity channels for $S$ and $P$. One difficulty is that the velocity decrease with depth ought to start at the depth of the Moho; but beneath the oceans the upper limit of the low velocity channel for $S$ is thought to be at a depth of about 50 km. Nor does the effect of temperature well below the melting point appear adequate to account for the regional variations in delay times. HALES and DOYLE (1967) found that a difference of temperature of the order of 1250°C spread over 500 km vertical extent would be needed to explain the variation in delay times in the United States, which is excessive and unrealistic.

Let us look at other possible ways of interpreting the seismic velocity pattern in the upper mantle. A good starting point is to assume that the composition is pyrolite consisting of three parts of average dunite to one part of average basalt. GREEN and RINGWOOD (1963) found that pyrolite can crystallize in four different mineral assemblages under conditions of temperature, pressure and water-vapour pressure which may occur in the upper mantle. These assemblages are as follows:

| | |
|---|---|
| ampholite | olivine, amphibole, accessory chromian spinel (3·25–3·28 g/cm³) |
| plagioclase pyrolite | olivine, plagioclase, enstatite, clinopyroxene, accessory chromite (3·24 g/cm³) |
| pyroxene pyrolite | olivine, aluminous enstatite, aluminous clinopyroxene, spinel (3·30–3·32 g/cm³) |
| garnet pyrolite | olivine, pyrope garnet, pyroxene(s) (3·37 g/cm³) |

The densities are stated for surface temperature and pressure. According to Green and Ringwood, these same assemblages would be expected for a wide range of possible upper mantle compositions ranging from basalt-dunite ratios of 1 : 1 to 1 : 10.

*Pyroxene pyrolite* is the mineral assemblage one would expect to occur under normal conditions down to about 100–200 km depth. This assemblage is characteristic of the peridotite nodules in basalts. At about 100–200 km depth it would be expected to give way to the high pressure form *garnet pyrolite*, with an increase in density of about 0·06 g/cm³ and a corresponding increase in seismic velocity of about 2½–3% according to Birch's velocity-density relationship (p. 65). The transition from pyroxene to garnet pyrolite may explain the Lehmann discontinuity near the base of the low velocity zone.

The *ampholite* assemblage would be expected to occur in a water-rich environment below the dehydration temperature for the amphibole, which is probably between 500 and 1000°C depending on the water vapour pressure. Ampholite may therefore occur in the uppermost part of the sub-oceanic mantle. *Plagioclase pyrolite* is possibly stable down to about 50 km depth provided that the temperature is above about 600–700°C and water vapour pressure is low. Such conditions might be expected in the uppermost part of the mantle beneath regions of high heat flow.

The density of plagioclase pyrolite is about 0·07 g/cm³ lower than that of pyroxene pyrolite. According to Birch's relationship this corresponds to a difference in the $P$ velocity of about 0·2 km/s. Regional differences in seismic velocity down to about 50 km would be expected to occur wherever pyroxene pyrolite passes laterally into plagioclase pyrolite or ampholite. However, this effect is not adequate to explain the observed range of $P$ velocities below the Moho (7·4–8·6 km/s) nor $P$ and $S$ delay times greater than a fraction of a second.

Ringwood has suggested a compositional model of the upper mantle in an attempt to explain the low velocity layer and its regional variations, partly as a temperature effect and partly by chemical inhomogeneity (CLARK and RINGWOOD, 1964). His model (Figure 4.30) is based on the different equilibrium assemblages of pyrolite. He considered that the primitive pyrolite has differentiated beneath continents to give the sial, leaving a residual layer of dunite with eclogite pockets. The differentiated layer is assumed to be thickest beneath the Precambrian shields. The differentiation has not been so strong beneath oceans. This model is tied to the idea that continents and oceans are permanent.

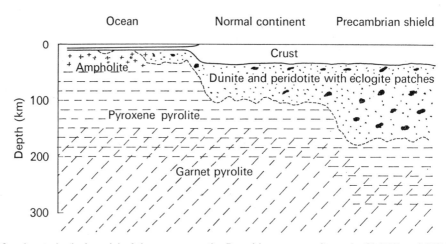

**Fig. 4.30** A petrological model of the upper mantle. Based in part on a figure in CLARK and RINGWOOD (1964), *Rev. Geophys.,* **2**, 57.

The melting point of basalt is probably most closely approached and in some regions reached between depths of about 50 and 200 km. This zone approximately coincides with the low velocity layer for $S$ in the upper mantle, which encourages us to investigate whether the low velocity zone can be explained by partial fusion or by weakening of the intercrystalline constraints not far below the melting point.

Complete fusion reduces the rigidity modulus to zero but only causes a small change in the bulk modulus and density. The result is that fusion causes the $S$ velocity to be reduced to zero and the $P$ velocity to be reduced much more than it would be in a phase reaction between mineral assemblages involving the same reduction in density (Table 4.2). A similar relationship between the

**Table 4.2**    The influence of fusion and solid-solid phase transition on density and seismic velocity for a rock of basic or ultrabasic composition.

| | Reduction in density | Reduction in $P$ velocity | Reduction in $S$ velocity |
|---|---|---|---|
| Complete fusion | 10%[1] | 21%[2] | 100% |
| Solid-solid phase transition | 10% (assumed) | 12%[3] | 12%[4] |

*Notes:* 1 as observed for basalt.
2 computed assuming the bulk modulus is the same in solid and liquid phases and that Poisson's ratio is 0·25 in the solid.
3 computed from Birch's velocity-density relationship assuming a 10% decrease in density.
4 computed from change in $P$ velocity assuming Poisson's ratio is the same in both phases.

reduction in velocities and density probably also applies to partial fusion. If true, this gives us a possible method for investigating whether seismic anomalies in the upper mantle are caused by proximity to the melting point by comparing the anomalies in $P$ and $S$ with each other and with the density anomalies as revealed by regional gravity surveys.

This sort of approach has been used by HALES and DOYLE (1967) to interpret the $P$ and $S$ delay times in the United States (p. 117). They found that the $S$ delay times had a range of nearly 8 seconds and that the observed ratio of $S$ delay to $P$ delay is $3·72 \pm 0·43$. They point out that if Poisson's ratio in the normal and anomalous parts of the upper mantle is the same, then the ratio of $S$ delay to $P$ delay would be 1·7–1·8, which is significantly different from the observed value. The observed ratio of 3·72 can be accounted for if the rigidity modulus alone changes and the bulk modulus is not affected. This may suggest that the delays are related to partial fusion or approach to it rather than to composition changes or solid-solid phase changes.

A similar argument can be applied to the $P$ delay of about 1·3 seconds observed in Iceland (p. 115); this needs a reduction of 0·6 km/s in $P$ velocity over a vertical range of about 200 km. If Birch's relationship between velocity and density applies, then the corresponding density reduction would be 0·2 g/cm$^3$ extending over a vertical distance of 200 km. This would cause a negative gravity anomaly of more than 800 mgal, which is certainly not observed on Iceland, even if allowance for a thin crust is made. Clearly Birch's relationship does not apply, and the simplest interpretation is that there is a fused fraction in the upper mantle beneath Iceland which causes a greater reduction in seismic velocity than in density.

We have seen that proximity to the fusion temperature is a simple way of explaining the low velocity zone in the upper mantle, and that large regional variations in the $P$ and $S$ velocity structure can be accounted for by variations in the depth extent of the layer where the melting point is approached. This seems to be the only explanation of the large regional variations of velocity which does not need unrealistically large variations in the density of the upper mantle.

Partial fusion may be the most important factor, but this does not rule out other less extreme variations in velocity caused by phase changes within the pyrolite assemblages, anisotropy of velocity, or even some degree of chemical inhomogeneity. However, there appears to be no need to postulate major differences in the chemical composition of the upper mantle beneath oceans and continents. Rather, the temperature and its regional variation seems to be the controlling factor. The temperature has a larger effect than in other parts of the Earth simply because it is in the upper mantle that fusion temperatures can be reached.

## The transition zone and the lower mantle

The main problem of the transition zone is to explain the rapid increase in seismic velocity with depth. The increase in velocity is now known to occur in two steps. As early as 1936, Bullen found that the seismic velocities of the mantle could not be explained just by self-compression and that extensive chemical or mineralogical inhomogeneity must occur at relatively shallow depth. Later work has shown that the inhomogeneity essentially occurs in the transition zone and that it is almost certainly caused by major phase transitions.

BULLEN (1936) used the Adams-Williamson differential equation for his computations. This relates increase in density with depth caused by adiabatic self-compression under hydrostatic pressure to the seismic velocities. It is derived as follows. A seismic parameter $\phi$ which can be computed from the $P$ and $S$ velocities is defined as follows:

$$\phi = k/\rho = V_p^2 - \tfrac{4}{3} V_s^2.$$

For adiabatic compression $k = \rho \, dp/d\rho$ by definition, where $p$ is pressure. If the pressure is hydrostatic then $dp/dr = g\rho$ where $g$ is acceleration due to gravity. Eliminating $p$ and $k$ gives

$$d\rho/dr = g\rho/\phi$$

which is the Adams-Williamson equation. Step-by-step integration of this equation makes it possible to compute the density distribution and gravity throughout a homogeneous shell of the Earth provided their values are known at the top. The following modified form of the equation can be applied to regions where the temperature distribution is not adiabatic:

$$d\rho/dr = g\rho/\phi + \alpha\rho\tau$$

where $\tau$ is the difference between actual and adiabatic temperature gradients and $\alpha$ is the coefficient of thermal expansion.

Starting with a density of 3·32 g/cm³ just below the Moho, Bullen used the Adams-Williamson equation to compute the density distribution from observed values of $\phi$ throughout a homogeneous self-compressed mantle. He then computed the moment of inertia of the mantle. By subtracting the computed moment of inertia of the crust and mantle from that of the Earth, he was able to show that the moment of inertia of the core, if associated with a self-compressed mantle, would be 0·57 $M_c R_c^2$ where $M_c$ is its mass and $R_c$ its radius. For comparison, the moment of inertia of a sphere of uniform density is 0·4 $MR^2$ and of a shell of radius $R$ is 0·67 $MR^2$. If the density increases towards the centre it is less than 0·4 $MR^2$. The value of 0·57 $MR^2$ implies that the density within the core must *decrease* strongly towards the centre, which is completely incompatible with the known fluid state of the outer core. Even if the impossibly high density of 3·7 g/cm³ is assumed to occur just below the Moho, the moment of inertia of the core associated with a self-compressed mantle must still be as high as 0·4 $MR^2$. A temperature gradient steeper than the adiabatic would only accentuate the problem. One is forced to the conclusion that a substantial increase in density with depth (of the order of 1·0 g/cm³) caused by inhomogeneity must occur at a relatively shallow depth within the mantle.

BIRCH (1952) took the argument one stage further by demonstrating that this major inhomogeneity does occur within the transition zone. He made use of the observation that the increase in isothermal bulk modulus with pressure $dk_T/dp$ for most solids is about 4. By differentiating $k = \phi\rho$ with respect to $p$ and by substituting $dp/dr = -g\rho$, one obtains

$$\frac{dk}{dp} = 1 - \frac{1}{g}\frac{d\phi}{dr}$$

for a self-compressed solid. The equation can be modified to take temperature gradient into account. Since isothermal and adiabatic bulk moduli differ by only a few per cent, the equation can be used to estimate $dk_T/dp$ within the Earth from the seismic parameter $\phi$ and the acceleration due to gravity. Birch used the equation to show that $dk/dp$ above a depth of about 250 km is less than 4 and even reaches a negative value; much of this can be explained by the steep temperature gradient. Between depths of about 400 and 750 km the estimate of $dk/dp$ is about 6 to 7; such a high value cannot be explained away as a temperature effect and it implies that the seismic velocities increase with depth much more rapidly than could be expected for self-compression. Between depths of about 1000 km and 2700 km the estimate is close to 4, suggesting that most of the lower mantle behaves as if it is homogeneous self-compressed material.

Experimental and theoretical high pressure studies have shown that a drastic re-organization of the crystal structure of olivine and pyroxene occurs at about the pressure range of the transition zone. This is the main reason for believing that change in phase rather than in chemical composition is the main cause of the large increase of seismic velocities and density between the upper and lower mantle. As knowledge has improved, it has been remarkable how well the mineralogical predictions have been found to fit into the seismic velocity framework. In particular, the expected mineralogical transformations can be divided into two stages:

(1) a breakdown of pyroxenes to give olivine and the high pressure form of quartz called stishovite; and of olivine to give the cubic mineral spinel;
(2) breakdown of spinel to give the more closely packed oxides periclase (MgO), wustite (FeO) and stishovite, and possibly also metasilicates with ilmenite structure.

These stages probably correspond to the two steps in the seismic velocity profiles.

BERNAL (1936) first suggested that the '20° discontinuity' may be caused by phase transition from olivine to spinel, by analogy with the two known forms of magnesium germanate ($Mg_2GeO_4$) one having an olivine structure and the other a spinel structure. In magnesium germanate the transition occurs at 810°C at atmospheric pressure and at 940°C at 5·7 kbar. High pressure techniques have now improved to such an extent that it has been possible to observe the transition in olivine for compositions ranging from pure fayalite to a 60% molecular ratio of forsterite (AKIMOTO and FUJISAWA, 1968); the data can be extrapolated to forsterite without difficulty. The transition involves an increase in density of about 10% as the orthorhombic structure of olivine changes to cubic structure of spinel with the silicon atoms tetrahedrally co-ordinated. At higher pressure, a second stage of breakdown to a more closely packed structure with octahedrally co-ordinated silicon atoms with a further density increase of about 8% is to be expected. This would involve transition of spinel to the oxides periclase (MgO), wustite (FeO) and stishovite ($SiO_2$). This transition has not been observed experimentally, but a phase diagram for it can be predicted from the thermochemical data which is available (ANDERSON, 1967b).

Anderson has constructed a phase diagram for the system $Mg_2SiO_4$–$Fe_2SiO_4$ at 800°C from available experimental and thermochemical data, showing the transitions from olivine to spinel and from spinel to the post-spinel phase. It is shown in Figure 4.31. The phase transitions occur at a discrete pressure for the end members forsterite and fayalite, but the effect of solid-solution is to spread the transition out over an interval of pressure. For olivine consisting of about 80% forsterite, both transitions are spread over a pressure range of about 20–30 kbar corresponding to a depth range of about 60–90 km. This is in excellent agreement with the two steps in seismic velocity, both of which are spread out over about this range of depth. Since the transition pressure increases with temperature at about 4 kbar per 100°C, a temperature gradient in the mantle will tend to spread out the transition zone slightly more. Presence of other phases such as pyroxenes would be expected to have the same effect.

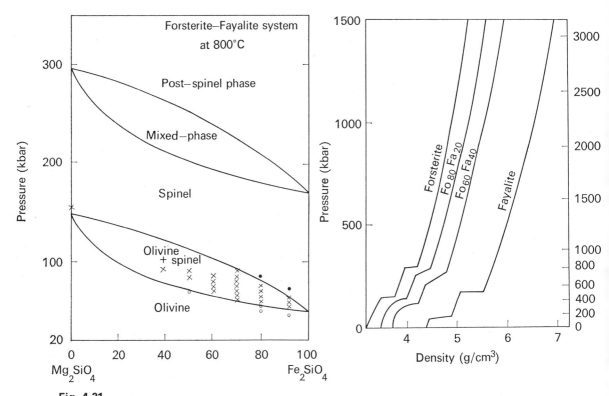

**Fig. 4.31**
*Left*: estimated phase diagram for the forsterite-fayalite system at 800°C according to Anderson based on the old pressure scale. Increase in temperature would cause the phase changes to occur at higher pressure.
*Right*: density of olivines at 800°C as a function of pressure and depth.
Redrawn from ANDERSON (1967b), *Science, N.Y.*, **157**, 1166.

ANDERSON (1967b) has also computed the depths at which the transition would occur if the temperature is in the range 1400–1800°C at 400 km depth and 1600–2000°C at 600 km. The predicted depths depend on whether the standard pressure scale (I) used in the experimental work is adopted, or whether a recent revision of the scale (II) is used. Anderson's predictions are shown in Table 4.3. These predicted depths are in excellent agreement with the seismological observations.

One important question remains. Is there a change in composition, in addition to the phase changes, on passing downwards through the transition zone to the lower mantle? ANDERSON (1968) has studied this problem by comparing the observed change in *P* velocity and density (using Birch's relationship

**Table 4.3**  Predicted depths of the olivine/spinel and spinel/post-spinel transitions in the mantle, computed for the standard pressure scale (I) and the revised pressure scale of JEFFERY and others (1966), (II).

| Composition | Olivine/spinel | Spinel/post-spinel |
|---|---|---|
| 80% forsterite ⎱<br>20% fayalite  ⎰ | 426±60 km (I)<br>388±60   (II) | 546±60 km (I)<br>670±60    (II) |
| 60% forsterite ⎱<br>40% fayalite  ⎰ | 368±60 km (I)<br>342±60   (II) | 473±60 km (I)<br>620±60    (II) |

to connect them) with the theoretically predicted changes for an olivine mantle. Anderson found that the $P$ velocity change is too small by $0 \cdot 4$ km/s and the density change is too large by $0 \cdot 24$ g/cm$^3$ for a mantle of uniform chemical composition. This suggests that the lower mantle is substantially richer in iron than the upper mantle, the FeO/MgO ratio increasing from $0 \cdot 1$ above to $0 \cdot 27$ below the transition zone. PRESS (1968) reached a similar conclusion. It should be pointed out that these conclusions are based on unproved empirical relationships and simplified mineralogical models of the mantle; however, Anderson considered that the presence of pyroxenes would probably not invalidate his conclusion. Such a chemical inhomogeneity would have the important implication that it would inhibit mantle-wide convection (p. 276).

## 4.7  Density, elastic moduli and allied properties

### Radial distribution of density

Knowledge of the $P$ and $S$ velocities at a specified depth within the Earth gives two equations connecting three important physical properties, namely density and the bulk and rigidity moduli. A further equation connecting them is needed if they are to be determined uniquely. It is important to seek such an equation because once the density distribution is known then the elastic properties and the gravity and pressure within the Earth can be estimated as a function of radius.

Any acceptable model of the density distribution of the Earth must reproduce the observed values of the Earth's mass and moment of inertia. Further constraint on allowable density distributions can be provided by assuming a specific petrological model for the upper mantle and computing the density at the relevant temperatures and pressures. These constraints are useful in that they set bounds to allowable distributions, but they do not permit a unique solution. Further hypotheses and assumptions are still needed.

In theory, the combination of the radial distributions of body wave velocities with surface wave dispersion curves as extended by free oscillation data opens up the possibility of directly determining the density distribution of the mantle. In practice, the $S$ velocity structure of the upper mantle and the detailed body wave structure near the core-mantle boundary are not well enough known to make full use of this approach. Nevertheless any satisfactory density model should be in agreement with the observed periods of free oscillations. Most of the well known models of density within the Earth pre-date the exploitation of free oscillations and do not use this constraint.

In the homogeneous shells of the Earth which are subject solely to adiabatic self-compression, the Adams-Williamson equation (p. 140) provides the additional relationship needed to determine the density distribution. Allowance can be made for known deviation from the adiabatic temperature distribution provided the coefficient of thermal expansion is assumed. The Adams-Williamson equation is almost certainly applicable to the fluid outer core (E) and it has been widely used for the lower mantle (D) with justification. It has been used for the upper mantle (B) with less justification. It cannot be used for the transition zones of the mantle (C) or core (F) or for the crust. The Adams-Williamson approach cannot determine the increases in density across transition zones and discontinuities; some of these jumps in density can be determined by using the constraints imposed by the Earth's mass and moment of inertia, but it is still necessary to make some further assumptions to obtain a unique model of density.

Models of the Earth's radial density distribution have been presented by Bullen, Bullard, Bolt, Birch, Clark and Ringwood, Landisman Satô and Nafe, and others. Each of these models depends on a different set of assumptions. Bullen's models and three other recent models are described briefly below.

Bullen's two well-known models are described in his book (1963). Model A is based on the following assumptions. The density just below the Moho was taken as $3.32 \text{ g/cm}^3$, which is appropriate for a rock of dunitic composition at 30 km depth. The Adams-Williamson equation was used for zones B, D and E and the outer core was taken to have zero rigidity modulus. The density distribution of

**Table 4.4**  Some estimates of the density distribution of the mantle and core.

|  | Depth (km) | Bullen Model A (g/cm³) | Clark & Ringwood pyrolite model (g/cm³) | Birch* (g/cm³) |
|---|---|---|---|---|
|  | 33 | 3·32 |  | 3·59 |
|  | 100 | 3·38 | 3·285 | 3·49 |
|  | 200 | 3·47 | 3·404 | 3·54 |
|  | 300 | 3·55 | 3·481 | 3·70 |
|  | 400 | 3·63 | 3·556 | 3·86 |
|  | 500 | 3·89 | 3·682 | 4·02 |
|  | 600 | 4·13 | 3·906 | 4·17 |
|  | 700 | 4·33 | 4·132 | 4·30 |
|  | 800 | 4·49 | 4·360 | 4·43 |
|  | 900 | 4·60 | 4·587 | 4·54 |
| Mantle | 1000 | 4·68 | 4·688 | 4·58 |
|  | 1200 | 4·80 | 4·805 | 4·68 |
|  | 1400 | 4·91 | 4·920 | 4·77 |
|  | 1600 | 5·03 | 5·031 | 4·85 |
|  | 1800 | 5·13 | 5·140 | 4·93 |
|  | 2000 | 5·24 | 5·247 | 4·99 |
|  | 2200 | 5·34 | 5·351 | 5·06 |
|  | 2400 | 5·44 | 5·454 | 5·12 |
|  | 2600 | 5·54 | 5·555 | 5·20 |
|  | 2800 | 5·63 | 5·657 | 5·27 |
|  | 2900 | 5·68 / 9·43 |  | 5·27 |
|  | 2920 |  | 5·716 / 9·851 |  |
|  | 3000 | 9·57 | 9·98 |  |
|  | 3200 | 9·85 | 10·28 |  |
|  | 3400 | 10·11 | 10·56 |  |
|  | 3600 | 10·35 | 10·82 |  |
| Outer core | 3800 | 10·56 | 11·04 |  |
|  | 4000 | 10·76 | 11·26 |  |
|  | 4200 | 10·94 | 11·46 |  |
|  | 4400 | 11·11 | 11·62 |  |
|  | 4600 | 11·27 | 11·79 |  |
|  | 4800 | 11·41 | 11·92 |  |
|  | 4982 | 11·54 | 12·04 |  |
|  | 5121 | 14·2 / 16·8 |  |  |
| Inner core | 6371 | 17·2 | 12·86 |  |

* Using a velocity distribution close to but not identical with that of Gutenberg.

the mantle transition zone (C) was taken to be a quadratic function of radius, tied to the density values at the top and bottom of the zone and to the density gradient at the top of D. The constraints imposed by the Earth's mass and moment of inertia were used to determine the increase in density across C and at the core-mantle boundary. The density at the Earth's centre was arbitrarily defined as 17·3 g/cm³, involving an additional increase of 5 g/cm³ across the core transition zone (F). The resulting density distribution is shown in Table 4.4.

One implication of model A is that the bulk modulus $k$ and its pressure gradient $dk/dp$ only change slightly across the core-mantle boundary. This led Bullen to formulate his 'compressibility-pressure hypothesis', that $k$ is a smoothly varying function of pressure, irrespective of composition or state, below about 1000 km. Model B is based on the use of the compressibility-pressure hypothesis for the lowest 200 km of the mantle (D″) and below the outer core (F and G); the Adams-Williamson equation is used for D′ (1000–2700 km) and E. Model B determines the density at the Earth's centre as about 18 g/cm³ and implies a greater increase in density with depth in D″ than would be expected for self-compression alone.

The observed periods of the Earth's free oscillations are in better agreement with model A than model B, although the agreement is not perfect for either model.

BIRCH (1961b) derived a density distribution for the mantle by using his empirical relationship between $P$ velocity, density and mean atomic weight (p. 65). For a constant composition this empirical relationship is

$$\rho = a + bV_{\mathrm{p}}$$

where $a$ and $b$ are constants. Birch assumed the moment of inertia of the core to be $0·39\ M_{\mathrm{c}}R_{\mathrm{c}}^2$ and used the two constraints imposed by the mass and moment of inertia of the Earth to determine $a$ and $b$ from the $P$ velocity distribution of the mantle. He thus determined the density distributions corresponding to both the Jeffreys and Gutenberg (Table 4.4) velocity-depth profiles. The result corresponds roughly to a mean atomic weight of 22·5 and is consistent with a mantle of uniform composition. Birch's density distributions show higher density than Bullen's models near the top of the mantle and lower values through the lower mantle.

CLARK and RINGWOOD (1964) constructed density models for the Earth which were based on:

(1) assumed petrological models for the upper mantle;
(2) uniform increase of density with depth across the mantle transition zone (C);
(3) application of the Adams-Williamson equation to D and E;
(4) a uniform density for the inner core (G);
(5) a moment of inertia of $0·388\ M_{\mathrm{c}}R_{\mathrm{c}}{}^2$ for the core;
(6) Gutenberg's velocity-depth distribution.

The density distribution of the upper mantle was computed for both a pyrolite and an eclogite composition. The density for each relevant assemblage of minerals can be calculated for surface conditions and corrected for pressure and temperature assuming a temperature gradient midway between the continental and oceanic ones. Using the Earth's mass and moment of inertia as constraints, the density distribution can be computed without ambiguity for both petrological models of the mantle. Table 4.4 shows the result for the pyrolite model.

The pyrolite model of Clark and Ringwood is closely similar to Bullen's model A in many respects, but differs in having a decrease in density with depth through the top 100 km of the mantle. The density must decrease with depth in the upper mantle if the temperature gradient is steeper than about 6–7°C/km, which it almost certainly is, unless chemical composition changes with depth.

The reason is that the effect of thermal expansion outweighs self-compression wherever the geo-thermal gradient is sufficiently steep.

LANDISMAN, SATÔ and NAFE (1965) have made a combined use of body waves and free oscillations to determine directly the density of the mantle between depths of 500 and 2900 km. They used a conventional approach for the rest of the Earth. They obtained the extraordinary result that the density appears to be approximately constant between depths of 1600 and 2800 km, suggesting that Bullen's model A strongly overestimates the density of the lower mantle. However, DORMAN, EWING and ALSOP (1965) have shown that the free oscillation data can be equally well interpreted on a more conventional density model provided that the depth to the core is about 10 km less than the usually accepted value and the velocities are about 25% lower than specified by Gutenberg and Jeffreys in the bottom 30 km of the mantle.

*Radial distribution of gravity, pressure and elastic moduli*

Once the density distribution within the Earth has been specified, the value of gravity can be computed at each depth and the pressure-depth distribution evaluated without complication. The rigidity and bulk moduli can be computed from the $P$ and $S$ velocity distributions and the density distribution, using the basic formulae for $P$ and $S$ velocities (p. 6). All these distributions are shown graphically in Figure 4.32.

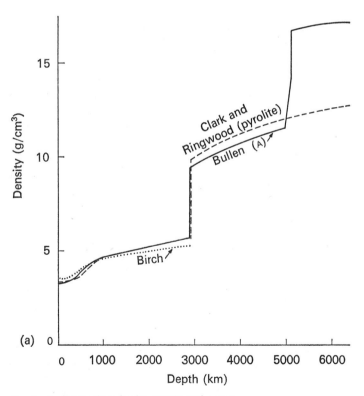

**Fig. 4.32** (a) Density-depth distributions in the mantle and core;

## Lateral density variations in the mantle

An exciting discovery of modern geophysics is that large lateral variations of density exist in the mantle. One type of density anomaly has been discovered through the combined use of gravity observations and seismic crustal structure studies to investigate the cause of isostatic equilibrium (p. 51). Such investigations have shown that the ocean ridges are underlain by a mass deficiency in the topmost mantle beneath, which is large enough to support the ocean ridge in isostatic equilibrium

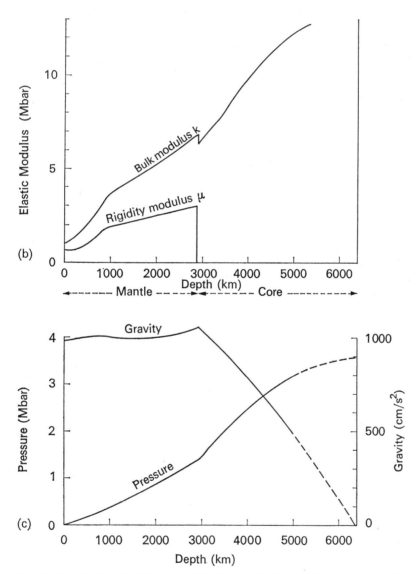

**Fig. 4.32** (continued) (b)   Bulk and rigidity moduli distributions in the mantle and core based on the CLARK and RINGWOOD (1964) pyrolite model of density and the modified Gutenberg velocity distribution; (c) Gravity and pressure distributions within the Earth according to the pyrolite model of density.

and to cause a Bouguer anomaly of $-150$ mgal. Another example comes from the western United States, where the regional plateau uplift of 1500–2000 m is isostatically supported by a mass deficiency in the upper mantle which causes a negative Bouguer anomaly of about $-200$ mgal.

A striking feature about the low density rocks of the upper mantle beneath ocean ridges and the western United States is that they are probably not a permanent feature. MENARD (1964) has shown that ocean ridges grow and decay, and it is also known that the plateau uplift mentioned above occurred during the Tertiary after a period from the Cambrian with relatively low relief. It is inconceivable that such large scale vertical movements occur without changes in the underlying density pattern such that approximate isostatic equilibrium is maintained. It is difficult to avoid the conclusion that these density anomalies in the upper mantle are a transitory feature, that they can appear and disappear in time. This suggests that they are caused by changes in temperature affecting the physical properties, rather than by chemical inhomogeneity.

Large scale changes in the density of the upper mantle could occur as a result of temperature change in three ways:

(1) *Thermal expansion and contraction*: Taking the coefficient of thermal expansion as $3 \times 10^{-5}/°C$, a rise in temperature of 200°C would cause a reduction in density of about 0·02 g/cm³ in the upper mantle. If this were spread over a vertical range of 200 km, it could cause a negative gravity anomaly as large as $-150$ mgal.

(2) *Phase transition*: At a fixed pressure, the mineral assemblage of a rock of fixed composition depends on the temperature, and the density may depend on the mineral assemblage. For instance, plagioclase pyrolite is about 0·07 g/cm³ lower in density than pyroxene pyrolite. Plagioclase pyrolite may occur locally in the topmost mantle where temperature is high. A 30 km thick layer of plagioclase pyrolite, replacing pyroxene pyrolite laterally, would give rise to a negative gravity anomaly of up to $-80$ mgal.

(3) *Partial fusion*: When basalt melts the density is reduced by about 0·3 g/cm³. If partial fusion in the upper mantle produced a 5% fraction of basalt magma, the consequent reduction in density would be 0·015 g/cm³. If 5% partial fusion occurred over a vertical range of 200 km, the resulting mass deficiency could cause a negative gravity anomaly of up to $-120$ mgal.

The simple calculations above show that quite large lateral variations in upper mantle density, sufficient to support ocean ridges and plateau uplifts, can result from raised temperature. If convection currents do occur in the upper mantle, then one would expect the highest temperature to be in the vicinity of the rising current, and consequently there would be an associated density anomaly.

It was mentioned in Chapter 1 that study of the orbits of artificial satellites has revealed broad variations in the Earth's gravity field. The variations are usually depicted as a map of the deviation of the geoid from a reference ellipsoid of revolution, but they could be presented equally well as a map of gravity anomalies. These anomalies may be partly caused by delay in isostatic recovery from the ice-loading and depletion of the ocean associated with the Pleistocene glaciation, but they are probably caused mainly by more permanent lateral variations of density within the Earth.

The satellite gravity anomalies (Figure 1.2) do not appear to be related systematically to the distribution of continents and oceans, or to the Earth's major linear surface features (except island arc-trench regions which are generally positive), or even to the density anomalies in the upper mantle described above. This is partly to be expected, because the major surface features are all in approximate isostatic equilibrium so that the gravitational effect of the surface feature at a point above the Earth's surface is approximately cancelled out by the opposite effect of the 'root'. But in fact a small residual gravity anomaly would be expected above the Earth even for a feature in perfect isostatic equilibrium. This is because the compensating mass must have a smaller areal extent than the surface feature as a result of the decrease of the Earth's circumference with depth, so that the gravitational effect of the surface feature should slightly outweigh that of the compensation. It is therefore all the more surprising that the satellite gravity anomalies do not reflect the surface features in an obvious way. This may be partly because insufficiently high harmonics are available, but the lack of correspondence cannot be entirely explained away like this.

Both the absence of an obvious relation between the surface features and the satellite anomalies, and isostatic considerations, suggest that the anomalies are not caused by variation in crustal thickness or density. Irregularity in the depth of the core-mantle boundary, at which the density jump is about 4 g/cm³, is a possible source of the harmonics lower than the 10th degree. COOK (1967) has shown that a fluctuation of about 38 km in the depth to this boundary would be needed to explain the $n = 10$ harmonic (which has a wavelength of the order of 4000 km or more), and even larger fluctuations would be required to explain the higher harmonics. These would cause unacceptably high stress differences and would be known to seismologists if they existed. It is unlikely that a large enough density contrast can exist in the lower mantle to contribute significantly to any of the harmonics.

Thus the higher degree harmonics, and possibly also the lower ones, must be caused by lateral density variation within the upper mantle or the transition zone. The density anomalies could be caused either by chemical inhomogeneity or by lateral variation in the temperature. Possible ways in which temperature can influence the density in the upper mantle have been described above. However, quite large density variations in the uppermost mantle are associated with isostatic equilibrium and are unlikely to be the main cause of the satellite anomalies for reasons discussed above.

One possible cause of some positive regional gravity anomalies revealed by artificial satellites is the sinking of cool oceanic lithosphere into the upper mantle. This occurs at the Benioff zones of deep earthquakes according to the theory of ocean-floor spreading and plate tectonics (p. 264). An individual sinking sheet of lithosphere has a relatively high density because of its low temperature, eventually returning to normal density as it warms up. If the process continues over a long period of time while the position of the sinking slab of lithosphere migrates, there would be a broad regional cooling of the upper mantle. If the dense, cool part of the mantle remained uncompensated or only partially compensated isostatically, then it would result in a regional positive gravity anomaly and a corresponding region of relatively high geoid. A belt of positive gravity anomalies which can be interpreted in this way is particularly conspicuous over the island arc region of the south-west Pacific and extends around the northern and eastern margins of the Pacific.

Another interesting possibility is that part of the broad gravity anomaly field revealed by satellites (and the corresponding warping of the geoid) is caused by lateral temperature variation affecting the mantle transition zone. The increase in density between depths of about 350 and 450 km caused by the olivine/spinel and associated transitions is about 0·35 g/cm³. According to experimental work of AKIMOTO and FUJISAWA (1968) the pressure-temperature relationship for the olivine/spinel transition at appropriate composition is

$$p(\text{kbar}) = 78 + 0 \cdot 062 \ T(°C).$$

Thus a rise in temperature of 11°C would cause the transition zone to migrate downwards by at least 2 km, the exact value depending on the geothermal gradient at the transition depth. Taking the $n = 14$ degree harmonic, this would cause a gravity anomaly of about $-12$ mgal at the surface, which is of the right order to account for the appropriate harmonic revealed by satellite observations.

On the other hand, theoretical predictions made by ANDERSON (1967b) and predictions based on lattice measurements made by MAO and others (1969) suggest that a rise in temperature would cause the deeper step in the transition zone (spinel to oxides) to migrate upwards causing a positive gravity anomaly.

Thus quite small regional variations in temperature affecting either the lower or the upper part of the transition zone (but possibly not both) could readily explain the bulk of the satellite gravity observations. These temperature variations could be caused by regional variation in the radioactive heat sources in the upper mantle or below, or they could be related to a pattern of convection currents, past or present, either in the whole mantle or restricted to the upper mantle. This explanation requires the transition zone to be significantly more viscous than the overlying asthenosphere, because otherwise the temperature variations would be dissipated more rapidly than they could be produced. The possible relationship between the satellite gravity observations and heat transfer within the Earth is taken up again in later chapters.

# 5 The core

The core extends from a depth of about 2900 km to the centre of the Earth. The core-mantle boundary has a radius of about 3470 km and the core forms 16% of the Earth by volume and 31% by mass. One of the most interesting aspects of the core is that the main magnetic field originates in it. It is the magnetic field that has made it possible to use palaeomagnetism to reveal past movements of the continents and magnetic surveys to show up the pattern of ocean-floor spreading. Thus the core has direct relevance to the problem of the origin of the Earth's surface features.

## 5.1 The structure of the core

The radial structure of the core is known from the travel-times of $P$ waves in the range of epicentral distances from 105° to 180°. Bullen used Jeffreys' velocity distribution to subdivide the core into three zones as follows (Chapter 1):

| | | |
|---|---|---|
| zone E | 2900–4980 km | outer core |
| zone F | 4980–5120 km | transition zone |
| zone G | 5120–6370 km | inner core |

Although the exact position of the upper and lower boundaries of the transition zone are subject to doubt, Bullen's subdivisions are generally accepted by seismologists.

Only $P$ waves are known definitely to propagate within the core. $S$ waves may possibly traverse the inner core but they have not yet been detected with certainty. However, rays which travel through the mantle as $S$ can give rise to, or result from, $P$ rays in the core by conversion at the boundary.

The usual nomenclature for the many types of seismic ray which are associated with the core is built up from the following symbols:

| | |
|---|---|
| $P$ | compression wave in the mantle |
| $S$ | shear wave in the mantle |
| $K$ | compression wave in the outer core |
| $I$ | compression wave in the inner core |
| $\mathcal{J}$ | shear wave in the inner core (undetected yet) |
| $c$ | incident mantle ray reflected at the core-mantle boundary. |

These letters are strung together in sequence to describe the complete path of a given ray through the Earth. For example, a ray which traverses the mantle as $P$ both downwards and upwards and which penetrates the outer core is $PKP$; if it also penetrates the inner core it is $PKIKP$. Other possible rays which undergo conversion at the core-mantle boundary but are not reflected are $SKS$, $PKS$, $SKP$, $SKIKS$, $PKIKS$ and $SKIKP$. $PcP$, $ScS$, $PcS$, $ScP$ are mantle rays reflected once at the core boundary, and $PKKP$ represents one type of ray reflected at the boundary after passage through the outer core.

The depth to the core-mantle boundary can be estimated from the velocity-depth distribution for the mantle by using the travel-time and epicentral distance (105°) of the deepest ray which penetrates the mantle, just grazing the boundary. The estimate can be refined by comparing the observed and computed travel-times of *PcP* and *ScS*. Using this approach, JEFFREYS (1939a and b) obtained a depth of $2898 \pm 2 \cdot 5$ km which is usually rounded to 2900 km. As mentioned in the last chapter (p. 146), interpretation of the periods of the free oscillations suggests that this estimate may be about 10–20 km too high.

The estimate of the depth to the core-mantle interface must be treated as a mean value for the following reason. The Earth's rotation causes surfaces of equal density within the Earth as well as the external surface to be distorted very nearly into ellipsoids of revolution. The flattening at depths within the Earth can be calculated provided the internal density distribution is known. BULLEN (1963) gave an estimate of about 1/390 for the flattening at the core-mantle boundary, which implies that the polar radius of the core is about 9 km less than the equatorial radius and that the corresponding difference in depth to the core-mantle interface is about 12 km. It is also possible that there are further undulations in the core-mantle boundary as some seismological investigations are beginning to suggest. If these are of the order of a few kilometres they might contribute significantly to the low degree harmonics of the Earth's gravity field, but they would also cause quite large stress-differences in the lower mantle.

The core-mantle interface gives rise to strong reflected phases such as *PcP*, *ScS* and *PKKP* and because of this it is known to be a sharp discontinuity rather than a gradational boundary. However, two quite independent pieces of evidence suggest that there may be a transitional layer at the bottom of the mantle with reduced seismic velocities and a high density. These are (1) the study of the periods of free oscillations of the Earth (p. 146), and (2) a study of mantle *P* waves which have been diffracted at the core-mantle interface to emerge beyond 105°, which suggest a layer 30–160 km thick with low *S* velocity and high density (PHINNEY and ALEXANDER, 1967).

Turning to the structure of the core itself, the ray paths for *PKP* and *PKIKP* according to GUTENBERG (1959, p. 104) are shown in Figures 5.1(a) and (b) respectively. Other phases such as *SKS* are also used to help determine the velocity-depth distribution, but *PKP* and *PKIKP* adequately illustrate the principles. Figure 5.2(a) shows the Gutenberg time-distance graph for these phases and Figure 5.3 shows three interpretations of the velocity-depth distribution in the core.

The *PKP* ray which has the shallowest penetration into the outer core emerges at the surface at an epicentral distance of about 188° at A, which is about 8° beyond the anticentre (Figures 5.1(a) and 5.2). With increasing depth of penetration into the outer core only, the epicentral distance of the emerging ray progressively decreases from about 188° to 143° (B in the figures) and then it increases again to about 169° (C in figures). The cusp in the time-distance curve at B is a geometrical effect of the total ray path and does not indicate a discontinuity of any sort. The time-distance graph is consistent with a steady rise in velocity with depth in the outer core, although the interpretation cannot be unique because no rays have their lowest point in the topmost part of the core. According to Jeffreys, the velocity increases from 8·10 km/s at 2900 km depth to 10·44 km/s at 4980 km, and Gutenberg's estimates are closely similar.

The rays which penetrate the inner core form the branch DE of the time-distance curve (Figure 5.2). The ray with the shallowest penetration emerges at an epicentral distance of about 110° and is represented by D. It was MISS LEHMANN (1936) who discovered the inner core by recognizing the significance of the branch DE, which gives rise to weak arrivals within the 'shadow zone'. The accepted interpretation is that (1) there is an overall increase in velocity of about 0·8 km/s between depths of about 4980 and 5120 km, and (2) below 5120 km the velocity is approximately constant at

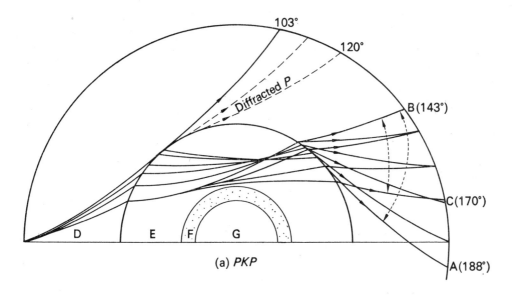

(a) *PKP*

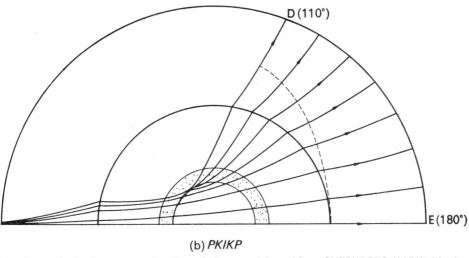

(b) *PKIKP*

**Fig. 5.1**    Ray paths for *P* waves passing through the core. Adapted from GUTENBERG (1959), *Physics of the Earth's Interior*, p. 104, Academic Press.

11·2 km/s to the centre of the Earth. The segment CD of the time-distance curve represents rays which reach their deepest point within the transition zone, but such arrivals are not observed because of confusion with the branch DE.

The main difference of opinion between seismologists who have studied the core concerns the

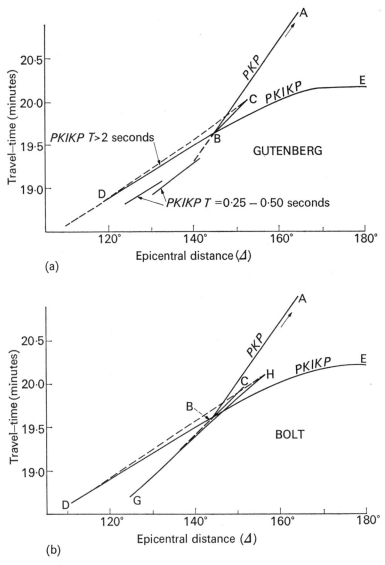

**Fig. 5.2**   Time-distance graphs for *PKP* and *PKIKP* according to (a) GUTENBERG (1959), *Physics of the Earth's Interior*, p. 111, Academic Press, and (b) BOLT (1962), *Nature, Lond.*, **196**, 122.

detailed structure of the transition zone. Three interpretations are shown in Figure 5.3. Jeffreys interpreted the structure as a *decrease* in velocity from 10·44 km/s at 4980 km depth to 9·47 km/s at 5120 km where there is an abrupt increase to 11·16 km/s. In contrast, Gutenberg claimed to be able to interpret the observations by a continuous increase in velocity with depth through the transition zone, although he interpreted some early short-period arrivals between epicentral distances of about 120° and 140° as caused by dispersion in the transition zone, the shorter periods travelling at a higher velocity. A new interpretation by BOLT (1962, 1964) suggested that these early arrivals of Gutenberg represent an additional branch of the time-distance curve as shown in Figures 5.2(b) and 5.3. He suggested that the observations can be best explained by two abrupt increases in velocity separated by a shell of uniform velocity.

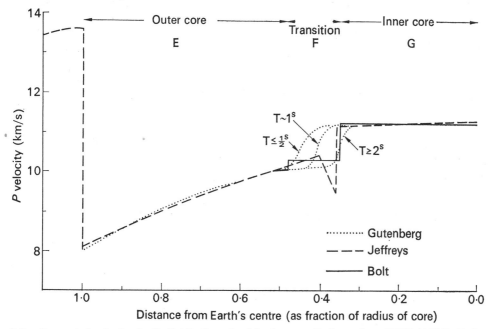

**Fig. 5.3**   Suggested velocity-depth distributions for P in the core. Redrawn from BOLT (1964), *Bull. seism. Soc. Am.,* **54**, 192.

## 5.2  The physical state of the core

### *The outer core*

The straightforward interpretation of the lack of observed shear waves passing through the core is that the outer core is liquid and cannot therefore transmit S. This argument is not quite conclusive because shear waves might be cut out by strong attenuation in the core. However, the large amplitude of the phase SKS corroborates the interpretation of the outer core as liquid. If it were solid the amplitude would be much smaller. This is because a much larger fraction of the incident energy is converted from S to P, and then back to S on return, at a solid-liquid interface than if both media are solid. The observed periods of the lower modes of the Earth's free oscillations also suggest that the outer core has a negligible rigidity modulus and is therefore liquid.

There are two other geophysical phenomena which when taken with the seismology of the core make the interpretation of the state of the outer core as liquid virtually conclusive. One of these is the Earth's magnetic field and its secular variation which are described later in the chapter and which could not be explained if the core were solid. The other is the response of the Earth to tidal forces and to impulses affecting the axis of instantaneous rotation which provide estimates of the rigidity of the core.

The *Earth tide* (MELCHIOR, 1966) arises as follows. The gravitational attractions between the Earth and the Sun and Moon only exactly balance the centrifugal forces at the centre of the Earth. Elsewhere within the Earth and at its surface there is a small residual gravity potential which causes both the ocean tide and a periodic deformation of the solid Earth which is known as the Earth tide. The tide-raising potential at a point on or within the Earth depends on its position and also that of the Sun and Moon in relation to a frame of reference fixed within the Earth and rotating with it. Semi-diurnal tides are the most prominent effect but longer period types such as diurnal, fortnightly and semi-annual tides do also occur both in the ocean and in the body of the Earth. The Earth tides are effectively in equilibrium with the tide-raising potential and are therefore in phase with it, but this is not true for the ocean tides especially in shallow water. Earth tides can be observed by recording the tidal variation of gravity, and tilting of the Earth's surface, and variation in linear strain using a strain seismometer (Figure 5.4). The amplitude of the Earth tide at the equator is about 20 cm.

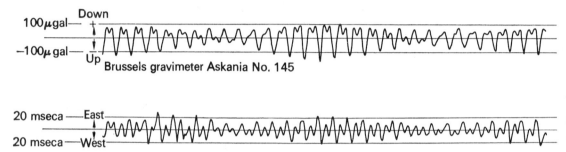

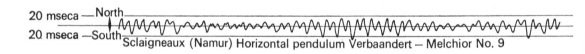

Fig. 5.4 The Earth tides simultaneously recorded *above* by Askania gravimeter at Brussels, and *below* by tiltmeters at Sclaigneaux (Namur). Reproduced from MELCHIOR (1966), *The Earth tides*, p. 18, Pergamon Press.

The response of the Earth to the tide-raising potential depends on the rigidity distribution in it. It is conveniently expressed in terms of two quantities $h$ and $k$ which are known as *Love's numbers* and which can be computed for specified distributions of the density and elasticity within the Earth*; $h$ is the ratio of the heights of the Earth tide to the corresponding theoretical ocean tide assuming equilibrium is reached; $k$ is the ratio of the gravitational potential caused by the tidal deformation of the Earth to the tide-raising potential. Shida introduced a further number $l$ which is the ratio of the horizontal displacement of the Earth tide to that of the corresponding equilibrium ocean tide. The numbers $h$, $k$ and $l$ can be determined from Earth tide observations. For instance, the quantity $(1 + h - \frac{3}{2}k)$ can be determined from the tidal variation in gravity and $(1 + k - h)$ from the tilting of the Earth's surface.

Another method of estimating Love's number $k$ is to study the natural period of the wobble of the Earth's instantaneous axis of rotation. The wobble of the Earth's axis is studied by recording the variation of latitude at a number of observatories. The average displacement of the poles from their

* On pp. 156 and 157, $k$ denotes one of Love's numbers; elsewhere in this book it denotes the bulk modulus unless otherwise specified

mean position is about 3 m and the corresponding angular disturbance to the axis of rotation is about 0·1″ (Figure 5.5). Part of the observed wobble is a forced oscillation with a period of about a year which is probably mainly caused by atmospheric phenomena. Most of the wobble, however, is a free oscillation with a period of 1·20 years which is called the *Chandler wobble*. This appears as a randomly excited damped oscillation with a decay time of the order of 12 years. It has been a long standing puzzle to know how the Chandler wobble is excited, but it has recently been suggested that the redistribution of mass associated with large earthquakes may be the principle cause (MANSINHA and SMYLIE, 1968).

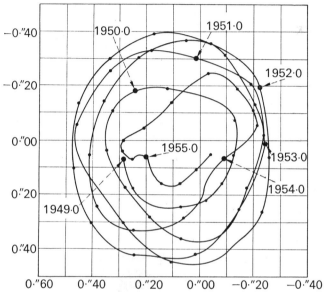

**Fig. 5.5**   Spirals described by the North Pole, showing time in years and months. Redrawn from MELCHIOR (1957), *Physics Chem. Earth*, **2**, 222, Pergamon Press.

The period of the Chandler wobble depends on the rigidity distribution within the Earth. If the Earth was an absolutely rigid body then the period $\tau_0$ would be 305 days. The effect of a finite rigidity is to increase the natural period of the wobble. Love's number $k$ can be estimated from the period of the wobble $\tau$ by the following relationship which was originally derived by Love (MELCHIOR, 1966)

$$k = (1 - \tau_0/\tau)\,(2g\epsilon/aw^2 - 1)$$

where $\epsilon$ is the ellipticity,

$a$ is the equatorial radius,

$g$ is gravity

and    $w$ is the angular velocity of rotation.

According to MELCHIOR (1966), the most satisfactory values of Love's numbers obtained from observations are

$$h = 0\cdot584$$
$$k = 0\cdot290$$
$$l = 0\cdot045.$$

There is still quite a lot of uncertainty, particularly as a result of anomalous observations of tidal gravity in Asia, but $h$ and $k$ are probably not in error by as much as 10%. The next stage is to interpret the observed values in terms of allowable distributions of the rigidity modulus within the Earth. TAKEUCHI (1950) did this by taking a distribution of density and elasticity similar to that of Bullen and assuming a range of possible values for the rigidity modulus of the core. He found that the observed values of Love's numbers required the rigidity modulus of the core to be between zero and $10^{10}$ dynes/cm². A rigidity modulus within this range would be characteristic of a liquid rather than a solid.

The only possible interpretation of the core-mantle boundary is that the melting point of the material forming the lowermost mantle is higher than that of the outer core, and that the actual temperature lies between the two melting points.

### The inner core

The inner core is believed to be solid for the following reasons (BULLEN, 1958). According to the Gutenberg velocity distribution, the compressional wave velocity increases by about 11% between the base of the outer core and the top of the inner core. According to Jeffreys there is an even larger jump in velocity at the base of the transition zone. The equation relating $P$ velocity to density and elasticity can be re-expressed as follows:

$$\rho V_P^2 = k + 4\mu/3$$

It is inconceivable that the density could decrease between the outer and inner core and it probably increases substantially. Therefore both sides of the equation must increase by at least 23% across the transition zone. If $\mu$ is zero in the inner core, then $k$ must also increase by at least 23%. According to Bullen's compressibility-pressure hypothesis, one would not expect $k$ to rise more rapidly through the transition zone than in the outer core. The compressibility-pressure hypothesis is probably not strictly correct, but it is unlikely that $k$ can jump by as much as 23%. The more acceptable explanation

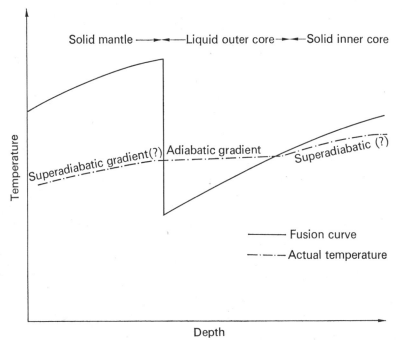

**Fig. 5.6**   Possible explanation of the outer-inner core boundary as solidification under increasing pressure.

of the steep increase in velocity is that the rigidity modulus takes a value of the order of $3 \times 10^{12}$ dynes/cm² in the inner core.

The transition from the liquid outer core to the solid inner core may be caused by solidification under increasing pressure. The temperature distribution in the outer core cannot be far removed from the adiabatic gradient. Almost certainly the increase in melting point with pressure is steeper than the adiabatic gradient. Provided that the temperature of the outer core is not greatly in excess of the melting point, the effect of increase in pressure with depth is to cause solidification below the depth at which the relevant adiabat of the outer core crosses the fusion curve (Figure 5.6). If the transition does represent the melting point of iron, then the temperature according to STRONG (1959) is probably about 3000°K (p. 128).

## 5.3 Composition of the core

Shortly after the discovery of the two main classes of meteorites in the mid-nineteenth century, it was suggested that the Earth has a core similar to the iron meteorites surrounded by a silicate shell analogous to the stony meteorites. Wiechert later adopted this hypothesis to explain the high mean density of the Earth but he over-estimated the radius of the core because he did not allow for self-compression. After the discovery of the core through the use of earthquake waves (OLDHAM, 1906), the concept of an iron core was taken over by seismologists and remained essentially unquestioned until 1941.

KUHN and RITTMAN (1941) suggested that the core consists of a condensed form of hydrogen. This hypothesis was based on the idea that the Sun and Earth have the same composition. It stimulated a lot of discussion but it is now known to be untenable because the pressure within the Earth is nowhere high enough to cause condensation of hydrogen.

A few years later RAMSEY (1949) suggested that the core may be formed of a high pressure modification of mantle material. He postulated that the silicate material is stripped of an outer electron to produce a metallic material with high density, low melting point and very high electrical conductivity. Ramsey's hypothesis gained serious support for a few years but has since lost favour for reasons outlined below. A few geophysicists still hold to the hypothesis, notably some of the supporters of the expanding Earth hypothesis (p. 270).

The point at issue since 1948 has been whether the core-mantle boundary represents a change in composition from silicate to metallic iron with impurities or a radical phase change as Ramsey postulated. The decisive evidence between them depends on the comparison of the known properties of the core such as density with the experimentally and theoretically derived values for silicates and metals at high pressure.

Core pressures have not yet been reached in static experiments but they have been momentarily attained in shock waves produced by explosives (e.g. McQUEEN, FRITZ and MARSH, 1964). A shock wave differs from a seismic wave in that the particle velocity $U_p$ is comparable to the velocity of propagation of the wavefront $U_w$ and the temperature rise associated with the passing wave is higher than would be caused by adiabatic compression. The velocities $U_p$ and $U_w$ can be observed in experiments. The pressure and the density within the wave can be deduced from $U_p$ and $U_w$ by the following equations which are based on the principles of conservation of mass and momentum (Birch; in CLARK, 1966, p. 99)

$$p = \rho_0 U_w U_p$$
$$\rho = \rho_0 U_w/(U_w - U_p)$$

where $\rho_0$ is the density at zero pressure. The resulting relationship between pressure and density is called the Hugoniot. By applying thermodynamic equations to correct for the temperature rise, the Hugoniot can be reduced to a more general relationship between pressure and density. The seismic parameter $\phi$ which is the square of the hydrodynamic sound velocity can be deduced from the adiabatic pressure-density curve using the following equation (p. 140)

$$\phi = k/\rho = dp/d\rho.$$

If $dp/d\rho$ is estimated from the Hugoniot it will give an overestimate of the parameter $\phi$.

According to the shock wave experiments, the density of iron at the core-mantle boundary at 2000°C would be about 11·2 g/cm³ and at the centre of the Earth it would be 13 g/cm³. The actual density of the outermost core is about 9·5 g/cm³ which is significantly lower even allowing for the reduction in density on melting. The presence of nickel would increase the disagreement, but the discrepancy of 15% can be explained by the presence of lighter elements such as silicon in solution.

Shock wave experiments also show that metals with lower atomic number such as aluminium and magnesium would have much too low a density to explain the core. Presumably this argument would also apply to the hypothetical metallic form of mantle silicate material suggested by Ramsey.

The shock wave results for metals can be conveniently compared with the seismic and density distribution within the Earth's interior by plotting $(dp/d\rho)^{\frac{1}{2}}$ against density for both (Figure 5.7). The quantity $(dp/d\rho)^{\frac{1}{2}}$ for metals can be determined from the Hugoniot; the error introduced by using the Hugoniot instead of an adiabatic equation of state is not serious. Within the Earth, $(dp/d\rho)^{\frac{1}{2}}$

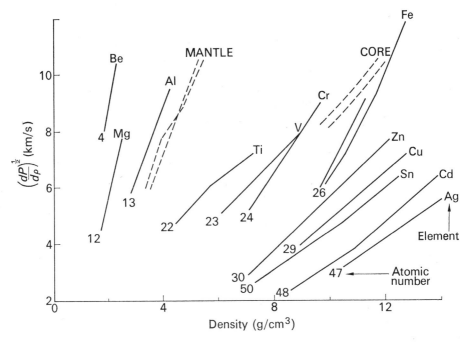

**Fig. 5.7** Hydrodynamic sound velocity as a function of density shown for a selection of metals as obtained by shock wave experiments (solid lines) and for the mantle and core as obtained from seismic observations and density models (dashed lines). The numbers shown are atomic numbers. Redrawn from BIRCH (1961b), *Geophys. J. R. astr. Soc.*, **4**, 309.

is the square root of the seismic parameter $\phi$, and in the outer core it is the $P$ velocity. The comparison is clearly consistent with a silicate mantle and a core of iron with some elements of lower atomic number in solution. It gives no support at all to Ramsey's hypothesis.

Most of the shock wave experiments on rocks and relevant silicate minerals reported by Birch (CLARK, 1966) do not quite reach the pressure of the outermost core. An exception is a single observation on dunite at 2·4 megabar made by Russian scientists yielding a density of 6·8 g/cm³, which is far too low to explain the core. KNOPOFF and MACDONALD (1960) also concluded that the core is unlikely to be a high pressure modification of silicates on theoretical grounds. They found from shock wave results that the equation of state at extremely high pressures could be represented by the Thomas-Fermi model, hitherto applied to gaseous stellar interiors. They extrapolated this down to lower pressures and found that the core has a density closer to iron than to silicates.

## 5.4  The Earth's magnetic field

*Introduction—the geomagnetic field in historical times*

Since about 1950 it has been recognized that the origin of the main geomagnetic field is related to fluid motions in the outer core, probably involving a dynamo process. A comprehensive theory has not yet been developed but what we do know adds considerably to our understanding of the structure of the core and processes which occur in it.

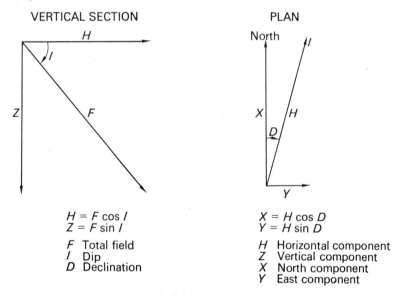

$$H = F \cos I$$
$$Z = F \sin I$$

$F$  Total field
$I$  Dip
$D$  Declination

$$X = H \cos D$$
$$Y = H \sin D$$

$H$  Horizontal component
$Z$  Vertical component
$X$  North component
$Y$  East component

**Fig. 5.8**   Components of the Earth's magnetic field.

The geomagnetic field at a point is a vector quantity and therefore it needs three quantities to describe it completely (Figure 5.8). It can be specified by the vertical component measured downwards ($Z$), the horizontal component ($H$) and the declination ($D$) which is the angle between true north and the direction of the horizontal component (taken as positive towards the east). Another common way of describing the field is to use the magnitude of the total field ($F$), the angle of its dip ($I$) and the declination. The variation with time of the elements of the geomagnetic field are studied

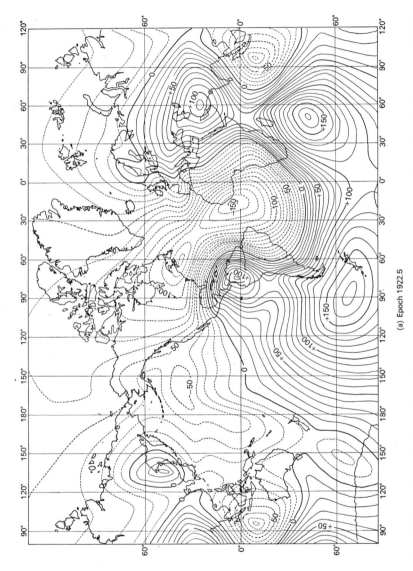

(a) Epoch 1922.5

**Fig. 5.9** (a)  Isoporic chart showing secular change in gamma per year, vertical intensity, for epoch 1922.5. Redrawn from VESTINE and others (1947), *Description of the main magnetic field and its secular variation, 1905-1945*, p. 380, Carnegie Institution of Washington.

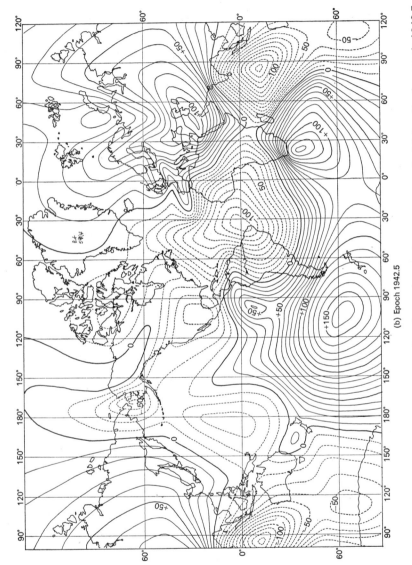

(b) Epoch 1942.5

**Fig. 5.9** (b) Isoporic chart showing secular change in gamma per year, vertical intensity, for epoch 1942.5 Redrawn from VESTINE and others (1947), *Description of the main magnetic field and its secular variation, 1905-1945*, p. 384, Carnegie Institution of Washington.

on a world-wide basis by recording them at permanently established magnetic observatories, such as the one run by the Institute of Geological Sciences at Eskdalemuir in south Scotland. Short-period variations can be studied using temporary observatories and the areal distribution of the geomagnetic field can be mapped by making magnetic surveys on land and at sea. The standard work on the description and analysis of the geomagnetic field is by CHAPMAN and BARTELS (1940).

About 90% of the present-day geomagnetic field can be represented by the field of a magnetic dipole at the Earth's centre which makes an angle of about 11·5° with the Earth's axis of rotation. An appreciable *non-dipole field* remains after the best-fitting *dipole field* has been subtracted from the observed field of the present day.

The mean annual values of all the magnetic elements vary in a regular way from year to year. This long-period change in the geomagnetic field from one year to the next is known as the *secular variation*. It was discovered by Gellibrand in 1634 when he recognized that the declination at London changes with time (Figure 5.10). The secular variation affects both the dipole and non-dipole parts of the field. Over the last century the dipole field has been decreasing by about 0·04% per year. The percentage annual change in the non-dipole field is on average somewhat larger but it varies from region to region and involves both increase and decrease of field strength.

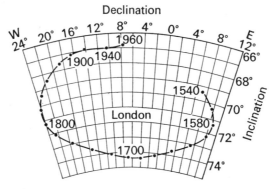

**Fig. 5.10**   Secular variation of the declination and inclination at London since 1540.

Spherical harmonic analysis has been used to show that both the main field and its secular variation originate within the Earth. This contrasts with the short-period variations of the magnetic elements (p. 117) which are primarily caused by electrical current systems above the Earth's surface.

A map showing the secular variation of a specific magnetic element for a given epoch is called an *isoporic chart*. Figure 5.9 is an example. This shows a series of centres where the secular variation is particularly large. These are known as *isoporic foci* and they are particularly associated with large changes in the non-dipole field.

Isoporic foci are conspicuously lacking from the Pacific Ocean, but whether this is an accident of chance or has more fundamental significance is not known. When isoporic foci for successive epochs are compared (Figure 5.9), a most interesting fact emerges. It is found that these foci are drifting westwards at about 0·2° of longitude per year. The secular variation and particularly its westward drift cannot be adequately explained by any process occurring entirely within solid subdivisions of the Earth, and this suggests that part of the Earth's interior, notably the outer core, must be fluid. It also implies that the outermost part of the core may be rotating at about 0·2° per year more slowly than the crust and mantle.

*Palaeomagnetism—the history of the geomagnetic field through geological time*

Palaeomagnetism is the study of the geomagnetic field during the geological past. It makes use of the permanent magnetization which may be picked up by a rock when it is formed or during some later geological event. The results of palaeomagnetic studies are of great importance in two ways. They show up important properties of the geomagnetic field which are not apparent in the relatively short historical record of it. They also provide an important tool for determining how continents have moved relative to each other and to the equator during geological time (p. 204). The great advances in palaeomagnetism over the past twenty years have been largely stimulated by Professors Runcorn and Blackett and scientists who have been associated with them. There are also active Japanese, French and other groups working on rock magnetism (NAGATA, 1961). A good general account of palaeomagnetism is given by IRVING (1964), and the techniques of the subject are described in a volume edited by COLLINSON, CREER and RUNCORN (1967).

An outline of the palaeomagnetic method is as follows. Orientated specimens of rocks from a locality of known age are collected. The direction of their permanent magnetization is measured using an astatic or spinner magnetometer. After certain precautions have been taken (see below), the resulting direction of magnetization for a sample is used as an estimate of the direction of the magnetic field (known as the ambient field) when the rock was formed or when its magnetization was picked up. The secular variation is averaged out by measuring several samples from slightly different geological horizons for each locality. The results are usually plotted on a stereographic projection such as is shown in Figure 5.11 and a statistical estimate of the true mean and of its 'circle of confidence' are obtained using a method especially developed for the purpose by R. A. Fisher. Recently a method of using the intensity of magnetization to determine the intensity of the ancient field has been applied to the Tertiary period (SMITH, 1967).

The permanent and the induced magnetization of rocks are effectively caused by relatively small fractions of ferromagnetic minerals, of which magnetite is the commonest. Rocks may pick up a permanent magnetization through several processes including the following. *Thermal remanent magnetization* (usually abbreviated to TRM) is the type picked up by igneous rocks when they cool through the Curie point, which is usually about 600°C or less, and well below the melting point. This is almost always in the direction of the ambient field but is occasionally in the opposite sense. As the igneous rock cools further the magnetization is strengthened. *Detrital remanent magnetization* (DRM) occurs in some sediments as a result of alignment of the magnetic grains in the direction of the ambient field, with some depositional bias, as deposition occurs. *Chemical remanent magnetization* (CRM) may occur after a rock has been formed by the production of new magnetic minerals through chemical reactions associated with diagenetic alteration, weathering or metamorphism. The newly formed magnetic minerals pick up permanent magnetization in the direction of the field at the time. Another type is *viscous magnetization* (VRM) which may be picked up when a rock lies in a weak magnetic field at relatively low temperature for a very long period of time.

If a permanent magnetization can be retained without change through a long period of geological time, it is said to be *hard*. If it is easily lost or its direction is changed, then it is said to be *soft*. VRM is usually soft. One of the objects of some of the more recently developed palaeomagnetic techniques is to remove the unwanted soft components of magnetization by 'washing'. Two standard ways of washing are: (1) demagnetizing the rock in an A.C. field of decreasing intensity, which removes the soft components but does not affect the hard magnetization; and (2) heating the rock to an appropriate

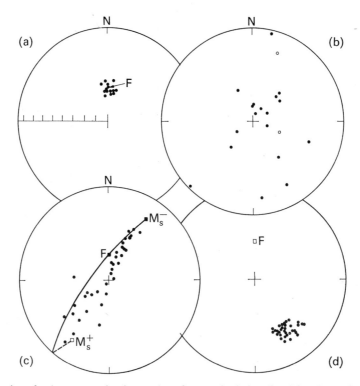

**Fig. 5.11** Examples of palaeomagnetic observations from a single locality. Directions of positive inclination are shown as solid circles and negative inclination as open circles. F is the present geocentric axial dipole field.
  (a) Payette formation (Neogene sediments), Idaho, U.S.A., showing palaeomagnetic directions consistent with present field;
  (b) Arikee formation (Miocene sediments), South Dakota, U.S.A., which are widely scattered;
  (c) Triassic marls from Sidmouth, England, showing scatter along great circle between the stable normal and reverse directions for the Triassic and the present field, caused by unstable magnetization (VRM), reproduced from CREER (1957);
  (d) Tertiary lava flow, Australia, showing consistent reversed directions differing significantly from present field.
  Redrawn from IRVING (1964), *Paleomagnetism*, p. 55, John Wiley.

temperature, thereby removing the components of magnetization picked up at low temperature but leaving the high temperature component.

An important discovery of palaeomagnetism relating to the past history of the magnetic field is that some rocks have picked up a permanent magnetization in the opposite sense to the present field. This is known as *reverse magnetization*. For instance, HOSPERS (1951) showed that the late Tertiary and Pleistocene lavas of Iceland could be stratigraphically subdivided into groups of normally and reversely magnetized rocks. Later work showed that reverse magnetization is about as common as normal magnetization through the geological column and that reversals occur in sediments as well as in igneous rocks. Figure 5.12 shows an example of a reversal within a pile of lavas. Either the reversely magnetized rocks have somehow picked up a magnetization in the opposite direction to the ambient field when they formed, or the Earth's field itself has suffered periodic reversal.

In this respect, it has been shown that a dacitic pitchstone from Mt. Asio in Japan picked up a

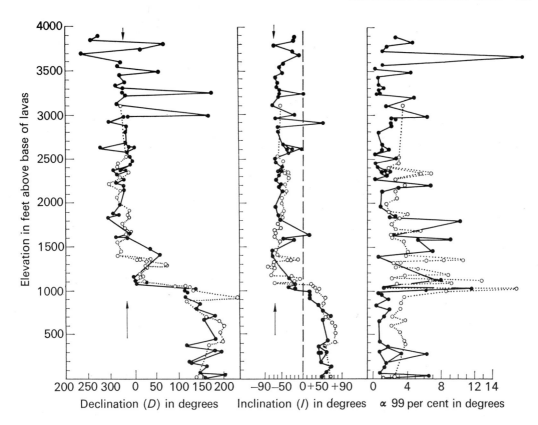

**Fig. 5.12** The palaeomagnetically determined declination (*D*), inclination (*I*) and the radius of the cone of 99% confidence (α 99 per cent) for sections of the Stormberg lava succession of Triassic-Jurassic age in Basutoland, from Sani Pass (solid circles) and Maseru area (open circles). The lavas at the bottom of the succession are reversely magnetized and there is a change to normal polarity at an elevation above the base of about 1000 ft. The arrows indicate the declination and inclination of the present field. Redrawn from VAN ZIJL, GRAHAM and HALES (1962), *Geophys. J. R. astr. Soc.*, **7**, 170.

reverse magnetization in the present normal Earth's field. This rock and other dacites from Haruna in Japan also undergo self-reversal in laboratory experiments (NAGATA, 1953). This self-reversal can be explained theoretically by several processes (NÉEL, 1951). The simplest of these is as follows. Two magnetic phases with different Curie points may be intergrown in a rock. As the phase with the higher Curie point cools, it picks up a normal magnetization, but the resulting lines of force within the mineral intergrowths are in the opposite direction and may exceed the Earth's field strength. As the second phase cools through its Curie point, it therefore picks up a magnetization in the reverse direction. As further cooling occurs, the reverse magnetization of the phase with the lower Curie point may become dominant, imparting an overall reverse magnetization to the rock. This simple mechanism is probably unimportant in relation to the other processes suggested by Néel, but it illustrates the principle that self-reversal is possible.

However, the dacites of Asio and Haruna are exceptional. Most igneous rocks contain only a single important magnetic phase and they do not undergo self-reversal in laboratory experiments. There

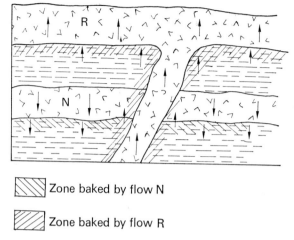

◣ Zone baked by flow N

◢ Zone baked by flow R

**Fig. 5.13**    Magnetic directions in the baked zone adjacent to a dyke, showing that the polarity of the normally magnetized lava has been reversed during thermal metamorphism. Redrawn from COX and DOELL (1960), *Bull. geol. Soc. Am.*, **71**, 736.

is circumstantial evidence from depositional magnetization, and also from the observation that normally magnetized lavas adjacent to a reversely magnetized dyke become reversely magnetized in the vicinity of the dyke as a result of thermal metamorphism (Figure 5.13). It is nowadays generally agreed that most observed reversals represent true changes in polarity of the geomagnetic field. This has important implications for theories of the origin of the main field.

The history of reversals of the Earth's magnetic field over the last 100 my is being worked out from three main sources of evidence, as described in Chapter 7. These are: (1) combined palaeomagnetic and K-A age dating studies of thick lava sequences such as that in Iceland; (2) study of palaeomagnetism of deep sea cores; and (3) study of oceanic magnetic anomalies. Observed changes in polarity of the field occur on average every 0·4 my, although the interval is not regular. In addition, short periods of opposite polarity known as 'events' occur within the main polarity periods. More conventional palaeomagnetic studies carry the record much further back although in much less detail. It has been found that reversals occur during most geological periods except that none have been observed for the Permian. Any theory of the Earth's magnetic field must take the existence and observed pattern of reversals into account.

Another important 'geomagnetic' result from palaeomagnetism is that the field averaged over a few thousand years appears to have remained within a few degrees of the axis of rotation during the Pleistocene and late Tertiary. The angle of dip is also the value which would be expected for the axial dipole field. This suggests that over a period of time represented by the specimens in a sample, the non-dipole field averages out to zero and the axis of the dipole field coincides with the axis of rotation rather than being ten or more degrees out of alignment as at the present. This is what would be expected on the modern theories of the origin of the field. It provides the justification for using the palaeomagnetic results from the Tertiary and before to estimate palaeolatitudes and azimuths; and the good agreement between ancient latitudes determined by palaeomagnetic and by geological methods further supports the hypothesis that the past magnetic field can be approximately represented by an axial dipole when secular variation is averaged out.

If the Earth's field in the past averaged over a few thousand years can be represented by an axial dipole, then the palaeomagnetic dip $I$ is related to the palaeolatitude $\theta$ by

$$\tan I = 2 \tan \theta.$$

The horizontal direction of the remanent magnetization gives the palaeoazimuth which can show up past rotations if they have occurred. The main restriction is that longitude cannot be obtained and therefore the past position cannot be uniquely deduced. The result of applying palaeomagnetism to the problem of continental drift is discussed in Chapter 7.

### *Origin of the main field and the secular variation*

In 1600 William Gilbert published his famous treatise *De Magnete*, in which he pointed out that the dip of the Earth's field resembles that of a sphere of the magnetized material lodestone. Between then and the early 20th century it was widely held that the geomagnetic field is caused by a strongly magnetized region within the Earth. For instance, a dipole at the centre with a magnetic moment of $8 \cdot 05 \times 10^{25}$ e.m.u. gives a reasonable approximation to the Earth's field as it was in 1955. Exactly the same effect would be caused by a uniform magnetization of $0 \cdot 075$ e.m.u./cm$^3$ affecting the whole Earth, or $0 \cdot 49$ e.m.u./cm$^3$ affecting the whole core, or about 10 e.m.u./cm$^3$ affecting the inner core only (Figure 5.14). The dipole field could also be produced by a current system such as is shown in Figure 5.14(d).

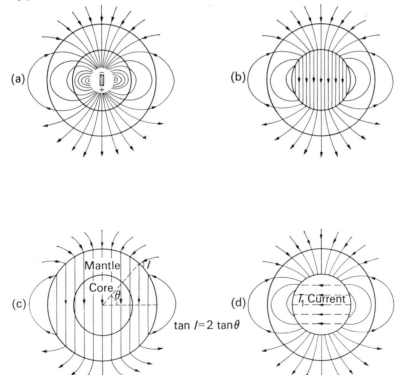

**Fig. 5.14**   Possible types of origin of the main dipole field: (a) a dipole at the Earth's centre; (b) a uniformly magnetized core; (c) a uniformly magnetized Earth; and (d) an east-west electric current system flowing along the core-mantle boundary with current density proportional to the cosine of the latitude (toroidal type $T_1$).

About the most highly magnetized common rocks of the crust form layer 2 of the oceanic crust (p. 82). Their magnetization is about 0·01 e.m.u./cm³ and they cause local disturbances of the magnetic field amounting to about 1%. A much stronger magnetization is needed to account for the main field. One of the obvious difficulties of the large magnet theory is that the Curie point of iron is about 770°C and this temperature is probably exceeded at and below a depth of 200 km at most. The effect of increased pressure is to reduce the Curie temperature slightly, although it is just possible that the effect is reversed at exceptionally high pressures.

Several new theories have been suggested during this century. One of the more exotic of these was the idea of BLACKETT (1947) that massive rotating bodies have an inherent magnetic field associated with them. This theory was disproved when (1) it was shown that the magnetic field increases with depth in mines, contrary to Blackett's predictions (RUNCORN, BENSON, MOORE and GRIFFITHS, 1951), and (2) a laboratory experiment made by Blackett failed to reveal the effect. Nowadays, the most decisive evidence contrary to most of the early theories is the realization that the Earth's field has periodically reversed itself in the geological past.

The only modern theory which seems capable of explaining the past and present features of the main field is the dynamo theory. It was originally suggested by LARMOR (1920) as a mechanism to explain the Sun's magnetic field, but COWLING (1934) showed that Larmor's mechanism would not work. Cowling's objection applies to any dynamo based on axially symmetrical fluid motions. The theory was revived again in a modified form as an explanation of the geomagnetic field by W. M. Elsasser and E. C. Bullard.

The dynamo theory attributes the observed geomagnetic field to a system of electric currents in the core and lower mantle. The electric currents must be maintained because otherwise they would die out in less than one million years. This is accomplished by fluid motions in the outer core which cause it to act as a dynamo. It is usually assumed that the fluid motion is caused by thermal convection. The outer core needs to be a good electrical conductor.

To illustrate the basic principle, a disc dynamo is shown in Figure 5.15. A rotating circular disc which is a conductor cuts a weak axial magnetic field. An e.m.f. is produced between the centre and

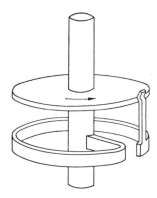

**Fig. 5.15**   A simple disc dynamo. After BULLARD and GELLMAN (1954), *Phil. Trans. R. Soc.,* **247A**, 214.

edge of the disc and this is used to drive an electric current through the attached coil. Provided that the coil is wound in the correct sense, the current causes an axial magnetic field which reinforces the original field. If the disc is rotated fast enough, a very slight stray magnetic field at the start can be amplified so that the dynamo maintains its own magnetic field. This is the principle of the self-exciting dynamo, although in practice the disc is replaced by an armature with many windings and the coil also has many turns, both being wound on soft-iron cores. A self-exciting dynamo of this type differs fundamentally from the core in one respect. The ability of the disc dynamo to work at all depends on its asymmetry; but the core is geometrically homogeneous (singly-connected) and cannot have the same type of asymmetry as the disc. The crucial question is whether a homogeneous self-exciting dynamo can exist at all.

The branch of applied mathematics dealing with the interaction of fluid motions and electromagnetic fields is called *magnetohydrodynamics*. Thorough investigation of the dynamo theory would involve looking for relevant solutions to the following simultaneous set of partial differential equations subject to specified boundary conditions:

(1) Electromagnetic equations connecting the magnetic field with the velocity field;
(2) The Navier-Stokes hydrodynamic equation connecting the velocity field in a viscous rotating fluid with boundary and body forces of electromagnetic, thermal and mechanical origin;
(3) The equation of heat flow appropriate to a fluid in motion;
(4) The equation of continuity for an incompressible fluid.

These equations are non-linear and the problem of solution is more than formidable. Furthermore, the boundary conditions and the relevant physical properties are not known. What BULLARD and GELLMAN (1954) did was to assume a pattern of fluid flow, thereby reducing the problem to solution of the electromagnetic equations (1) alone. They solved these by replacing the partial differential equations by finite-difference approximations and solving the resulting set of algebraic equations on a computer. Bullard and Gellman were able to show that certain plausible patterns of fluid motion could cause the core to act as a homogeneous dynamo, but their method was not rigorous and left some loose ends untied. Later on both HERZENBERG (1958) and BACKUS (1958) gave rigorous mathematical demonstrations of the possibility of homogeneous dynamo mechanisms, but the models they used are unlikely to resemble true fluid motions in the core. More recently still, LOWES and WILKINSON (1963) have constructed an experimental model of the Herzenberg dynamo which works and which has the added interest of being capable of reversing its field periodically.

The velocity of the fluid at a point within the core is a vector quantity. The complete pattern of fluid motion can be represented by a vector field which specifies the velocity vector at each point. A vector field such as the velocity field can conveniently be divided into two complementary types of field as follows: (1) a *toroidal field* in which the vector at every point is perpendicular to the radial direction; and (2) a *poloidal field* which contains radial and tangential components. These are represented by the symbols $T$ and $S$ respectively.

Here is a broad outline of the Bullard dynamo process. Each stage of the process involves the electromagnetic interaction of a magnetic field with a fluid motion to produce an induced electric current system. The electric currents produce a further magnetic field system. To simplify the account, we omit reference to the electric current systems. Each stage of the process can therefore be described as interaction of magnetic field $X$ with fluid motion $Y$ to produce a new magnetic field $Z$. Looking at it in a slightly different way, the fluid motions in the highly conducting core have the effect of dragging the lines of magnetic force with them, thereby increasing the energy of the magnetic field at the expense of the fluid motions, and changing its pattern (Figure 5.16). The Bullard process in its simplest form involves four such interactions which enable a weak initial dipole field to be amplified and maintained.

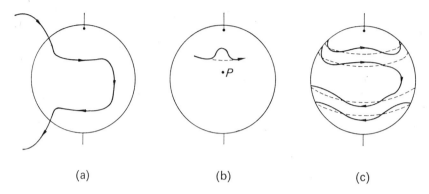

(a)                              (b)                              (c)

**Fig. 5.16**   The stretching of a magnetic line of force in a conducting fluid by (a) a $T_1$ motion, (b) $S_{2c}$ motion, and (c) the combined result. Reproduced from BULLARD and GELLMAN (1954), *Phil. Trans. R. Soc.*, **247A**, 260.

Two superimposed types of fluid motion in the core are required by the Bullard dynamo. These are: (1) a differential rotation of the core so that the inner parts rotate more rapidly than the outer part (this is to be expected if angular momentum is to be conserved in a convecting and rotating fluid); and (2) thermal convection in the core, with uprising and sinking currents near the equator but with a three-dimensional pattern of flow involving north-south and east-west movement (this is essential to remove axial symmetry—otherwise the dynamo could not work, as Cowling showed). (1) is the simplest possible type of toroidal velocity field and is referred to as $T_1$. (2) is a poloidal motion and the type referred to here has two sinks and two sources on the equator and will be called $S_2^{2c}$. The two types of velocity field and the four types of magnetic field involved are shown in Figure 5.17. The four stages in the process are as follows:

*Stage 1*: A weak initial dipole field (type $S_1$) interacts with the $T_1$ velocity field associated with differential rotation to produce a toroidal magnetic field of type $T_2$, which wraps itself around the core along lines of latitude, in different directions in opposite hemispheres;

*Stage 2*: The $T_2$ field interacts with the convection cells $S_2^{2c}$ to give a more complicated type of toroidal magnetic field of type $T_2^{2c}$;

*Stage 3*: The $T_2^{2c}$ field interacts with the $T_1$ motion to produce a $T_2^{2s}$ magnetic field; this is similar to $T_2^{2c}$ apart from a rotation of 90° about the axis;

*Stage 4*: The $T_2^{2s}$ magnetic field interacts with the $S_2^{2c}$ motion to produce a dipole field of type $S_1$, reinforcing the original field.

An interesting feature of this and other dynamo hypotheses is that a very strong toroidal field of type $T_2$ must be present in the outer core and lowermost mantle. Toroidal magnetic fields can only exist within conductors and for this reason it does not penetrate to the Earth's surface. All we can observe is the relatively weak poloidal fields such as $S_1$. Another interesting corollary of Bullard's theory is that the outermost part of the core must rotate more slowly than deeper parts do.

The dynamo theory depends on effective conversion of thermal energy into electromagnetic energy through the agency of convection currents in the outer core. According to the second law of thermodynamics, the ratio of heat converted into electromagnetic energy to total heat transported upwards by the convection must be less than 1%. Thus the dynamo theory can only work if there is an adequate source of heat in the lower parts of the core—perhaps the inner core—and an efficient mechanism for

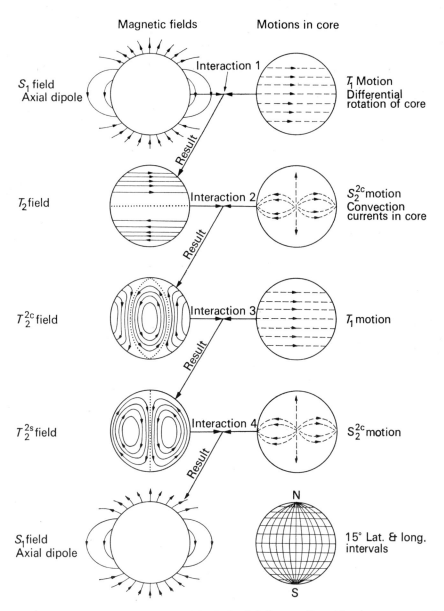

Magnetic fields        Motions in core

Interaction 1

$S_1$ field
Axial dipole

$T_1$ Motion
Differential
rotation of core

Result

$T_2$ field

Interaction 2

$S_2^{2c}$ motion
Convection
currents in core

Result

$T_2^{2c}$ field

Interaction 3

$T_1$ motion

Result

$T_2^{2s}$ field

Interaction 4

$S_2^{2c}$ motion

Result

$S_1$ field
Axial dipole

N

15° Lat. & long.
intervals

S

**Fig. 5.17** The electromagnetic interactions of the Bullard-Gellman self-exciting homogeneous dynamo, as described in the text. The diagram showing the $S_1$ magnetic field is a diametrical section through the core, passing through the poles. The other diagrams depict magnetic lines of force and flow lines on the surface of the core as they would be viewed from a point vertically above the equator. $S_2^{2c}$ motion represents a convection current in the core rising at two diametrically oposite points on the equator and sinking between them. Partly after BULLARD and GELLMAN (1954), *Phil. Trans. R. Soc.*, **247A**, 222.

disposal of the excess heat transported when it reaches the core-mantle boundary. Most geophysicists think that heat produced by radioactive decay in the lower parts of the core is the main source of energy and that the mechanism of heat transfer in the lower mantle (radiative heat transfer) is adequate to remove the heat convected upwards through the outer core. Other possible heat sources are (1) downward migration of dense phases towards the Earth's centre, thereby releasing gravitational energy as heat, or (2) energy released by phase transitions within the core occurring as a result of steady reduction of pressure associated with a decreasing value of the gravitational constant (see p. 270). Another possibility is that the lower part of the mantle and the core have been cooling systematically through geological time; this would depend on an efficient mechanism of upward transfer of heat through the whole mantle.

The secular variation is believed to be caused by electromagnetic induction in the core within the framework of the dynamo theory. The most spectacular aspect of the secular variation is the periodic reversal of the main field. Both theory and experiments relating to homogeneous self-exciting dynamos suggest that reversal may occur either because of an inherent instability in the mechanism or because of slight changes in the pattern of fluid motion. Smaller fluctuations in the intensity of the dipole field and its orientation would also be expected.

It used to be thought that the secular variation of the non-dipole field was caused by varying magnetization within the crust and upper mantle. This explanation was discounted when the westward drift was discovered and nowadays nearly everyone accepts a core origin depending on electromagnetic interactions between the magnetic field and fluid motions there. LOWES and RUNCORN (1951) showed that the centres of maximum secular variation could be explained by horizontal electric current loops at the surface of the core. The westward drift is attributed to a slightly slower rotation of the outer core as it rotates. It has been suggested that the non-drifting part of the non-dipole secular variation is caused by local uplifting of the strong toroidal field in the outer core by fluid motion.

*Irregular changes in the Earth's rate of rotation*

The rate at which the main field has drifted westwards appears to have changed several times over the past hundred years. The rate of westward drift also varies conspicuously from one harmonic component to another, but all components tend to show irregular changes in rate of drift at about the same times. The effect is well shown by the apparent motion of the eccentric dipole which best fits the main field (Figure 5.18). The rate of drift was fairly steady until about 1900. Just before 1900 the rate decreased and it increased again at about 1910. These changes in drift rate suggest that the relative rotation rates of outermost core and mantle have changed rather abruptly. This would be expected to affect the rate of rotation of the crust and mantle and hence cause small changes in the length of day.

In Chapter 1 we saw that records of ancient eclipses and banding in fossil corals indicated that the Earth's rate of rotation is progressively slowing down mainly as a result of tidal friction. Superimposed on this regular change are irregular variations which have been revealed by astronomical studies over the last 200 years or so. The irregular changes in rotation affect the Earth only. A particularly sharp increase in the length of day, corresponding to a slower rate of rotation, occurred in 1899, and a corresponding decrease occurred between about 1910 and 1925 (Figure 5.18). These changes in length of day correspond in a remarkable way to the changes in the rate of westward drift of the main field.

Of all the possible explanations of these irregular fluctuations in the length of day, the only adequate one is that angular momentum is periodically exchanged between the core and mantle. If the crust and mantle speed up, then the core must slow down so as to maintain constancy of angular momentum, and vice versa. The only feasible known mechanism for the transfer of the angular momentum is electro-magnetic coupling between the outermost core and mantle. This theory seems to be able to

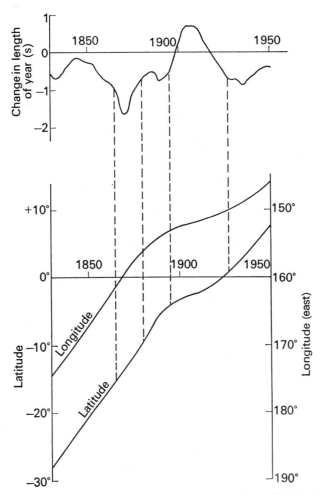

**Fig. 5.18** Correlation between the irregular change in length of the day with latitude and longitude of the eccentric dipole which best fits the Earth's observed field. Redrawn from VESTINE (1962), *Proc. Benedum Earth Sci. Symp.,* University of Pittsburgh Press.

explain both the changes in length of day and the westward drift and its fluctuations comprehensively. The changes in the rate of westward drift are smoother than the fluctuations in length of day because the relatively high electrical conductivity of the lower mantle filters out the higher frequency changes in field from our observations.

# 6 Terrestrial heat flow

## 6.1 Introduction

The study of thermal processes within the Earth is one of the most speculative branches of geophysics. This is because the available evidence on heat flow at the surface and on temperatures within the Earth can be interpreted in many different ways. Below a depth of about 100 km, the temperature distribution is highly uncertain and the distribution of heat sources and the mechanism of heat transfer are unknown. However, the subject is important because the process of heat escape from the Earth is probably the cause, directly or indirectly, of most tectonic and igneous activity.

Let us look at the major energy transactions which affect the Earth (Table 6.1). The largest item

**Table 6.1**  The major energy transactions which affect the Earth.

|  | erg/y |
| --- | --- |
| solar energy received by Earth (and re-radiated) | $10^{32}$ |
| geothermal loss of heat | $10^{28}$ |
| energy lost by slowing of the Earth's rotation | $3 \times 10^{26}$ |
| elastic wave energy released by earthquakes | $10^{25}$ |

is the heat received from the Sun, but this is mainly re-radiated back into space and only a minute fraction penetrates below a depth of a few hundred metres. It is the main source of energy for processes on and above the solid Earth's surface. It also controls the Earth's surface temperature, aided by the blanketing effect of the atmosphere. Nevertheless its influence on the interior of the Earth is negligible in comparison with that of heat generated within the Earth. Energy released by earthquakes and by the tidal slowing down of the Earth's rate of rotation is also small in comparison with the geothermal heat loss. The main source of heat energy within the Earth at the present time is believed to be radioactive decay of long-lived isotopes but other sources of heat such as release of gravitational energy may have been substantial early in the Earth's history. As heat escapes through the Earth, a small fraction of it is converted to other forms of energy which drive tectonic processes, cause igneous and metamorphic activity and give rise to the Earth's magnetic field.

## 6.2 The measurement of terrestrial heat flow

Heat flow per unit area is equal to the temperature gradient multiplied by the thermal conductivity. Applying this to the Earth, the heat escaping from the interior at a place on the surface can be determined by measuring (1) the temperature gradient just below the Earth's surface, and (2) the thermal conductivity of the rocks.

On land, the temperature gradient is usually measured in a borehole using maximum mercury thermometers of special design or more conveniently by lowering a thermistor probe (BECK, 1965). The measurements should be made below a depth of about 200 m to avoid the transient effects of the

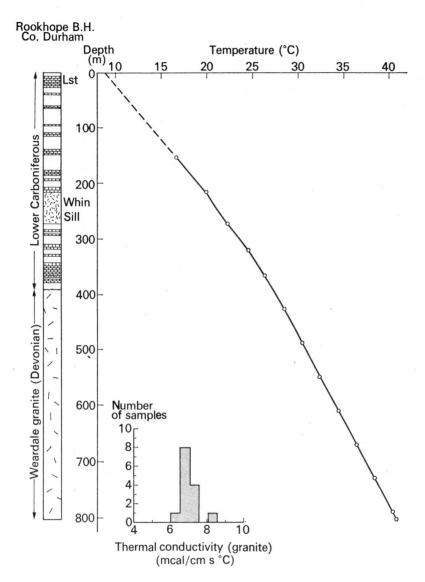

**Fig. 6.1** Heat flow determination in the Rookhope borehole, Stanhope, north England. The temperature-depth profile was obtained by Dr. G. A. L. Johnson and the author using a thermistor probe, three years after completion of drilling. Thermal conductivity measurements on thirteen granite specimens from regularly spaced intervals between 427 and 792 m depths were kindly made by Dr. J. Wheildon, and these are plotted on a histogram shown as an inset. The results are:

observed temperature gradient (427–792 m) = (32·45±0·10) °C/km
correction for topography = – (1·55±0·50) °C/km
corrected temperature gradient = (30·90±0·51) °C/km
estimated thermal conductivity = (0·0070±0·0002) cal/cm s °C
heat flow corrected for topography = (2·16±0·07) $\mu$cal/cm$^2$ s

If a glacial correction is applied, the heat flow estimate becomes 2·3 to 2·4 $\mu$cal/cm$^2$ s.

Pleistocene glaciations. In non-porous strata, the borehole needs to be left for several times as long as the drilling took in order to establish thermal equilibrium; alternatively temperature may be measured at the bottom of the borehole at intervals during drilling provided it can be left to settle for about a day before each measurement. The most reliable measurements are obtained from boreholes in non-porous basement rocks because the flow of water in porous sedimentary strata can carry away heat, thus disturbing the true geothermal gradient. A correction can be applied for disturbances to the geothermal gradient caused by irregular topography in the vicinity of the borehole. The thermal conductivities of samples from the borehole are measured in the laboratory, or alternatively the thermal conductivity is measured *in situ* by putting a heating coil in the borehole and measuring the temperature response. An example of a heat flow determination in the Rookhope borehole, Stanhope, northern England, is shown in Figure 6.1. Here, the heat flow is estimated to be $2 \cdot 16 \pm 0 \cdot 07$ $\mu$cal/cm² s.

The method used for measuring heat flow at sea was devised by Sir Edward Bullard in 1950. A probe two or more metres in length is dropped into the soft sediments of the ocean-floor and the temperature gradient is measured between two or more thermistors attached to the probe. A sample of the sediment is collected in the hollow barrel of the probe for later measurement of the thermal conductivity. One recent version of the probe is shown in Figure 6.2. The probe is left in position for a few minutes to ensure that thermal equilibrium has been reached. The method works because the temperature of the ocean-floor remains almost constant and has probably done so in the past. Some small errors may be caused by heat exchange between the ocean-floor and the water near the bottom, by movement of interstitial water in the sediments and by sedimentation (LUBIMOVA, VON HERZEN and UDINTSEV, 1965). The best confirmation of the reliability of the method comes from close agreement between heat flow measurements in the preliminary Mohole and those made by a probe nearby (VON HERZEN and MAXWELL, 1964).

Although measurements at sea have only been possible since about 1950, there are now a larger number of reliable values at sea than on land. The reason is the expense of drilling suitable holes on land; many of the measurements have to be made with boreholes drilled for other purposes which are often in unsuitable regions such as sedimentary basins.

## 6.3 The pattern of terrestrial heat flow

By 1969 over 3000 heat flow measurements had been made. The reliable published measurements up to 1964, amounting to about 1150, have been summarized and analyzed by LEE and UYEDA (1965). Only 11% of the analyzed measurements are from continents and these are badly distributed, leaving large gaps in South America, Antarctica, Africa, and Asia apart from Japan. The oceanic observations are more evenly distributed but show important gaps in the Arctic and Antarctic. The available data is hardly ideal for statistical investigation.

LEE (1963) estimated the world's mean heat flow by conduction to be $1 \cdot 5 \pm 0 \cdot 15$ $\mu$cal/cm² s; a more recent analysis using additional data confirms this figure. This means that the total heat escaping from the Earth's interior by conduction at the surface is about $2 \cdot 4 \times 10^{20}$ cal/y, or $10^{28}$ erg/y, with an uncertainty of about 10%. The corresponding estimated heat loss through volcanic activity is smaller by a factor of about 400. Thus most of the heat escaping from the Earth reaches the surface by thermal conduction through the geothermal gradient and volcanic transport of heat is a relatively minor contributor.

The best method to compare the heat flow of continents and oceans is to use the average heat flow values for a grid of equal area partitions rather than for individual measurements. This counteracts

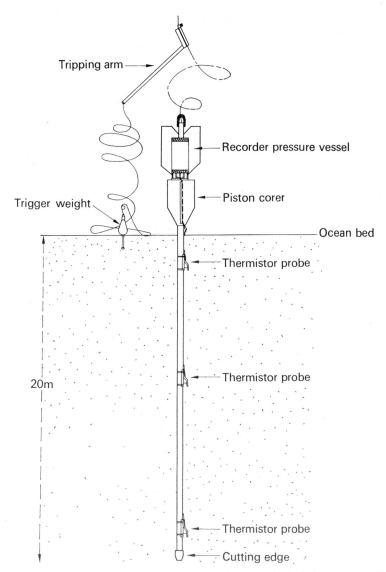

**Fig. 6.2**    An ocean-floor temperature gradient probe attached to a Ewing piston corer. When the trigger weight hits the seabed, the corer is released to fall freely into the soft sediments. The thermistor temperatures are recorded in the attached vessel. Redrawn from LANGSETH (1965), *Terrestrial heat flow*, p. 62, American Geophysical Union.

bias caused by the irregular geographical distribution of observations. Using grid partitions with an area of $9 \times 10^4$ square nautical miles, Lee and Uyeda found that the average for 51 continental partitions is $1 \cdot 41 \pm 0 \cdot 52$ s.d. $\mu$cal/cm$^2$ s (standard error of mean $= \pm 0 \cdot 07$), and the average for 340 oceanic partitions is $1 \cdot 42 \pm 0 \cdot 78$ (standard error of mean $= \pm 0 \cdot 04$). These average values do not differ significantly. However, HORAI (1969) has shown that the lowered surface temperature resulting from the Pleistocene glaciations may cause an underestimation of the continental heat flow values by about $0.2\ \mu cal/cm^2$ s. This suggests that the mean continental heat flow is higher than the oceanic by 10–15%, although this difference is hardly statistically significant. Although the mean values are closely similar to each other, the histograms showing the distribution of grid averages for continents and oceans are quite distinct from each other, as shown in Figure 6.3. The continental values form a stumpy bimodal distribution but the oceanic values form an asymmetrical distribution with a narrow peak and a long tail. Undoubtedly the shapes of these distributions do differ significantly.

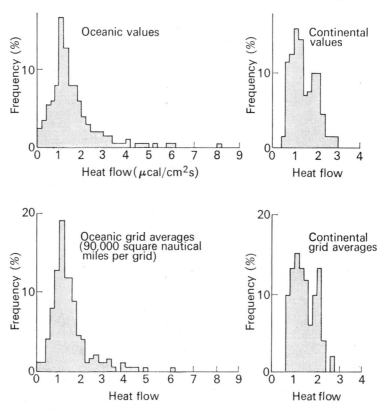

**Fig. 6.3**   Comparison of oceanic and continental heat flow distributions. *Above*: histograms of heat flow values; *below*: histograms of average values over 90,000 square nautical mile areas. Redrawn from LEE and UYEDA (1965), *Terrestrial heat flow*, pp. 139 and 140, American Geophysical Union.

## Regional variations of heat flow

Within continental and oceanic regions, the pattern of heat flow does show a close correlation with the major geological subdivisions (Table 6.2).

   Let us first look at the heat flow pattern of the continents, as shown in Figure 6.4. and in Table 6.2. Low heat flow values showing very little variation are characteristic of the Precambrian shields. Higher than average values are characteristic of post-Palaeozoic orogenic areas and also of Tertiary volcanic areas. There is a tendency for the younger orogenic belts to have the higher heat flow. There are, however, some other regional variations which do not correspond to geological character; one of the best known of these is the relatively high heat flow characteristic of south-eastern Australia. As would be expected from its volcanic associations, the Lake Baikal rift system of south-

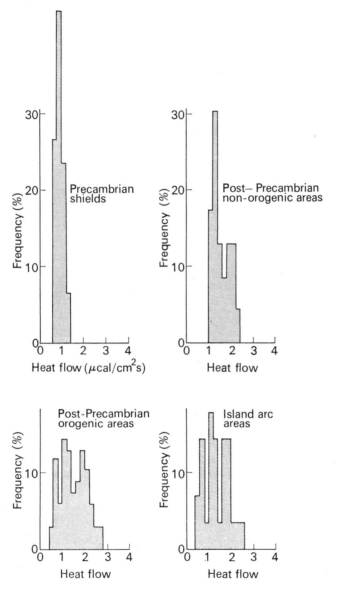

**Fig. 6.4**   Histograms of heat flow values for different types of continental region. Modified from LEE and UYEDA (1965), *Terrestrial heat flow*, pp. 148 and 149, American Geophysical Union.

central Siberia yields heat flow values ranging from $1 \cdot 2$ to $3 \cdot 4$ $\mu$cal/cm² s contrasting with the relatively low values of about $1 \cdot 0$ $\mu$cal/cm² s for the adjacent Siberian platform (LUBIMOVA, 1967).

The regional values for the oceans are shown in Table 6.2 and Figure 6.5. The ocean ridges show highly variable heat flow with a high average value; locally, abnormally high values of up to 8 $\mu$cal/cm² s have been observed. In contrast, lower than average values tend to occur on the flanks of the ridges.

**Table 6.2** Statistics of heat flow observations for the major geological subdivisions of continents and oceans (after LEE and UYEDA, 1965)

| Geological feature | Number of values | Mean ($\mu$cal/cm² s) | Standard deviation | S.E. of mean |
|---|---|---|---|---|
| *Continents* | | | | |
| Precambrian shields | 26 | 0·92 | 0·17 | 0·03 |
| post-Precambrian non-orogenic areas | 23 | 1·54 | 0·38 | 0·08 |
| Palaeozoic orogenic areas | 21 | 1·23 | 0·40 | 0·09 |
| Mesozoic-Tertiary orogenic areas | 19 | 1·92 | 0·49 | 0·11 |
| Tertiary volcanic areas (excluding geothermal areas) | 11 | 2·16 | 0·46 | 0·14 |
| *Oceans* | | | | |
| ocean basins | 273 | 1·28 | 0·53 | 0·03 |
| ocean ridges | 338 | 1·82 | 1·56 | 0·09 |
| ocean trenches | 21 | 0·99 | 0·61 | 0·13 |
| other oceanic areas | 281 | 1·71 | 1·05 | 0·06 |

The ocean basins show a relatively low and uniform heat flow pattern. The trenches show the lowest average value of the oceans. The pattern does not vary greatly from one ocean to another, although the average values for the Atlantic are the lowest by a small margin.

As further observations are added for the poorly covered regions, it is to be expected that there will be changes in our statistical estimates of the regional pattern of heat flow. However, the broad regional characteristics discussed above can be regarded as fairly well established.

*Global analysis of heat flow observations*

A more general way of studying the global variation of heat flow is to make a spherical harmonic analysis of the observations, although the results of such an analysis should be treated with caution. The main practical difficulty is that the observations are not evenly distributed over the Earth's surface, resulting in bias. This problem was overcome by LEE and MACDONALD (1963) by first expanding the heat flow in terms of orthogonal functions and then determining the maximum number of spherical harmonic coefficients which could be regarded as significant. They produced a map incorporating spherical harmonic coefficients up to the third order. A revised version of the map has been constructed by Lee and Uyeda and is shown in Figure 6.6. This gives a highly smoothed version of the true global heat flow pattern. The main criticism of this map is that observations are totally lacking from quite large areas, so that the existence of some of the peaks and troughs should be regarded with caution. For instance, there is a conspicuous lack of observations over north-east Africa where the most prominent maximum appears, this being a throw-off from observations over the rest of the Earth and without real significance.

If we take the map at its face value, there are 'highs' over the eastern Pacific and East Africa, and 'lows' over the central Pacific and the Atlantic. There is no obvious correlation between the heat flow 'highs' and 'lows' and the major division of the Earth's surface into continents and oceans. Nor is there any obvious correlation with major surface features. But perhaps this is to be expected if harmonics higher than the third are not included.

It is particularly interesting to compare the global heat flow map with the map of the geoid as revealed by satellite gravity observations (Figure 1.2). One might expect some correlation between the two maps because high temperature in the upper mantle would be expected to cause high heat flow, and also a lowering of the geoid as a result of thermal expansion reducing the density and causing a

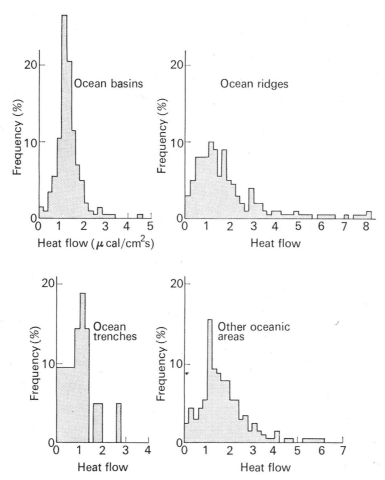

**Fig. 6.5** Histograms of heat flow values for the four main types of oceanic region. Modified from LEE and UYEDA (1965), *Terrestrial heat flow*, pp. 149-151, American Geophysical Union.

negative gravity anomaly. There is some indication that gravity is low (i.e. the geoid is warped downwards) where the heat flow is high and vice versa, but the agreement is not good enough to be convincing. We look forward to better analyses in a few years time when the global heat flow distribution is more complete.

## 6.4 Thermal properties of rocks

At low pressure and low to moderate temperature, experimental measurements of most thermal properties of common rocks have been made. At high temperature and pressure, it is necessary to rely on the theoretical predictions of solid state physics and thermodynamics.

The *thermal conductivity* $(K)$ at low temperature is caused by lattice vibrations. Its value for most types of non-porous rock measured at room temperature lies between 0·004 and 0·014 cal/cm s °C

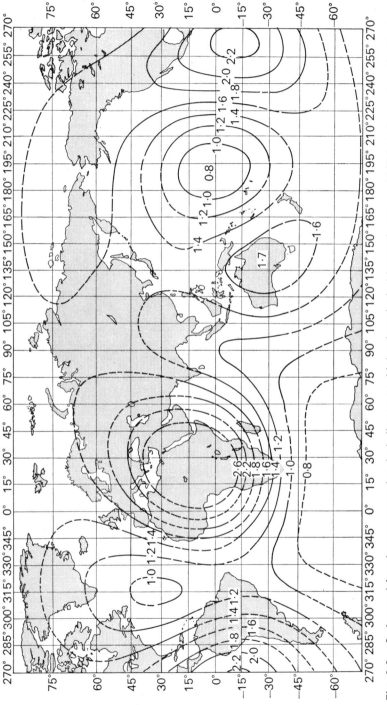

**Fig. 6.6** Orthogonal function representation, including up to third-order spherical harmonic terms, of 987 heat flow values. Contours are $\mu cal/cm^2$ s and are dashed where no data exist. Redrawn from LEE and UYEDA (1965). *Terrestrial heat flow*, p. 152, American Geophysical Union.

(BIRCH and CLARK, 1940). The range narrows to 0·004–0·009 at 200°C, mainly because the values for poorly-conducting quartz and feldspar rise with increasing temperature and the values for the other minerals, which are better conducting, fall with increasing temperature. Above 200°C the average thermal conductivity of rocks tends to decrease slightly with temperature up to 600°C which is the limit of most measurements. The effect of pressure is less well known but an increase appears to cause a slight rise in thermal conductivity. Above temperatures of 1000–1500°C, it is possible that the thermal conductivity is effectively increased by a substantial amount by the transfer of heat through solid rocks by radiation (see below p. 192). To summarize, a good estimate of the thermal conductivity down to about 50 km depth is 0·006 cal/cm s °C, but below this depth there is uncertainty.

The *specific heat at constant pressure* ($c_p$) is best computed from the observed values of the constituent minerals. The value of $c_p$ for most crystalline rocks rises from 0·15–0·20 cal/g °C at 0°C to 0·27–0·32 at 800°C. At higher temperatures $c_p$ can be fairly accurately predicted by theory. Assuming that the mean atomic weight of the mantle is 21 and that the specific heat is principally due to lattice vibrations, solid state theory predicts that the specific heat at constant volume ($c_v$) approaches a maximum value at high temperature of 0·3 cal/g °C. Thermodynamic relationships show that $c_p$ should be a few percent higher than $c_v$ (about 10%). The specific heats are the most accurately predictable thermal property of the mantle. A good estimate for $c_p$ throughout the mantle is 0·3 cal/g °C, which is probably accurate to better than 50%.

The *thermal diffusivity* ($\kappa$) is needed for solving problems of heat conduction in which the temperature changes with time. It is defined as $\kappa = K/\rho c_p$. In the crust and top 50 km of the mantle $\kappa$ is probably about 0·006 cm²/s, but it may increase substantially below this depth if radiative heat transfer is effective.

The *volume coefficient of thermal expansion* ($\alpha$) of common types of crystalline rock is calculated from observed linear coefficients of expansion and lies in the range $1·5–3·3 \times 10^{-5}$/°C. The value of $\alpha$ for olivine and pyroxenes between 20°C and 1000°C is about $3 \times 10^{-5}$/°C which is a good estimate for the upper mantle. ANDERSON (1965) has shown that estimates of $\alpha$ at all depths in the mantle can be obtained indirectly from knowledge of the seismic velocities, using the theory of solid state physics. He first showed that the Grüneisen ratio $\gamma$ can be estimated from the increase in bulk modulus with pressure ($dk/dp$) which can be obtained from the body wave velocities (p. 141). The estimate of $\alpha$ can then be obtained from the thermodynamic relationship $\alpha = \gamma c_v/k\rho$. Anderson's calculations yield values of $\alpha$ which decrease from $5·5 \times 10^{-5}$/°C at 33 km depth to $0·5 \times 10^{-5}$/°C at the core-mantle boundary. Compared with the experimental values, the theory yields rather high values for the upper mantle, but this is to be expected because phase changes must influence the estimate of $dk/dp$. A good estimate is probably 2 to $3 \times 10^{-5}$/°C in the upper mantle and about $10^{-5}$ in the lower mantle, with an uncertainty of the order of 50%.

## 6.5  The Earth's internal sources of heat

Before radioactivity was discovered, the flow of heat out of the Earth was believed to be the result of the cooling of an initially hot body. On this supposition Lord Kelvin deduced that the Earth could not be older than about 80 my (p. 12). After the discovery of radioactivity it was recognized that radioactive decay within the Earth provides a source of heat adequate to explain the observed heat flow without recourse to the cooling hypothesis.

The modern view is that the Earth formed as an initially cold body by accretion (p. 18). Assuming this to be true, sources of heat within the Earth are needed to explain both the present-day heat flow and the relatively high internal temperatures. Let us compare the relative magnitudes of the heat loss and the heat used in raising the temperature. The observed annual heat loss is about $2 \cdot 4 \times 10^{20}$ cal/y, and assuming that the same rate has applied over the whole age of the Earth the total heat loss would be $1 \cdot 1 \times 10^{30}$ cal. Taking the average specific heat at constant pressure to be $0 \cdot 3$ cal/g °C and the Earth's mass to be $5 \cdot 977 \times 10^{27}$ g, then the heat required to raise the average temperature by 3000°C would be $5 \cdot 4 \times 10^{30}$ cal. Thus a significantly larger amount of internal heat generation is needed to explain the high internal temperature than to produce the present heat flow. This puts the energy requirements into perspective. The possible sources of heat are discussed below.

## Long-lived radioactive isotopes

There is a loss of molecular binding energy when a radioactive isotope decays. This provides the energy of the $\gamma$-rays and imparts kinetic energy to $\alpha$- and $\beta$-particles. This energy is dissipated as heat in the immediate vicinity of the decaying isotope.

The radioactive isotopes which contribute significantly to the present heat production within the Earth are $U^{238}$, $U^{235}$, $Th^{232}$ and $K^{40}$. These have half-lives comparable to the age of the Earth and hence they are still sufficiently abundant to be important heat sources. Uranium consists essentially just of these two isotopes, the present-day proportion of $U^{235}$ being $0 \cdot 71\%$. $K^{40}$ forms $0 \cdot 0118\%$ of present-day potassium.

The rate of heat production for each of these four isotopes has been determined experimentally, and is shown together with the half-life in Table 6.3. An isotope with decay constant $\lambda$ was more abundant in the Earth by a factor of $e^{\lambda t}$ at time $t$ before the present. This means that the radioactive heat production from these four isotopes was larger in the past and has progressively decreased since the Earth's formation, simply because they were more abundant in the past. The abundances of the isotopes at various times in the past, relative to the present value, are shown in Table 6.3. The rate at which heat has been produced for each gramme now present of U, Th and K can then be computed from the known isotopic composition of each element, and is shown in Table 6.4. The past and present heat production in rocks can be estimated using this table, provided that the average content of the radioactive elements can be estimated. Estimates for granite, basalt and chondritic meteorite are shown in Table 6.5.

According to MACDONALD (1965), intermediate igneous rocks have an average present-day rate of heat production of $340 \times 10^{-8}$ cal/g y. Estimates for eclogite vary between about 8 and $34 \times 10^{-8}$. The average value for peridotite is $0 \cdot 91 \times 10^{-8}$, and for dunite it is $0 \cdot 19 \times 10^{-8}$. These estimates taken with those given in Table 6.5 provide the quantitative data for assessing the importance of long-lived radioactive isotopes as a heat source within the Earth, past and present.

Let us start by investigating whether a radioactive content equivalent to that of chondritic meteorites could explain the present heat loss from the Earth. Taking the Earth's mass to be $5 \cdot 977 \times 10^{27}$ g,

**Table 6.3**   The half-lives, rate of heat production and abundances in the past (relative to present) for long-lived radioactive isotopes

| | Half-life (1000 my) | Heat production (cal/g y) | Abundances in past relative to present | | | | | |
|---|---|---|---|---|---|---|---|---|
| | | | 0 my | 1000 my | 2000 my | 3000 my | 4000 my | 4500 my |
| $U^{238}$ | 4·50 | 0·71 | 1·00 | 1·17 | 1·36 | 1·59 | 1·85 | 2·00 |
| $U^{235}$ | 0·71 | 4·3 | 1·00 | 2·64 | 6·99 | 18·5 | 48·8 | 80 |
| $Th^{232}$ | 13·9 | 0·20 | 1·00 | 1·05 | 1·11 | 1·16 | 1·22 | 1·25 |
| $K^{40}$ | 1·3 | 0·21 | 1·00 | 1·70 | 2·89 | 4·91 | 8·35 | 10·9 |

**Table 6.4**    Rate of heat production per gramme of U, Th, and K now present (in cal/y). Computed from the data in Table 6.3 and the isotopic ratios of uranium and potassium as stated in text.

|  | 0 my | 2000 my ago | 4000 my ago | average over 4500 my |
|---|---|---|---|---|
| U | 0·74 | 1·18 | 2·81 | 1·57 |
| Th | 0·20 | 0·22 | 0·24 | 0·22 |
| K | $2·6 \times 10^{-5}$ | $7·4 \times 10^{-5}$ | $21·4 \times 10^{-5}$ | $10·6 \times 10^{-5}$ |

**Table 6.5**    Average content of long-lived radioactive elements in granite, basalt and chondritic meteorites and the resulting past and present heat production of these rocks. Compositions taken from MACDONALD (1965).

| | Composition (ppm) | | | | Heat production ($10^{-8}$ cal/g y) | | | |
|---|---|---|---|---|---|---|---|---|
| | U | Th | K | K/U | 0 my | 2000 my | 4000 my | average over 4500 my |
| granite | 4·75 | 18·5 | 37900 | 8000 | 820 | 1250 | 2600 | 1550 |
| basalt | 0·60 | 2·7 | 8400 | 14000 | 120 | 192 | 414 | 243 |
| chondrite | 0·012 | 0·0398 | 845 | 70000 | 3·9 | 8·6 | 22·4 | 11·7 |

the annual radioactive heat produced by a chondritic Earth would be $2·3 \times 10^{20}$ cal. The present annual heat escape is $2·4 \times 10^{20}$ cal. This suggests that the chondritic model for the Earth can explain the heat flow provided that there is an approximate balance between heat production and heat escape. This does not necessarily mean that the Earth is chondritic—as mentioned on p. 135 there are serious geochemical difficulties with this hypothesis—but it shows that it is plausible to attribute the present heat flow mainly to radioactive decay of long-lived isotopes.

Perhaps a more realistic geochemical model of the Earth would be a rock producing equivalent heat to a chondritic model but having a K/U ratio more typical of igneous rocks than of chondrites (Table 6.5). This would mean that the heat production shortly after the origin of the Earth would be about 30% less than on the chondritic model, because of the higher proportion in chondrites of $K^{40}$, which has a relatively short half-life.

Another important factor is the strong concentration of heat-producing isotopes in rocks of granitic and intermediate composition, such as are usually assumed to form the continental crust. If it is assumed that the continental crust has an average composition equivalent to intermediate igneous rock and a density of $2·8$ g/cm$^3$, then radioactive decay within a 33 km thick crust would contribute $1·0$ $\mu$cal/cm$^2$ s towards the average heat flow of $1·5$ $\mu$cal/cm$^2$ s. If this estimate is accurate, only about a third of the continental heat flow comes from the upper mantle, the rest being produced in the crust itself. Looked at in another way, radioactive decay in the continental crust may provide about 16% of the total heat flowing out of the Earth, which is quite out of proportion to its relatively small volume. In contrast, the thin oceanic crust would be unlikely to contribute more than about $0·1$ $\mu$cal/cm$^2$ s towards the average oceanic value of $1·5$ $\mu$cal/cm$^2$ s, so that over 93% of the oceanic heat flow must come from the mantle. The problem raised by this highly significant difference was touched on earlier (p. 126) and will be discussed later in the chapter (p. 194).

What part did the long-lived radioactive isotopes play in the initial heating up of the Earth, especially during the first 1000 my of its life? Taking a rate of heat production of $21 \times 10^{-8}$ cal/g y appropriate to the chondritic model and the specific heat to be $0·3$ cal/g° C, the resulting rise in temperature over the first 1000 my (assuming no heat escapes) would be 700°C. If all the heat produced by a chondritic Earth over its life of 4500 my was trapped within the Earth, then the average rise in temperature would still only be 1800°C. Thus the long-lived radioactive isotopes can account for part of the initial heating up of the Earth, but other heat sources are needed to explain most of it.

## Short-lived radioactive isotopes

The decay of three known short-lived radionuclides would significantly heat up accreting planetary bodies for a period of about 5–15 my after the termination of nucleosynthesis in the primitive solar system (FISH, GOLES and ANDERS, 1960). These are $Al^{26}$, $Cl^{36}$ and $Fe^{60}$. $Al^{26}$ is the most important of these; it has a half-life of 0·73 my and would last as a significant heat source for about 10 my. The other two isotopes have half-lives of about 0·3 my and would consequently lose their effectiveness as heat sources after about 4 my. These three extinct isotopes are the only known source of energy available to have caused the heating up of the parent meteorite bodies; consequently the meteorite bodies must have formed within a few million years of the nucleosynthesis of these isotopes (see p. 18).

Whether or not these extinct radionuclides were also effective in heating up the Earth's interior depends on how long it took for the Earth to form by accretion. If it only took 20 my after termination of nucleosynthesis, then heat evolved through decay of $Al^{26}$ might be the dominant cause of the Earth's high internal temperature. On the other hand, if it took about 100 my then the only effect of the decay of the extinct radionuclides would be to raise the temperature of the innermost 10% of the original Earth.

## Heat produced during the Earth's formation

Two physical processes other than radioactivity would raise the internal temperature of the Earth as it grew by accretion. These are: (1) kinetic energy released on the impact of protoplanetary bodies as they collided with the growing Earth; and (2) rise in temperature caused by progressive adiabatic compression.

As small bodies hit the growing Earth, kinetic energy of the order of 9000 cal/g would be released by the collisions (VERHOOGEN, 1960). This greatly exceeds the heat needed to raise the internal temperature to its present value. Nearly all of it would be dissipated as heat in the immediate vicinity of the impact and would be radiated into space without affecting the deep interior. However, a small fraction of the incident energy would be transformed into the elastic wave energy of seismic waves, and this would be dissipated as heat throughout the growing Earth through anelastic processes such as are known to occur (p. 247). If the trapped energy amounted to 0·1% of the incident energy, then the internal temperature would be raised by about 30°C altogether.

A more significant rise in temperature would be caused by adiabatic compression as the internal pressure progressively increased during the process of accretion. The adiabatic gradient expressed in terms of pressure (p. 129) is

$$dT/dp = \alpha T/\rho c_p.$$

To get a very rough estimate of the magnitude of this effect we can use average estimated values of the quantities on the R.H.S. of the equation as follows: $\alpha = 2 \times 10^{-5}/°C$, $T = 750°K$, $\rho = 4$ g/cm³ and $c_p = 0·3$ cal/g °C $= 12 \times 10^6$ erg/g °C. Substituting these quantities gives $dT/dp = 0·3°C/kbar$. The pressure at the Earth's centre after accretion would probably be about 3000 kbar, suggesting an adiabatic temperature rise during accretion of 900°C at the centre but progressively less going outwards towards the surface.

## Conversion of the Earth's rotational energy to heat

Another possible source of internal heat is the dissipation of the Earth's rotational energy as it slows down through tidal interaction with the Moon, and less significantly with the Sun (p. 20). It has been suggested that the Earth's rotational period after formation was less than 10 hours and possibly between 2 and 4 hours. Part of the energy lost is dissipated by tides in the oceans, particularly in the shallow seas. Part of it is dissipated in the interior by the Earth tides (p. 156). A relatively small fraction is used to increase the Moon's orbital energy as it recedes from the Earth. MACDONALD (1964) has shown that a total of $1 \cdot 5 \times 10^{38}$ erg would be released on an overall slowing down of the rotation from a 3 hour to a 24 hour period. This amounts to a release of heat of 600 cal/g for the whole Earth. If all this heat was dissipated within the Earth by the Earth tides, then internal temperature would rise by about 2000°C. On the other hand, if 90% of the energy was dissipated in the shallow seas then the internal rise in temperature would be 200°C. The great uncertainty is to know how the energy loss is divided between the ocean tides and the Earth tides.

The present-day loss of rotational energy is a small fraction of the total heat flow. The main period in the Earth's history when rotational energy possibly might have been converted to internal heat in significant quantities would be when the Earth and Moon may have been relatively close to each other about 1500 my ago (p. 22).

## Formation of the core

It is usually assumed that the Earth formed by accretion from approximately homogeneous material and that it has been differentiated subsequently into crust, mantle and core. The formation of the core would release a large amount of gravitational energy as a result of the concentration of the high density nickel-iron phase nearer to the Earth's centre. There would also be a small but significant release of energy accompanying the differentiation of the crust from the mantle. Part of the energy released during core formation would be converted to elastic strain energy, because the concentration of dense material towards the centre increases gravity within the Earth and thus increases the pressure and causes further compression. Most of the remaining gravitational energy released would be dissipated as heat through friction and viscous flow.

TOZER (1965b) has estimated the total change in gravitational and strain energy during the process of core formation from an originally undifferentiated Earth. He found that the release of heat amounted to an average of about 500 cal/g for the whole Earth. About 6% of this would be used to melt the nickel-iron phase; the remaining 94% would raise the average temperature of the Earth by about 1500°C. The total heat evolved during core separation is about the same as the total amount of radioactive heat evolved by long-lived isotopes during the whole life of the Earth. Evidently core separation is a major factor in raising the temperature within the Earth.

How did the core form? By analogy with meteorites, the primitive Earth may have consisted of an admixture of silicate minerals with high melting point and low density, and a metallic iron-nickel phase with lower melting point and higher density. As the primitive Earth gradually warmed up, eventually the melting point of the nickel-iron phase would be reached at a depth just below that part of the outer shell affected by cooling through thermal conduction, possibly about 400 km below the surface. A thin shell consisting of liquid nickel-iron phase and solid silicate would form at this depth (Figure 6.7(b)). The denser liquid phase would then drain towards the Earth's centre, by processes akin to magma intrusion, but in the opposite direction.

The time when the core started to form would depend on whether short-lived or long-lived

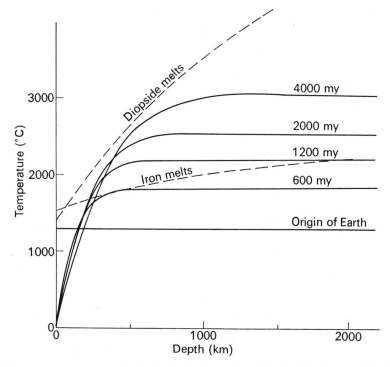

**Fig. 6.7** (a)    Computed temperature-depth distribution in a chondritic Earth having an initial temperature of 1300°C computed at various times after formation. Iron first melts according to this model at a depth of about 400 km, 600 my after origin of the Earth. Adapted from MACDONALD (1959), *J. Geophys. Res.*, **64**, 1991.

radio-isotopes caused the main initial heating of the primitive Earth. If the Earth was formed by accretion over a period of 1 my or thereabouts after termination of the nucleosynthesis of $Al^{26}$ and other short-lived isotopes, then decay of these may have produced sufficient heat to initiate core formation within a few million years of the Earth's formation. Alternatively, if the Earth took longer to form, these short-lived isotopes would have decayed before they could effectively heat the Earth's interior; then the long-lived radio-isotopes of uranium, thorium and potassium would cause the initial heating, and initiation of core formation would be delayed for a period of the order of 400–1000 my after the Earth's formation, as shown in Figure 6.7(a).

BIRCH (1965) and TOZER (1965b) are of the opinion that the process of core formation, once started, would accelerate and continue to completion over a relatively short period of time compared to the age of the Earth. However, an alternative hypothesis that the core has grown steadily over the Earth's geological history was put forward by UREY (1952) and has been taken up by Runcorn as an explanation for changing patterns of convection in the mantle (p. 285).

Tozer has put forward convincing criticisms of the steady growth hypothesis for core formation. He showed that the present rate of heat production within the Earth would greatly exceed the present heat flow if the core were still forming. He also showed that if the metallic phase is still being carried down from the mantle to the core by convection currents, then the convection pattern would cause much larger global gravity anomalies than are observed. The balance of evidence is in favour of the hypothesis that the core formed relatively rapidly at an early stage in the Earth's history.

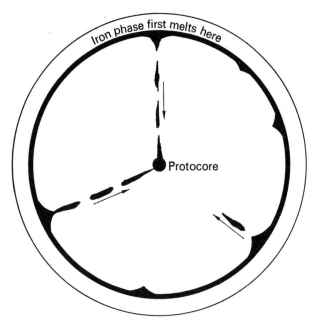

**Fig. 6.7** (continued) (b)    The start of the process of core formation.

The result of core formation by differentiation would probably be a doubling of the temperature at most depths. This would establish a new thermal regime, which might give rise directly or indirectly to more effective methods of heat transfer in the Earth. This might initiate the formation of the crust and start the geological record as we know it.

The above speculations on core formation should be treated with caution. Another view is that the accretion of a metallic core is the first stage in the formation of a planet such as the Earth. OROWAN (1969) has suggested that the metallic particles in the protoplanetary nebula may stick together on collision because they are plastic-ductile at high and low temperature and can loose kinetic energy by plastic deformation. On the other hand, the brittle silicate particles would break up on collision except near the melting point. Thus a metallic protoplanet would form first, and when it had grown to a big enough size, silicate particles would be attracted to it by gravity. Orowan considered that the boundary between metallic core and silicate mantle would be sharpened by later melting of the metallic phase.

## 6.6  Transfer of heat within the Earth

*Thermal conduction*

One of the central problems of terrestrial heat flow is to explain how heat escapes from the deep interior of the Earth. Most of the lost heat reaches the surface by thermal conduction through the rocks of the crust. This raises the following problem. Taking the heat flow to be $1 \cdot 5\ \mu$cal/cm$^2$ s and the thermal conductivity to be $0 \cdot 006$ cal/cm s °C, the temperature gradient is 25°C/km. If this gradient is extrapolated to a depth of 100 km then the temperature would be 2500°C which would cause wholesale melting. On the other hand, the temperature at 60 km depth must locally reach 1200°C to produce basalt magma. Therefore there must be a decrease in the temperature gradient by a factor of about ten before a depth of 100 km is reached. The flattening of the temperature-depth curve could be caused either by concentration of the Earth's internal heat sources near the surface, or by a more effective mechanism of heat transfer than normal thermal conduction, or by a combination of these.

If the transfer of heat below 100 km depth occurs mainly through thermal conduction in rocks possessing thermal conductivities of about 0·006 cal/cm s °C, then 80–90% of the radioactive heat sources of the Earth must be concentrated above a depth of about 100 km. Otherwise there would be wholesale melting of the mantle. It is believed that about 60–70% of the radioactive heat sources in continental regions are concentrated in the crust. But the problem is more acute beneath the oceans; in the absence of more effective mechanisms of heat transfer nearly all the heat-producing isotopes would have to be concentrated in the uppermost 100 km of the mantle which seems unlikely. We are forced to conclude that some other mechanism of heat transfer probably occurs in the mantle beneath the oceans and possibly also beneath the continents.

## Transfer of heat by radiation and by excitons

At temperatures above about 800–1500°C it is likely that a significant amount of heat can be transferred through rocks by radiation. The result of radiative heat transfer is to increase the thermal conductivity by an additional amount $K_r$. It has been known for some time that the thermal conductivity of glasses and ceramic materials may be increased at high temperature by radiation. It has recently been verified that the thermal diffusivity of some common minerals, including quartz and olivine, starts to increase at about 450°C (KANAMORI, FUJII and MIZUTANI, 1968). This is presumed to be caused by the increase in radiative heat transfer at high temperature. At 750°C the radiative heat transfer in single crystals of olivine is probably about equal to the normal conduction caused by lattice vibrations, and the radiative effect would be expected to become dominant at higher temperature. Radiative heat transfer would be impeded by crystal boundaries; because of this the relative importance of lattice conduction and radiative heat transfer in the mantle would depend on the size of crystals.

The effectiveness of radiative heat transfer depends on the transparency of the appropriate silicate minerals to heat carrying radiation at the red end of the visible spectrum and in the infra-red. An approximate expression for the radiative contribution to the thermal conductivity at temperature $T$ is given by

$$K_r = 16n^2sT^3/3e$$

where $n$ is the refractive index, $s$ is the Stefan-Boltzmann constant and $e$ is the opacity (CLARK, 1957). The quantities $n$ and $e$ are understood to be weighted means over the appropriate band of wavelengths. The first point to note is that $K_r$ increases strongly with increase in temperature. The second point is that both $e$ and $n$ probably depend on temperature and pressure, the dominant influence being the influence of temperature on opacity. The opacity is caused by scattering and absorption of radiation. Both the absorption of radiation and the electrical conductivity depend directly on the density of free electrons, which in turn depends on temperature. The electrical conductivity increases by three orders of magnitude between depths of 500 and 1000 km in the mantle (p. 120). Thus an increase in opacity would also be expected between 500 and 1000 km depths and this would tend to counteract any increase of radiative heat transfer from the $T^3$ term above.

Yet another mechanism of heat transfer which may possibly increase the effective thermal conductivity at high temperature is by 'excitons'. The idea is that neutral atoms can be excited by radiation which has not enough energy to cause free electrons. The energy of an excited atom can be transferred to a neighbouring atom, and heat can flow by this mechanism. It is possible that exciton thermal conductivity equals or even outweighs radiative thermal conductivity in some parts of the mantle below 100 km depth. A useful review of exciton conductivity is given in the comprehensive paper by LUBIMOVA (1967).

To summarize, it is possible that the effective thermal conductivity of the upper mantle below about 100 km depth is an order of magnitude higher than the surface value because of radiative and exciton conduction and this would therefore reduce the temperature gradient at these depths. The increase in conductivity with depth may be inhibited in the mantle transition zone and below because of increase in opacity.

## Transfer of heat by thermal convection

The process of thermal convection can transfer substantial quantities of heat upwards through fluids in the presence of relatively small temperature gradients. Thermal convection is probably the mechanism of upward transfer of heat through the outer core. There is also a lot of evidence based on the origin of the major surface features to suggest that convection of some kind occurs in the upper mantle (p. 273). The mechanism of thermal convection is described in Chapter 9, where it is shown that the estimated viscosity of the upper mantle may be low enough to permit thermal convection in the presence of a relatively small temperature gradient in excess of the adiabatic gradient.

An important geothermal implication of convection is that heat produced at depth can be transported towards the Earth's surface much more rapidly than by conduction. The temperature gradient within the convection cell would not be expected to be more than a few times the adiabatic gradient which is about 0·3°C/km. Thus the upper mantle convection hypothesis may explain why the temperature gradient decreases substantially with increasing depth at about 50–100 km below the surface. This is illustrated in Figure 6.8 and in Figure 6.9 which contrasts the types of temperature-depth profiles to be expected in the upper mantle in the presence and absence of convection.

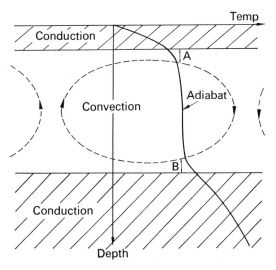

**Fig. 6.8** Schematic diagram showing the temperature-depth distribution associated with a convecting zone in the upper mantle, overlain and underlain by layers in which heat is transported by conduction. A is the boundary zone affected by cooling of the upper surface of the convection cell, and B is the zone affected by heating near the lower surface. Based partly on Elder (1965), *Terrestrial heat flow,* p. 235, American Geophysical Union.

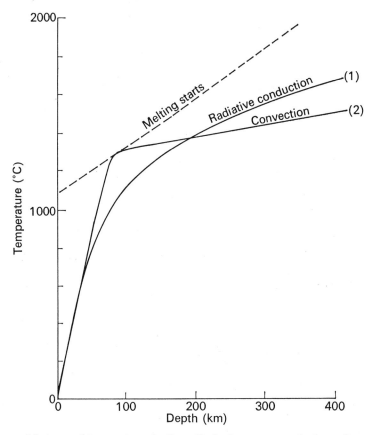

**Fig. 6.9**  Two possible types of temperature-depth profile in the upper mantle depending on:
(1)  Heat transfer by conduction, including radiative heat transfer or concentration of radioactive heat sources towards the surface or both;
(2)  Heat transfer below 75 km depth by convection.

Penetrative convection (ELDER, 1965) is another mechanism which can rapidly transfer heat upwards towards the Earth's surface. It is the name given to the upward flow of hot, low density fluids, such as magma and hydrothermal solutions, through a denser porous medium. It is an irreversible process in that the magma or hot water is deposited at or near the Earth's surface where it remains. This is probably the principle method of upward transfer of heat in geothermal regions including ocean ridges.

## 6.7 The problem of equivalent continental and oceanic heat flow

Another central problem of terrestrial heat flow is to explain why the average heat flow values of continents and oceans are about equal. The solutions which have been suggested carry with them far-reaching implications relating to the permanence of continents and oceans.

The problem arises as follows. It is believed that radioactive decay in the continental crust accounts for about two-thirds of the observed continental heat flow. The situation for oceans is different because the oceanic crust is too thin and too basic to produce more than a small fraction of the heat produced in the continental crust. Consequently before heat flow measurements had been made at sea it was anticipated that the oceanic heat flow would be much less than the continental value. The observations (p. 179) showed to everyone's surprise that the oceanic heat flow was about the same as the average continental value. The consensus of opinion is that nearly all the oceanic heat flow originates below the Moho. This can only mean that the heat flow in the uppermost sub-oceanic mantle is two to three times greater than in the sub-continental mantle. The associated temperature differences in the upper mantle have already been discussed (p. 126). The problem here is to explain why the heat flow in the uppermost mantle differs by such a large amount beneath continents and oceans.

Two types of explanation have been suggested. One of these suggests that roughly the same amount of radioactivity exists beneath oceans and continents, and that most of it has been concentrated in the crust in continental regions but has not been separated from the upper mantle beneath the oceans. The other idea appeals to mantle convection currents to provide more rapid upward transfer of heat through the mantle beneath the oceans. These two ideas are not necessarily mutually exclusive, although they have tended to develop that way in the literature. They are discussed in turn below.

## Explanation based on different radioactive heat source concentrations

The most complete attempt to explain the similarity of oceanic and continental heat flow in terms of different sub-oceanic and sub-continental distributions of radioactivity has been made by MACDONALD (1963, 1965), who tackled the problem as follows. He assumed an initial temperature distribution 4500 my ago and a specified distribution of radioactive heat sources within the Earth, differing beneath continents and oceans. He then solved the equation of heat conduction in a sphere by computer. This yielded the internal temperature distribution at subsequent times and the surface values of heat flow. He sought for internal distributions of radioactivity which could explain the present-day values of heat flow, and worked out the corresponding temperature distributions in the upper mantle beneath continents and oceans.

Two of Macdonald's models which can explain the present heat flow are shown in Figures 6.10 and 6.11. Both models assume that the initial temperature 4500 my ago rose from 0°C at the surface to 1000°C at 600 km depth and was constant below 600 km. They have the same distribution of radioactivity, but they differ in that radiative heat transfer is negligible in Figure 6.10 but dominant at high temperature in Figure 6.11. The temperature distributions below about 50 km differ significantly between the two models. In particular, the sub-oceanic temperatures are unacceptably high in the absence of radiative heat transfer. Macdonald found this to be true for all models which do not include radiative heat transfer, unless the radioactivity is concentrated above 100–200 km depth which would be unlikely. It must be concluded that a strong increase in thermal conductivity above about 1000°C is an essential property of the mantle if the present-day heat flow is to be explained by conduction without incurring temperatures well above the melting point in the sub-oceanic mantle.

Macdonald's computations have shown that continental and oceanic heat flow can be explained by conduction provided that the distribution of radioactivity is similar to that shown in Figure 6.10 and that radiative heat transfer occurs in the upper mantle. The hypothesis can be extended without difficulty to explain the regional variations of heat flow such as the low average values of Precambrian shields, by postulating a suitable distribution of radioactivity in the crust and upper mantle below.

Macdonald's hypothesis has far-reaching implications. It requires that the upper mantle has

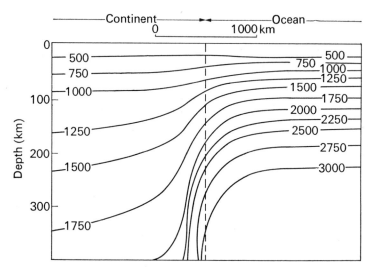

**Fig. 6.10**   The temperature distribution below continental and oceanic structures computed by MACDONALD (1965), *Terrestrial heat flow,* p. 208, American Geophysical Union, for a thermal conduction model of heat transfer, based on the following assumptions:
   (1)   average U concentration in mantle $3·3 \times 10^{-8}$ g/g;
   (2)   all radioactivity now concentrated above 1500 km depth;
   (3)   beneath *oceans*
         U (1500–465 km) : U (465–0 km)=1 : 18·2;
   (4)   beneath *continents*
         U (1500–465 km) : U (465–45 km) : U (45–0 km)=1 : 3·29 : 40·0;
   (5)   $K/U=10^4$; $Th/U=3·7$;
   (6)   initial temperature as in text;
   (7)   opacity $=1000$ cm$^{-1}$ (i.e. radiative conduction negligible).

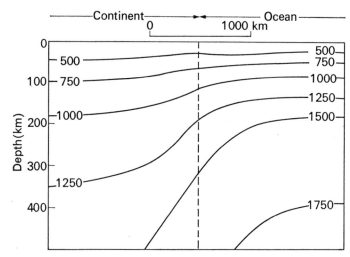

**Fig. 6.11**   The temperature-depth distribution below continental and oceanic structures computed by MACDONALD (1965), *Terrestrial heat flow,* p. 207, American Geophysical Union, for a thermal conduction model of heat transfer, based on the same assumptions as Figure 6.10 except that the opacity is 10 cm$^{-1}$ giving a highly significant radiative contribution to heat transfer at high temperature.

widely differing distributions of radioactivity beneath oceans and continents. A segregation of the Earth's radioactive isotopes into the upper mantle and crust is also a necessary feature of the hypothesis. These factors imply the permanence of continents and oceans, and suggest that the upper mantle underwent chemical differentiation early in its history. The hypothesis also implies that the continental crust formed by differentiation of the immediately underlying part of the mantle only, so that the radioactivity once present in the sub-continental mantle is now divided between crust and upper mantle; in contrast the sub-oceanic mantle still retains most of its original radioactivity.

The main criticism of the hypothesis comes from the implied permanence of the continents and oceans. This is particularly difficult to reconcile with the growing body of evidence for continental drift and ocean-floor spreading. For this reason, it is desirable to look for an alternative explanation of the continental-oceanic heat flow problem.

*Explanation based on the convection hypothesis*

The alternative explanation of the oceanic-continental heat flow problem depends on transfer of heat in the upper mantle by convection. This hypothesis supposes that most of the oceanic heat flow is carried through the upper mantle by convection. In the upper mantle beneath continents, convection is assumed to be absent or to carry a much smaller proportion of the heat flow reaching the surface. It will be shown in Chapter 9 that the pattern of oceanic heat flow places some stringent limitations on possible patterns of convection, although it is not inconsistent with the convection hypothesis.

The general idea is that the convection currents rise near ocean ridges and discharge heat as they flow towards the continents. Most of the observed oceanic heat flow is explained as heat carried through the upper mantle by convection. The outstanding problem is to explain exactly how the heat is lost. On the other hand, the continental heat flow is explained by radioactive heat sources in the crust and upper mantle, possibly supplemented by weak convection in the mantle or by cooling of it.

One important question remains unanswered by the convection hypothesis. Why are the mean values of oceanic and continental heat flow so closely similar to each other? This would not be expected, simply because the oceanic and continental heat flow are attributed to different and apparently unconnected processes. Is this a coincidence, or is there some underlying control of convection and continental drift which keeps the mean values about the same? The answer is not yet known. In this respect, Macdonald's hypothesis is more satisfactory in that equal continental and oceanic values are to be expected for it. However, the convection hypothesis fits much better with modern ideas on continental drift and ocean-floor spreading. The pattern of heat flow associated with convection is treated in more detail in Chapter 9.

## 6.8 High heat flow anomalies

Highly variable values of heat flow occur on ocean ridges. Locally values of up to $8 \mu cal/cm^2$ s have been observed (Figures 6.12 and 3.11). The observed pattern gives the impression of a series of hot spots situated at shallow depth superimposed on a more normal regional value of heat flow, which may be below or above average. Highly variable heat flow is also characteristic of volcanic and geothermal regions of the continents as would be expected (ELDER, 1965; McNITT, 1965).

The immediate source of heat flow as high as $8 \mu cal/cm^2$ s must be shallow for the following reason. Take the thermal conductivity to be $0.006$ cal/cm s °C; then the temperature gradient is 134°C/km

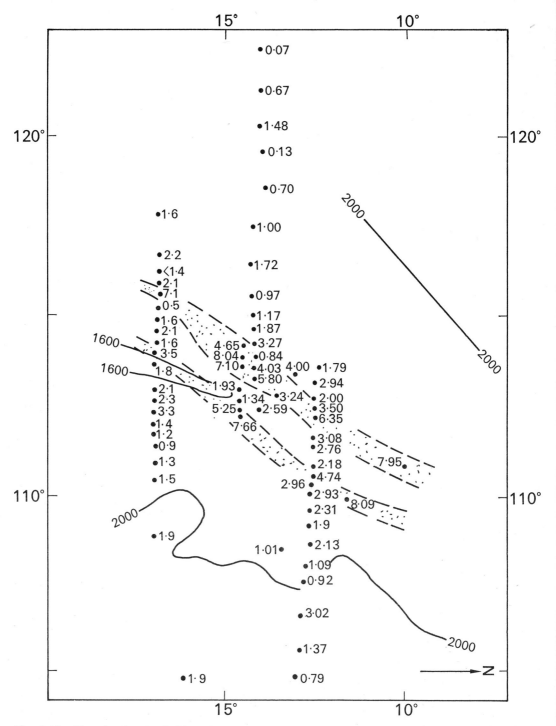

**Fig. 6.12**  Map showing detailed heat flow measurements across the crest of the East Pacific rise, obtained by scientists from Scripps Institution of Oceanography and Lamont Geological Observatory. The depth contours are shown in fathoms and the two parallel strips of exceptionally high heat flow are stippled. Redrawn from LANGSETH, GRIM and EWING (1965), *J. geophys. Res.*, **70**, 376.

and at 9 km depth the temperature would be 1200°C which is the melting point of basalt. The source of the heat flow could not be much deeper than this without widespread melting.

It is generally agreed that the high heat flow values of ocean ridges and of many geothermal areas on continents are caused by the presence of cooling magma below at relatively shallow depth. In geothermal areas, hot fumarolic gases from the cooling igneous body are evident and may also interact with groundwater to cause hot springs and geysers. Even without such hydrothermal activity, the presence of magma at a temperature of 800–1200°C at a depth of a few kilometres or less would cause heat flow values of the order of 8 $\mu$cal/cm$^2$ s. The igneous bodies could be in the form of shallow sheets or major intrusions.

The magma producing the hot spots is formed at much greater depth. Beneath oceans, basalt magma is probably produced by partial fusion of the mantle (in a rising convection current?) at a depth of 60 km or deeper. Beneath continents, the magma may form by partial fusion of the upper mantle or the lower part of the crust. Penetrative convection may cause the magma to rise to higher levels because it has a lower density than its host rocks.

## 6.9 Regional variations of continental heat flow

It was shown above (p. 180) that broad regional variations of heat flow do occur over continents, quite apart from volcanic areas. Young mountain ranges such as the Alps, and also some non-tectonic and non-volcanic regions, possess a heat flow which is about 50% above the average for continents. In contrast, the value for Precambrian shields is about 30% less than average. The most active tectonic regions show the highest values and the most stable regions show the lowest.

The high heat flow of young fold mountain ranges may be mainly a result of crustal thickening during mountain-building. If the crustal thickness is doubled, then the heat flow coming from radio-active decay in the crust would also be doubled. If only half of the normal continental heat flow is produced by crustal radioactivity, doubling the thickness would increase the total heat flow by 50%. The heat flow in active mountain-building regions may also be increased as a result of heat given off as strain energy is released by faulting and folding, and by the rise of magma into the upper crust.

Some regions show systematically high heat flow without evidence of crustal thickening. For instance, south-eastern Australia (Figure 6.13) has an average value of about 2·1 $\mu$cal/cm$^2$ s (SASS, 1964). This regional anomaly could be caused by an unusually high concentration of radioactive heat sources in the crust or mantle beneath. Alternatively, such high heat flow would be expected if a mantle convection current flows beneath the region or did so until recently.

Another interesting region of high regional heat flow is the Hungarian basin (BOLDIZSÁR, 1964). This is a basin of thick Tertiary sediments surrounded by the Carpathians and the Dinaric Alps. The average heat flow in the basin is about 2·4 $\mu$cal/cm$^2$ s. This is believed to be caused by volcanic activity. Normally such activity would cause a highly irregular pattern of heat flow, but if cooling magma and hydrothermal activity heat up the bottom part of a basin consisting of water-saturated porous rocks, the heat will be distributed more evenly over the area occupied by the basin. Hot water drawn from depths of 1000–2000 m in the basin is a significant source of thermal energy in Hungary.

The low values of heat flow of the Precambrian shields could be caused in various ways. According to Macdonald's hypothesis, it would be attributed to relatively low concentrations of radioactive heat sources in the underlying crust and upper mantle. According to the thermal convection hypothesis, the shields would overlie a relatively cold upper mantle which has probably not been subjected to convection currents since the Precambrian; the heat flow would represent the continental radio-active decay supplemented by a small contribution from cooling of the mantle below.

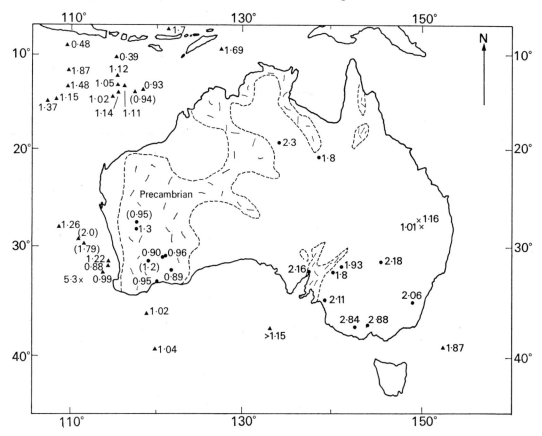

**Fig. 6.13** Heat flow values in and around Australia. Redrawn from LEE and UYEDA (1965), *Terrestrial heat flow*, p. 106, American Geophysical Union.

# 7 Continental drift by ocean-floor spreading

## 7.1 Introduction

The idea of continental drift was first taken seriously by geologists as a result of the work of TAYLOR (1910), BAKER (1911) and WEGENER (1912). In particular, it was Alfred Wegener, the German meteorologist and Greenland explorer, who developed the hypothesis in detail, seeking evidence from widely different disciplines. The good correspondence between opposite coastlines across the Atlantic had been commented on long before; one of the new mainstays of Wegener's argument was the Permo-Carboniferous glaciation which affected South America, South Africa, Australia and India, suggesting that these land masses were grouped around the south pole at that time.

Wegener suggested that during the Upper Palaeozoic there was a single continental mass which he called Pangaea. This mass broke into fragments which tended to drift away from the pole and towards the west during the Mesozoic and Tertiary. The last stage was the rapid westward drift of the Americas which according to Wegener occurred as late as the Pleistocene. Wegener believed that the Upper Palaeozoic south pole was situated just off the present coast of South Africa; Africa had remained fixed in position since then and the pole of rotation had wandered as well as the other continental masses.

Wegener suggested that centrifugal force acting on the relatively high-standing continents would cause them to migrate towards the equator. Tidal attraction of the Sun and Moon and precessional effects could cause them to move westwards. These mechanisms are quite inadequate and received severe criticism. Wegener bowed to these criticisms in the 1928 edition of his book *Die Entstehung der Kontinente und Ozeane* and inclined towards the modern theory of sub-crustal convection currents as the mechanism.

Since Wegener's original work, there have been three main stages in the development of the theory of continental drift:

(1) Up to about 1950 was a period of unresolved controversy. A large amount of evidence supporting drift was assembled, more especially by geologists working in the southern hemisphere. To geologists with first-hand knowledge of the evidence there appeared to be an impressive and almost conclusive case for drift. This is the viewpoint put forward in the book by DU TOIT (1937). Other geologists, notably those of the northern hemisphere, were more sceptical; the main criticisms centred on the inadmissability of some of Wegener's evidence which he had overpressed, and on the general inability to find an adequate mechanism.

(2) Between about 1950 and 1960, palaeomagnetism introduced a new quantitative approach to the determination of palaeolatitudes and palaeoazimuths. Palaeomagnetic results could be accepted by the scientific public without the need to visit the southern hemisphere. The palaeomagnetic evidence confirmed that the continents had drifted in much the same way as suggested by geologists such as Du Toit, thereby raising the hypothesis to the status of a valid theory. The theory became much more widely accepted although there were still many notable opponents.

(3) Since 1961, the development of the concept of ocean-floor spreading has confirmed continental

drift in a spectacular way. It has taken us one stage further in that now the exact history of the opening up of the oceans since the late Mesozoic can be worked out by studying oceanic magnetic anomalies and the stratigraphy of ocean-floor sediments. This exciting stage is still in progress.

Wegener's concept of a single continental mass during the Palaeozoic is not accepted nowadays. It is believed that there were two main continental masses at that time. The southern continent *Gondwanaland* consisted of South America, Africa, Madagascar, India, Australia and Antarctica. The northern continent *Laurasia* consisted of North America, Greenland, Europe and Asia excepting India. Wegener's late date for the opening up of the Atlantic has been generally replaced by a more gentle drift during late Mesozoic and Tertiary time.

## 7.2 The geological evidence

A short summary of the main geological arguments which support continental drift is given here. For a fuller account, the reader is referred to the book by DU TOIT (1937) and the short review article by WESTOLL (1965).

The *fit of the continents* across the Atlantic has been a longstanding argument for drift. CAREY (1958) has shown that there is an excellent fit between the edge of the continental shelves of South America and Africa provided that the Niger delta has been added to Africa since the start of drift. As would be expected, the coastlines themselves do not fit so well. There is also a good fit across the North Atlantic provided that Iceland is regarded as oceanic in origin and not as a fragment of a continent. The fit across the Atlantic calculated by computer is shown in Figure 7.1 (BULLARD, EVERETT and SMITH, 1965). The jig-saw pieces around the Indian Ocean have been fitted together at the 500 fathom depth contour by SMITH and HALLAM (1970); this fit is supported by good geological consistency between the reassembled fragments of the type summarized below.

The next stage is to investigate the *fit of tectonic features*, including fold mountain belts, major faults, sedimentary basins, radioactive age dates on intrusions and metamorphic belts, and dyke swarms. If these also fit together, the argument for drift becomes greatly enhanced. Examples include the excellent match of Caledonian and Hercynian fold belts across the North Atlantic, and the fitting together of the Samfrau orogenic belt of Gondwanaland. In general, there is a convincing fit of tectonic features across the Atlantic and between the fragments of ancient Gondwanaland.

A further extension of this approach is to study the *comparative stratigraphy* of the edges of continents which are now apart but may have been joined before drift. For instance, the Upper Palaeozoic successions of rocks in the Gondwana continents are closely similar to each other but are strikingly different in lithology and fauna from rocks of the same age in Laurasia. Across the North Atlantic the Cambrian and Ordovician rocks of north-west Scotland are remarkably similar to those of the St. Lawrence area of North America in rock type and fauna, although they are quite different from rocks of the same age in southern Scotland and further south.

*Palaeoclimatology* is the branch of geology which aims at reconstructing past climates. It provides some of the most convincing evidence for the drift hypothesis. The methods available include the use of:

(1) past glacial deposits to indicate arctic climates;
(2) evaporites suggesting high temperature and low precipitation;
(3) bauxite which is formed by tropical or subtropical weathering;

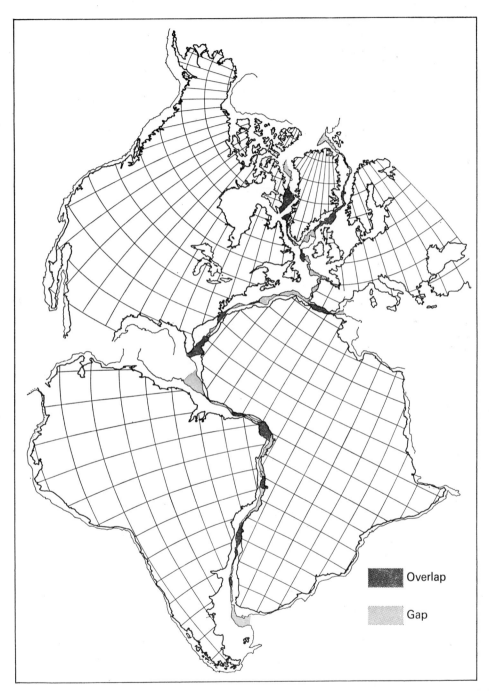

**Fig. 7.1**    Fit of all the continents around the Atlantic, obtained by least square fitting at the 500 fathom contour. Redrawn from BULLARD, EVERETT and SMITH (1965), *Phil. Trans. R. Soc.,* **258A**, opp. p. 49.

(4) reef-deposits suggesting tropical or subtropical climates;

(5) dune bedding indicating past wind directions;

(6) direct temperature measurements by oxygen isotope studies.

Caution is needed in interpreting some of the evidence, and also in attempting to deduce palaeo-latitudes from the evidence because climatic conditions over the globe may have been severely different at times in the past. The results are most convincing when the global pattern of climate at a given time can be assembled. For instance, during the Permo-Carboniferous time the Gondwana continents were all affected by widespread glaciation at the same time as reef deposits, coal and evaporites were being formed in Britain and U.S.A. There can be little doubt that the Gondwana countries were nearer to the south pole, and Britain and U.S.A. were nearer to the equator than they are now. The good agreement between the results obtained from palaeomagnetism and palaeoclimatology supports the general reliability of both methods. The results of palaeoclimatological studies are described at length in the book edited by NAIRN (1961).

The *past and present distributions of animals and plants* have been used widely as an argument for continental drift. To give an example, PLUMSTEAD (1965) has pointed out that the Gondwana continents, including Africa south of the Sahara and eastern South America, formed a single floral province from Lower Devonian to Lower Jurassic. This is in marked contrast to the present plant distribution in former Gondwana fragments, each of them forming an independent floral province with few species in common. The present distribution of some groups of animals and plants, in relation to their evolutionary history, is most easily interpreted if it is assumed that the continents have drifted apart from each other; the discredited alternative is that there used to be 'land-bridges' across the oceans which have now foundered.

The geological argument for drift has been aptly summarized by WESTOLL (1965): 'Purely geological evidence taken piece by piece cannot prove or disprove drift. Many kinds of evidence, individually supporting the drift theory with only a low probability, may collectively form a linked network that argues powerfully in its favour.'

## 7.3 The palaeomagnetic evidence

Palaeomagnetism (p. 165) has introduced a new quantitative approach into the testing of continental drift hypotheses. The method assumes that the direction of remanent magnetization in suitable rock specimens can be used, if precautions are taken, to estimate the dip and declination of the magnetic field when the rock was formed, to an accuracy of a few degrees. The secular variation is averaged out within statistically defined limits by using the mean direction for several samples of slightly differing age. The mean direction is used to estimate the ancient latitude (using the formula $\tan \theta = \frac{1}{2} \tan I$, p. 169) and the ancient north direction on the assumption that the magnetic field of the past, averaged over a few thousand years, can be approximately represented by a geocentric axially aligned dipole as at present. This is a basic assumption of palaeomagnetism, justified theoretically by the predictions of the dynamo theory and experimentally by the good agreement between palaeomagnetic and palaeoclimatic estimates of past latitudes (RUNCORN, 1965). The existence of reversals of the magnetic field does not interfere with the use of palaeomagnetism to determine past positions.

We need to know three quantities to determine the past position and orientation without ambiguity, such as latitude and longitude of a fixed point and orientation of a fixed direction. Palaeomagnetism yields estimates of two of these quantities to an estimated accuracy which is typically about 3–10°, but it cannot tell us the longitude. This limitation can be partially overcome by using other indications of the relative positions of continents in the past, such as the fit between the continental margins and the fact that they cannot overlap.

The use of palaeomagnetism to test continental drift has developed in two stages. The first stage was to demonstrate that the continents have drifted relative to each other. The second stage is to use the palaeomagnetic data together with the other evidence to reconstruct the positions of the ancient continents as far back through geological time as possible.

### Testing whether the continents have drifted

The basic palaeomagnetic approach is to determine the latitude and azimuth for each continental mass at as many points in the geological time scale as possible. Confidence in the method is given by the internal consistency between the results obtained for a given age at different locations within a single continental region which has not suffered internal deformation subsequently (Figure 7.2).

Palaeomagnetic results obtained from a single continent such as South America can be presented pictorially in two basic ways as shown in Figure 7.3. If it is assumed that the continent has remained

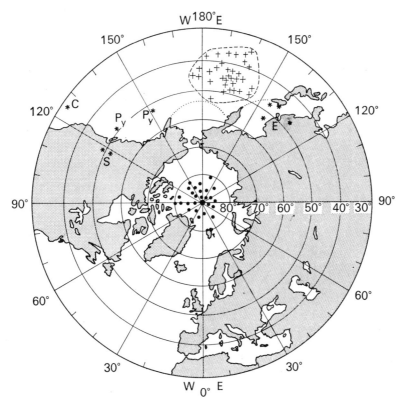

**Fig. 7.2** Palaeomagnetic determinations of the north pole (1) for the Pleistocene and Pliocene (black dots), (2) for the Permian based on determinations in Europe north of the Alpine belt (crosses), and (3) for the Permian based on European sites within the Alpine orogenic belt (stars); C= Corsica, Py= Pyrenees and S= southern Alps. The internal consistency between the undisturbed Permian poles gives confidence in the method. Redrawn from HOLMES (1965), *Principles of physical geology,* 2nd edition, p. 1210, Thomas Nelson and Sons.

fixed in position, then the past positions of the poles can be estimated from the ancient dip and declination; a convenient way of showing this is to construct a *polar wandering curve*, which marks the migration of one of the poles relative to the fixed continent through geological time (Figure 7.3(b)). The alternative method is to assume that the Earth's axis of rotation has remained fixed and to plot the past positions of the continent, assuming values of the longitude (Figure 7.3(a)). The palaeomagnetic results for a single continent which has not been internally deformed can be validly interpreted in either way or as a combination of both of them.

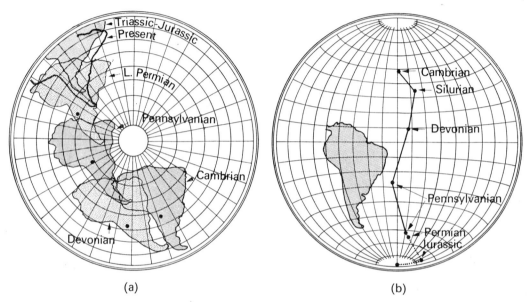

(a)                                    (b)

**Fig. 7.3** Palaeomagnetic results for South America since the Cambrian shown by (a) past palaeolatitudes and orientations assuming a fixed south pole, and (b) a polar wandering curve assuming a fixed continent. Redrawn from CREER (1965), *Phil. Trans. R. Soc.*, **258A**, 29.

The next stage is to compare the results obtained from different continents. A convenient way of doing this is to plot the apparent polar wandering curves for several continents on the same diagram, as shown in Figure 7.4. If all the continents had a common polar wandering curve, then we should know that they had remained stationary relative to each other and that the pole had wandered during geological time. Figure 7.4 shows that this is not so. The polar wandering curves diverge from each other going back in time. This means that the continents must have moved relative to each other. The pole may or may not have wandered—the palaeomagnetic evidence is not able to decide this issue.

It was RUNCORN (1956) who first recognized that the polar wandering curves for Britain and North America deviate from each other going back to the Triassic. As palaeomagnetic results became available from the southern hemisphere, it became apparent that the supposed Gondwana fragments had also drifted relative to each other and to the Laurasian continents.

### Reconstructing the continents of the Palaeozoic

Palaeomagnetic results support the idea of Du Toit and others that there were two main continental masses in the Upper Palaeozoic, Laurasia in the north and Gondwanaland in the south. They broke up in the early Mesozoic and drifted towards the present positions.

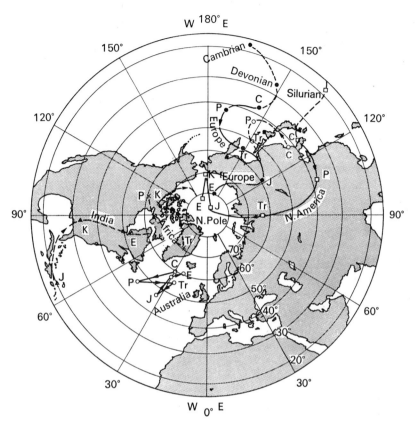

**Fig. 7.4** Apparent polar wandering curves for North America, Europe (with an offshoot for Siberia), Australia and India. E= Eocene, K= Cretaceous, J= Jurassic, Tr= Triassic, P= Permian, and C= Carboniferous. There is some scatter in actual results, and this has been averaged out to produce the curves. Redrawn from HOLMES (1965), *Principles of physical geology,* 2nd edition, p. 1211, Thomas Nelson and Sons.

In the *northern hemisphere* (RUNCORN, 1965), the polar wandering curves of Europe and North America can be reconciled if it is assumed that America has drifted about 30° westwards relative to Europe starting in about the Triassic period. Both the palaeomagnetic and palaeoclimatic results show that both continents were much closer to the equator during the Palaeozoic and Mesozoic. Both Britain and U.S.A. were in the equatorial belt from the Carboniferous to the Trias (RUNCORN, 1962). A small discrepancy is that the geological evidence suggests a closer proximity between the two continents during the Upper Palaeozoic.

In the *southern hemisphere* (CREER, 1965), the palaeomagnetic evidence shows that the Gondwana fragments were at high southern latitudes during the late Palaeozoic, as indicated by the Permo-Carboniferous glaciation. A summary of the movements of each main fragment from the Palaeozoic to the present, assuming no polar wandering, is as follows:

*South America*: near the south pole from the Cambrian to the Carboniferous, drifted rapidly northwards during the Permian and has remained near to its present latitude since the Trias;

*South Africa*: in south polar latitudes during the Upper Palaeozoic and drifted north to its present
latitude during the Permian and Trias;

*Australia*: possibly near the equator in the Devonian, close to the south pole during the Permo-
Carboniferous and drifted northwards towards its present position during the Mesozoic and
Tertiary;

*Antarctica*: drifted from middle latitudes to its present location since the Jurassic;

*India*: drifted rapidly northwards from middle latitudes since the Mesozoic, crossing the equator.

A reconstruction of the positions of South America, Africa and Australia during the Palaeozoic based
on palaeomagnetic results is shown in Figure 7.5. Although this does not agree in exact detail with
geological reconstructions, the broad principle that the Gondwana continents were in close proximity
to each other and near to the south pole has been verified by palaeomagnetism. As the number of
reliable palaeomagnetic measurements increases, we are to expect an improvement in our knowledge
of the past geography of Gondwanaland.

The author is of the opinion that anyone who has gone into the geological and palaeomagnetic
evidence for drift cannot but be impressed by the strength of the case. It is difficult to see how the
evidence could be interpreted otherwise.

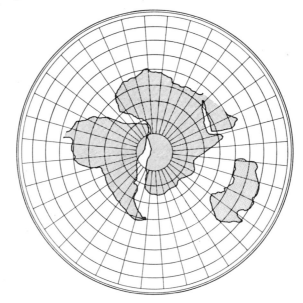

**Fig. 7.5**    The relative positions of South America, Africa and Australia for the Palaeozoic obtained by bringing
their polar wandering curves together. Redrawn from CREER (1965), *Phil. Trans. R. Soc.*, **258A**, 35.

## 7.4 Ocean-floor spreading and continental drift

The ocean-floor spreading hypothesis of Hess and Dietz was introduced in Chapter 3 as the process
by which new oceanic crust may be formed at the crests of ocean ridges. This hypothesis gives a new
dimension to the evaluation of the mechanism and history of continental drift. It has given rise to the
new concept of plate tectonics which explains how continents can move relative to each other. The
detailed study of oceanic magnetic anomalies according to the Vine-Matthews hypothesis makes it

possible to trace back to the Mesozoic the history of the opening up of the Atlantic and Indian Oceans.

## Transform faults

One of the important discoveries resulting from the magnetic anomaly surveys in the north-eastern Pacific was the large lateral displacements of the magnetic anomaly pattern across the Mendocino, Pioneer and Murray fracture zones (Figure 3.6). The Mendocino and Pioneer faults show a combined left-lateral displacement of the magnetic anomaly strips amounting to about 1400 km, and the

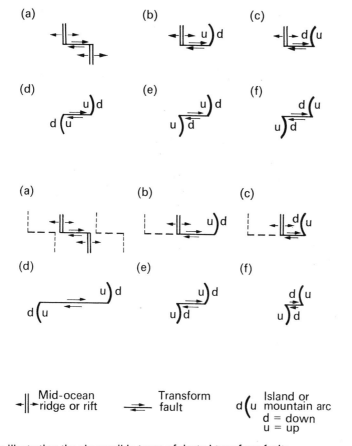

**Fig. 7.6** Diagram illustrating the six possible types of dextral transform fault:
(a) ridge to ridge;
(b) ridge to concave side of arc;
(c) ridge to convex side of arc;
(d)–(f) the three possible types of arc to arc connection.
The lower part of the figure shows the same faults at a later stage of development, with the now inactive parts marked as dashed lines. Note that the direction of motion in (a) is in the opposite sense to that required to offset the ridge. Redrawn from WILSON (1965), *Nature, Lond.,* **207**, 344.

right-lateral displacement at the Murray fault is 680 km at the western end and only 150 km at the eastern end (VACQUIER, 1965). The problem is to understand how such large horizontal movements can affect adjacent blocks of the oceanic crust, and how this lateral displacement can change drastically along the length of a single fault such as the Murray fracture zone. A similar problem is posed by the great continental strike-slip faults, such as the San Andreas fault of California and the Great Glen fault of Scotland. These faults must die out somewhere because they do not continue right round the Earth.

A simple yet profound solution as to how these large strike-slip faults can terminate has been suggested, within the framework of the ocean-floor spreading hypothesis, by WILSON (1965). He suggested that they terminate at the ends of mobile belts, which they meet, commonly, but not necessarily, at right angles. The lateral displacement on one side of the fault is taken up either by formation of new crust along a terminated segment of ocean ridge or by crustal shortening along a terminated segment of mountain range or ocean trench. Wilson calls this newly recognized class of strike-slip faults by the name *transform fault*. The concept of transform faults also explains the long-standing problem as to how mobile belts can be terminated. It leads to the idea that mobile belts are

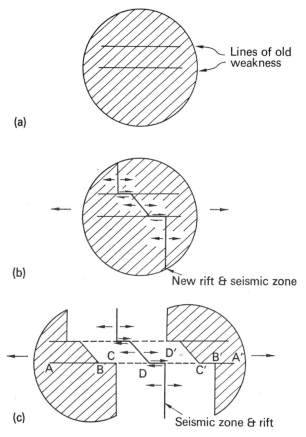

**Fig. 7.7**  Diagram illustrating three stages in the drifting apart of two continents such as South America and Africa, showing how transform faults have developed and played their part in the movement. The seismic activity would be mainly along the heavy lines. Redrawn from WILSON (1965), *Nature, Lond.*, **207**, 345.

linked by transform faults into an interconnected network which subdivides the Earth's surface into a series of 'rigid' plates which undergo relatively little internal deformation. This is the basic idea of plate tectonics.

Transform faults were grouped by Wilson into six basic classes depending on the type and orientation of the two mobile belts they join (Figure 7.6). The three possible types of junction at one end are (1) an ocean ridge, (2) a compression feature joined from the concave side, and (3) a compression feature joined from the convex side. Each of the six classes can be further subdivided into left-lateral (sinistral) and right-lateral (dextral) types.

Oceanic fracture zones are the best studied type of transform fault. Many of the fracture zones are known to interconnect segments of ocean ridge at both ends. A well-known series of parallel fracture zones displaces the crest of the mid-Atlantic ridge in equatorial latitudes as shown in Figure 8.24 (p. 264). These used to be interpreted as *left-lateral* transcurrent faults which displace a once-continuous crest. However, Wilson interpreted them as a series of *right-lateral* transform faults related to the opening up of the south Atlantic Ocean and the separation of South America from Africa (Figure 7.7). The apparent lateral displacement of the crest is not a true offset, but is the consequence of the shape of the original split between the continents. The shape of the original split must be maintained by the mid-ocean ridge if the crust is to behave as a pair of separating plates. Between the positions of the ridge crest on either side of a fracture zone, the crustal blocks on opposite sides move laterally relative to each other in the opposite sense to the displacement of the crests. Beyond the crests on both sides, the fractures cease to be active faults and the two adjacent plates move together. As described in Chapter 8, this interpretation is supported by studies of the distribution and mechanism of earthquakes.

To give another example, Wilson interpreted the northern termination of the mid-Atlantic ridge by transform faults (Figure 7.8(a)). The ridge splits south of Greenland and one branch passes on either side of it. The western branch terminates against the postulated Wegener transform fault passing between Greenland and Ellesmere Island. The eastern branch terminates against the De Geer fault which crosses from north Norway to Greenland passing just south-west of Spitsbergen. These faults show how a spreading ocean-floor can abut against less mobile parts of the crust.

Other examples of transform faults include (1) the fracture zones of the eastern Pacific Ocean, (2) the San Andreas fault joining the termination of the East Pacific rise in the Gulf of California to the short Juan de Fuca ridge off Vancouver Island, (3) the Owen fracture zone of the Indian Ocean connecting the end of the Carlsberg ridge to the Himalaya (Figure 7.8(b)), and (4) the Dead Sea rift zone transform fault connecting the north-west end of the expanding Red Sea to the Taurus mountains. The concept of tranform faults in relation to ocean-floor spreading is illustrated diagramatically in Figure 7.9.

*Magnetic stratigraphy using fossil reversals*

According to the Vine-Matthews hypothesis (p. 83), the past history of the Earth's main magnetic field is fossilized in the oceanic crust at least as far back as the late Mesozoic. The oceanic layer 2 has acted as a magnetic tape, recording the polarity of the magnetic field as new crust is formed at the crest of ridges (VINE, 1966). This crustal 'tape-recorder' can be replayed by observing the magnetic anomalies along profiles crossing the ocean ridges. The record is distorted by noise, mainly produced by variation in the depth to the upper surface of layer 2. Provided the noise does not swamp the signal, study of oceanic magnetic anomalies can be used to study the past character of the main

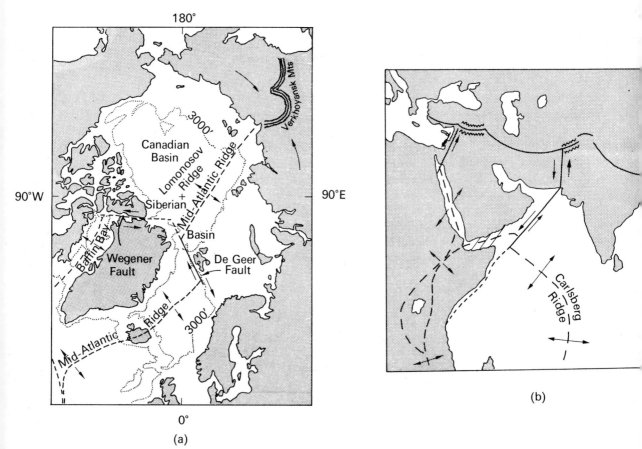

90°W           90°E

(a)

(b)

**Fig. 7.8**   Two examples of transform faults:
  (a) Termination of the mid-Atlantic ridge by two transform faults (De Geer and Wegener faults) and by transformation about a centre of rotation on the north Siberian coast into the Verkhoyansk Mountains.
  (b) Sketch map showing the sinistral transform faults of type (c) which join the end of the Carlsberg ridge to the Himalaya, and the north-western end of the Red Sea to the Turkish Mountains.
Note that the motions depicted must be treated as relative and not absolute. Reproduced from WILSON (1965), *Nature, Lond.*, **207**, 344 and 345.

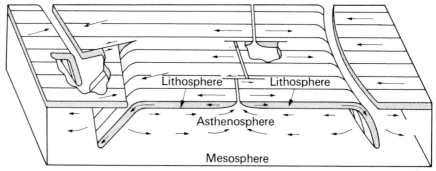

**Fig. 7.9**   Block diagram illustrating schematically the ocean-floor spreading hypothesis, and more particularly the relationship between ocean ridges, island arc-trench systems and transform faults of ridge-to-ridge and arc-to-arc types. The arrows indicate the relative motion between adjacent blocks. Adapted from ISACKS, OLIVER and SYKES (1968), *J. geophys. Res.*, **73**, 5857.

magnetic field as well as providing a tool for dating the formation of oceanic crust and working out the history of continental drift over the last 100 my or so (PITMAN and HEIRTZLER, 1966; VINE, 1966).

The first stage in the development of this method was to establish that magnetic anomalies near ridge crests could be produced by the known palaeomagnetic pattern of reversals over the last 4 my. This palaeomagnetic time-scale was originally obtained by measuring both remanent magnetization and potassium-argon age dates on the same lava specimens (COX, DOELL and DALRYMPLE, 1964; McDOUGALL and CHAMALAUN, 1966). The scale is as follows:

| epoch | magnetic polarity | age range (my) |
|---|---|---|
| Brunhes | normal | 0·7 – present |
| Matuyama | reverse | 2·4 – 0·7 |
| Gauss | normal | 3·35–2·4 |
| Gilbert | reverse | ?–3·35 |

Within the Matuyama epoch, there are short periods of normal polarity at 0·9, 1·6 and 2·0 my, and there are periods of reversed polarity at 2·8 and 3·2 my within the Gauss epoch. These are called the Jaramillo, Gilsa, Olduvai, Kaena and Mammoth events respectively. It is to be expected that further events may be found as further work is done.

Palaeomagnetic measurements on deep-sea cores have confirmed the details of the palaeomagnetic time-scale over the last 5 my and have allowed it to be accurately tied in to the faunal zones of the oceans (OPDYKE and others, 1966; HAYS and OPDYKE, 1967). Cores from high latitudes are most convenient to use because the magnetic field is nearly vertical so that accurate polarities can be obtained without need to orientate the core. Results obtained from some antarctic cores are shown in Figure 7.10, illustrating the excellent agreement between the different cores and with the faunal zones based on microfossils. Deep-sea cores have an advantage over the use of lavas in that the record is known to be continuous and relative ages can be resolved much more accurately. Using deep-sea cores, it has recently been reported that there is a short reverse polarity event between 114 000 and 120 000 years ago. Thus the magnetic time-scale has been well established and a new method of stratigraphical correlation has been introduced for the last 5 my or more.

Using this palaeomagnetic time-scale of reversals, PITMAN and HEIRTZLER (1966) and VINE (1966) theoretically computed the magnetic anomalies which would be produced near ocean-ridge crests for various spreading-floor rates. The agreement with the observed magnetic anomalies was found to be excellent provided appropriate spreading rates were chosen, as shown for the Juan de Fuca ridge and the East Pacific rise in Figure 7.11. The excellent agreement between the theoretical and observed patterns of anomalies confirms the validity of the Vine-Matthews hypothesis and it also shows that magnetic anomalies produced by the irregular topography of layer 2 do not swamp the anomaly pattern caused by the reversals of polarity of the field. Furthermore, the magnetic anomalies can now be used to work out the spreading rate over the last 4 my. The early results were as follows:

| ridge | spreading rate (cm/y) | reference |
|---|---|---|
| Juan de Fuca (46°N) | 2·9 | Vine |
| East Pacific rise (51°S) | 4·4 | Vine |
| N.W. Indian Ocean (5°N) | 1·5 | Vine |
| South Atlantic (38°S) | 1·5 | Vine |
| Pacific-Antarctic | 4·5 | Pitman and Heirtzler |
| Reykjanes | 1·0 | Pitman and Heirtzler |

These and other spreading rates are shown in Figure 7.15. The rate of crustal separation across a ridge is twice the spreading rate, and is about what would be expected if the separation of Europe and North America had occurred during the Mesozoic and Tertiary.

PITMAN and HEIRTZLER (1966) and VINE (1966) showed that a single sequence of normally and reversely magnetized blocks going further back in time could explain the magnetic anomaly profiles

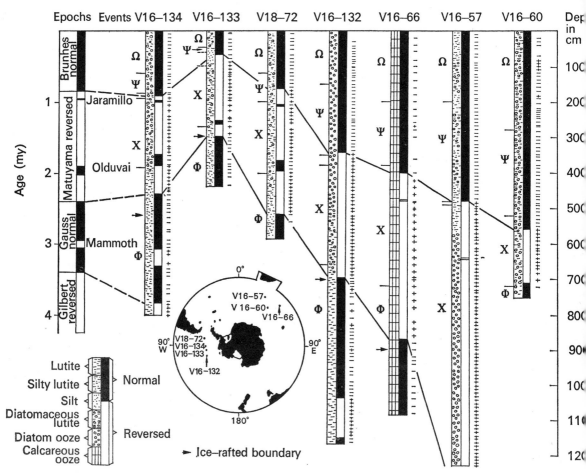

Fig. 7.10  Correlation of the magnetic stratigraphy in seven cores from the Antarctic, from locations shown in the inset. The magnetic polarity of the individual specimens is shown to the right of each section, minus signs indicating normal and plus signs indicating reversed polarity. Boundaries between the fossil zones are shown to the left of each section. Reproduced from OPDYKE and others (1966), *Science, N.Y.*, **154**, 350.

across the ridges they studied. Figure 7.12 shows the good agreement between the Pacific-Antarctic and Reykjanes ridges provided different spreading rates are assumed. The good agreement has now been shown to go back into the Mesozoic and to apply to most ridges (HEIRTZLER and others, 1968), although the pattern is difficult to recognize in some regions—notably the south-western branch of the Indian Ocean ridge system. This opens up the possibility of constructing a time-scale over the last 100 my or so based on reversals of the geomagnetic field.

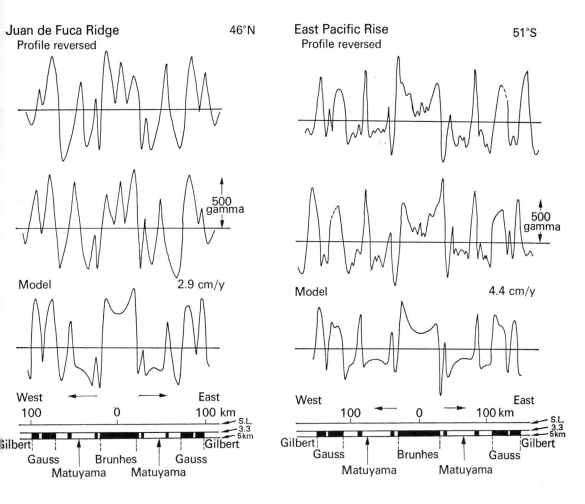

**Fig. 7.11**   Observed magnetic anomaly profiles across the Juan de Fuca ridge and the East Pacific rise compared with *above* the profile in the reverse direction to demonstrate the symmetry about the ridge crests, and *below* the theoretical magnetic anomaly according to the Vine-Matthews hypothesis assuming the palaeomagnetic time-scale of reversals and an appropriate spreading rate. It is assumed that the magnetic blocks are confined to layer 2. Black denotes normal magnetization, unshaded denotes reverse. Redrawn from VINE (1966), *Science, N.Y.,* **154**, 1409.

The practical problem in constructing a geomagnetic time scale is that the rate of ocean-floor spreading in some or all of the oceans has not been constant. This is shown in Figure 7.13 where the spacings of the magnetic anomalies patterns of different oceans are compared. In the absence of a better alternative, HEIRTZLER and others (1968) constructed a geomagnetic time scale by assuming that the South Atlantic spreading rate has remained constant over the last 80 my (Figure 7.14). This is a good working hypothesis until independent estimates of the ages of some of the conspicuous magnetic anomalies have been obtained (see below).

Conspicuous magnetic anomalies on the Heirtzler time-scale have been allocated numbers (Figure 7.14). For instance, anomaly 5 is about 10 my old and marks a change in the character of the anomalies.

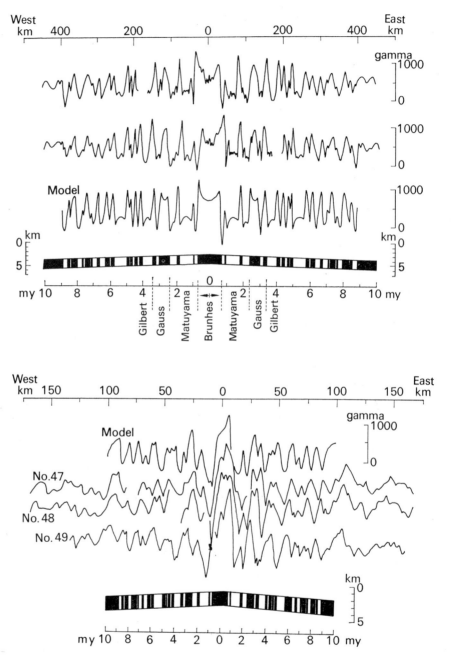

**Fig. 7.12** Observed magnetic anomaly profiles across the Pacific-Antarctic ridge (*above*) and the Reykjanes ridge (*below*) compared with computed models according to the Vine-Matthews hypothesis with the same sequence of normally and reversely magnetized blocks, but different spreading rates of 4·5 and 1·0 cm/y respectively. The Pacific-Antarctic profile is shown in the reverse direction at the top to indicate the symmetry. Reproduced from PITMAN and HEIRTZLER (1966), *Science, N.Y.*, **154**, 1166.

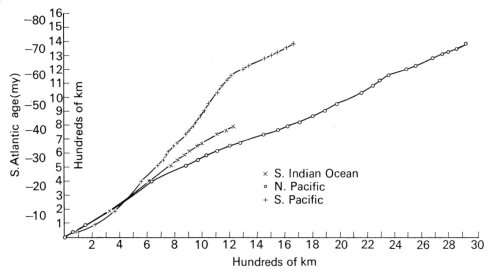

**Fig. 7.13**    The distance from the ridge crest to a given magnetic anomaly in the South Atlantic plotted against the distance to the same anomaly in the South Indian, North Pacific and South Pacific Oceans, demonstrating changes in the spreading rates. Redrawn from HEIRTZLER and others (1968), *J. geophys. Res.*, **73**, 2121.

Anomaly 31 also provides an easily recognizable marker horizon. If the positions of marker anomalies can be mapped on the ocean-floor by magnetic surveys, then the history of ocean-floor spreading and continental drift can be worked out as far back as the mapped marker-horizons are available. Figure 7.15 shows the extent of the opening up of the oceans estimated for the last 10 my by mapping the position of anomaly 5. A reconstruction of the position of the continents in the Palaeocene, obtained by closing the gap between the ridge crests and anomaly 31, is shown in Figure 7.16.

*Corroborating the geomagnetic time scale*

Tracing the past history of ocean-floor spreading beyond 10 my depends on the absolute reliability of the geomagnetic time scale. The main problem is to determine whether spreading has occurred relatively uniformly or in episodes. An independent check is becoming available as evidence on the age of the oldest sediments in layer 1 accumulates. One would expect the underlying layer 2 rocks to be slightly older. Even better, if the rocks at the top of layer 2 can be dated, the geomagnetic time-scale can be accurately tied in to the absolute time-scale.

   EWING and EWING (1967) showed that the layer 1 thickness increases abruptly at the position of anomaly 5, in the direction away from the ridges towards the adjacent ocean basins (Figure 7.17). They suggested that this represents a break in spreading and continental drift which ended about 10 my ago; the break was interpreted as being 30–40 my long and before it the spreading rate was interpreted as being faster than during the last 10 my. HEIRTZLER and others (1968) pointed out that the same pattern of magnetic anomalies occurs in all oceans, and therefore that any break in spreading must occur concurrently in all of the oceans. LE PICHON (1968) suggested that the geomagnetic time-scale can be reconciled with the abrupt thickening of sediments at anomaly 5 provided that a gap in the spreading lasting 10 my preceded anomaly 5 and that anomaly 32 has an age of 60 my rather than 77 my.

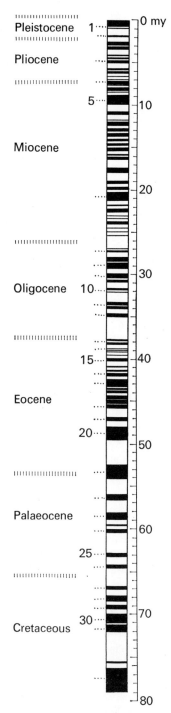

**Fig. 7.14** The geomagnetic time-scale based on a uniform spreading rate in the South Atlantic Ocean. Periods of normal polarity are shown in black and the numbers assigned to prominent magnetic anomalies are shown to the left of the column. Redrawn from HEIRTZLER and others (1968), *J. geophys. Res.,* **73**, 2123.

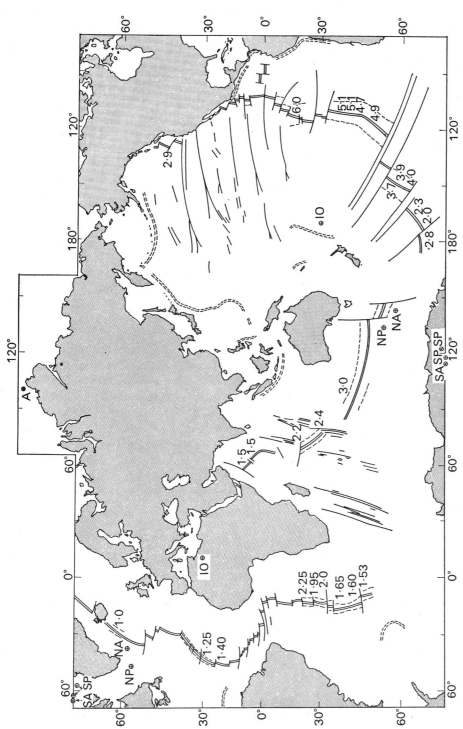

**Fig. 7.15** The position of anomaly 5 (dashed line) is shown in relation to the crests of ocean ridges (double line), indicating the amount of spreading during the last 10 my. The spreading rate deduced from the comparison of the magnetic anomalies with the palaeomagnetic time-scale is shown in cm/y. The poles of rotation about which spreading has occurred are: North Atlantic NA, South Atlantic SA, North Pacific NP, South Pacific SP, Arctic A, and Indian Ocean IO. A plus sign indicates pole deduced from geometry of fracture zone and a cross indicates pole deduced from spreading rates. Redrawn from HEIRTZLER and others (1968), *J. geophys. Res.*, **73**, 2130.

Results of the third leg of the Deep Sea Drilling Project (Chapter 3) have recently been announced (MAXWELL and others, 1970). Most of the drill holes on this leg, which is in the South Atlantic Ocean, penetrated through the whole of layer 1, reaching the top of layer 2. The positions of the holes are shown in Figure 7.18. The age of the oldest sediment of layer 1 encountered in each hole is plotted against distance from the ridge crest in Figure 7.19 and is compared with the age of the magnetic anomaly according to the time-scale of HEIRTZLER and others (1968) in Table 7.1.

**Table 7.1**   South mid-Atlantic ridge drilling sites (MAXWELL and others, 1970).

| Site No. | Sediment thickness (metres) | Hypothesized magnetic anomaly age (my) | Sediment age (my) | Distance from ridge axis (km) |
|---|---|---|---|---|
| 16 | 175 | 9 | 11 | 191 |
| 15 | 141 | 21 | 24 | 380 |
| 18 | 178 | —* | 26 | 506 |
| 17 | 124 | 34–38* | 33 | 643 |
| 14 | 107 | 38–39 | 40 | 727 |
| 19 | 141 | 53 | 49 | 990 |
| 20 | 72 | 70–72 | 67 | 1270 |
| 21 | 131† | — | 76† | 1617 |

* Location of these sites within the characteristic magnetic anomaly pattern is uncertain.
† Basement rock not reached at site 21.

These results from leg 3 of the Deep Sea Drilling Project are of exceptional scientific importance. They give the first direct confirmation of the predictions of the Vine-Matthews hypothesis and of the theory of ocean-floor spreading. Moreover, they confirm the reasonable accuracy of the Heirtzler geomagnetic time-scale (Figure 7.14), and give us confidence in the use of the scale back to an age of 70 my with an accuracy of about $\pm 5$ my. The results show that spreading in the South Atlantic Ocean appears to have been a continuous process over the last 70 my. This must also apply to the Pacific and Indian Oceans where the same uninterrupted sequence of magnetic anomalies has been observed; however, the relatively uniform spreading rate in the South Atlantic confirmed by the drilling suggests that the changes in spreading rate from the Pacific and Indian Oceans depicted in Figure 7.13 are real. The results from leg 3 do not support the idea of episodic spreading of EWING and EWING (1967), implying that we must seek some other explanation of the abrupt thickening of sediments reported to occur at anomaly 5.

*Plate tectonics*

The concept of plate (or block) tectonics has evolved from the hypotheses of ocean-floor spreading and transform faults. The overall concept was proposed almost simultaneously by McKENZIE and PARKER (1967) and by MORGAN (1968). The basic idea is that the outermost shell of the solid Earth which forms the lithosphere suffers strong deformation only along relatively narrow linear mobile belts. The mobile belts and the interconnecting transform faults divide the lithosphere into a series of 'rigid' plates which do not undergo any significant internal stretching, folding or distortion. Most of the Earth's tectonic activity is concentrated at the boundaries between adjacent plates.

It is a well-established facet of isostatic theory that the weak asthenosphere is overlain by a relatively strong lithosphere (or tectonosphere) about 50–100 km thick. This rheological model is supported by much modern observational, experimental and theoretical evidence as described in Chapter 8. Another important basis of plate tectonics is that the plates of lithosphere are capable of transmitting stress over large horizontal distances without buckling (p. 238).

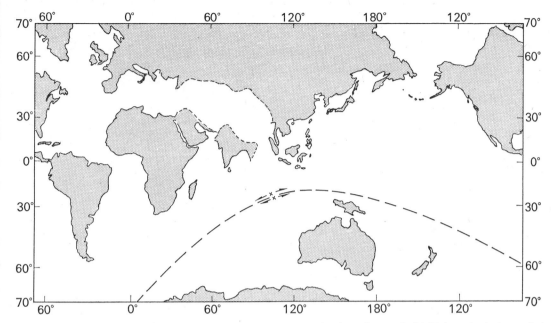

**Fig. 7.16** Reconstruction of the positions of the continents at the time of anomaly 31 (Palaeocene). Antarctica and Africa have been assumed to be in their present positions, and the relative position of the groups of continents on either side of the dashed line is not known. Redrawn from LE PICHON (1968), *J. geophys. Res.*, **73**, 3685.

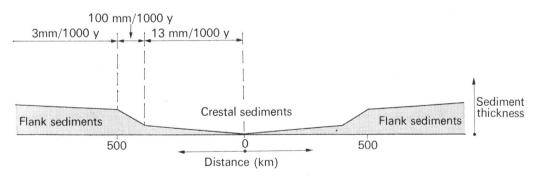

**Fig. 7.17** Generalized sketch of crestal and flank sediment thickness across the equatorial East Pacific rise. The rates of sedimentation assume a uniform spreading rate of 4·5 cm/y. Alternatively, if the rate of sedimentation is constant, then the abrupt increase in sediment thickness marks a break in spreading preceded by a period of more rapid spreading than the present rate. Adapted from EWING and EWING (1967), *Science, N.Y.*, **156**, 1592.

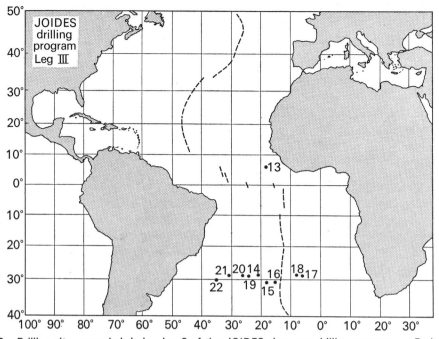

**Fig. 7.18**  Drilling sites occupied during leg 3 of the JOIDES deep-sea drilling programme. Redrawn from MAXWELL and others (1970), *Science, N.Y.*, **168**, 1048.

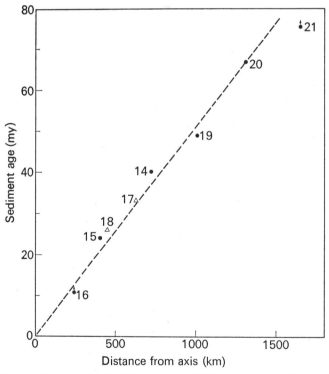

**Fig. 7.19**  Age of oldest sediments at leg 3 drilling sites of JOIDES plotted as a function of distance from the axis of the mid-Atlantic ridge. Numbers next to points indicate site numbers (Figures 7.18), the triangles showing sites on the eastern flank. Arrows at sites 16 and 21 indicate possibility of older sediments directly above basalt which were not recovered. Redrawn for MAXWELL and others (1970), *Science, N.Y.*, **168**, 1055.

The three basic types of plate boundary are as follows:

(1) *Extension boundaries*, where new crust is produced at the crests of ocean ridges;
(2) *Compression boundaries*, where the surface is being destroyed as plates approach each other;
(3) *Transform faults*, where plates move laterally relative to each other, and crust is neither produced nor destroyed.

Compression boundaries occur where the crust is shortened by thickening in young fold mountain ranges, and at the ocean trenches where oceanic crust is recycled into the upper mantle and where the continental blocks may progressively override the oceanic crust.

The beauty of plate tectonics is that it provides a geometrical explanation of how ocean-floor spreading and continental drift can occur on the surface of a nearly spherical Earth without deformation of the ocean-floor or the continents except at the well-known mobile belts. It relates most of the Earth's primary tectonic activity including continental drift and the formation of ocean ridges, young fold mountains, trenches, island arcs, plateau uplifts and rift valleys to the single process of ocean-floor spreading. Most of the implications of plate tectonics follow from purely geometrical reasoning without need to refer to the underlying cause of the movement of plates. In fact, discussion of the underlying mechanism is deferred entirely until later chapters.

Any conceivable displacement of a plate on a spherical surface can be produced by rotation about an appropriate axis passing through the centre of the sphere. If you do not believe this, then try to produce another type of displacement—you will find it is impossible. Any given displacement can be completely specified by one of the two poles where the rotation axis cuts the surface and by the angular rotation about the axis needed to cause the displacement. Similarly, the relative movement of two plates is defined by the pole of rotation and the angular velocity of rotation.

The pole of rotation for two adjacent portions of spreading ocean-floor on opposite sides of an oceanic ridge can be estimated from the geometry of the fracture zones which are active transform faults. Movement of the ocean-floor is ideally parallel to the transform faults, which must therefore form lines of latitude (small circles) relative to the axis of rotation. Another method of estimating the pole of rotation is to study the variation in spreading rate along an ocean ridge, using the values obtained from magnetic anomalies. Relative to the axis of rotation, the spreading rate must be a maximum at the equator and must fall-off with increasing latitude $\theta$ as $\cos \theta$. The reasonable agreement between the two methods gives confidence in their reliability (LE PICHON, 1968). The angular velocity of the separation of two plates can be estimated from the spreading rate at any known position relative to the pole of rotation.

As an approximation, Le Pichon divided the Earth's surface into six main plates as shown in Figure 7.20. Using fracture zones and spreading rates, he computed the pole of rotation and angular rate of separation associated with each pair of separating plates (Table 7.2). He then computed the relative motions of plates at the compression boundaries assuming that the Earth's surface area remains constant, obtaining the results shown in Figure 7.20. To do this, it was necessary to avoid specifying the spreading rate across the ridges of the southern part of the Indian Ocean, because otherwise the problem would be overdetermined; in fact the computation yielded estimates of the spreading rates across these ridge portions which agreed excellently with the observed rates.

Le Pichon's computations indicate that the creation of new plate in the Pacific is insufficient to account for the destruction of plate at its edges. We thus conclude that Eurasia and America are slowly converging at a rate of about 2–4 cm/y: at this rate the Pacific Ocean would vanish in about 300 my, although collisions between adjacent continents would probably stop the process before

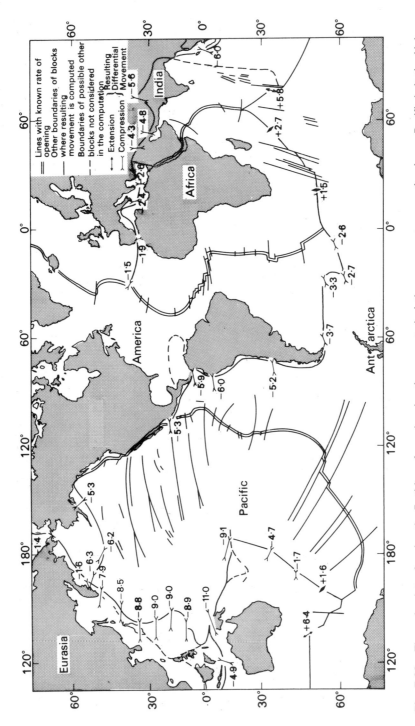

**Fig. 7.20** The main plates forming the Earth's surface, showing the computed relative movements across the circum-Pacific belt, the Alpine-Himalayan belt, the southern Indian Ocean ridges and the Scotia arc (shown in cm/y, positive signs indicate extension and negative signs indicate shortening), based on the observed spreading rates across the other ridges. Redrawn from LEPICHON (1968). *J. geophys. Res.*, 73, 3675.

Lines with known rate of opening
Other boundaries of blocks where resulting movement is computed
Boundaries of possible other blocks not considered in the computation
Extension
Compression
Resulting Differential Movement

Eurasia
America
Pacific
Africa
India
Antarctica

this happened. The computations suggest that the American, Eurasian and African plates are at present increasing in size, and that the Indian and Antarctican plates are not changing greatly in size.

**Table 7.2**   Poles of rotation for ocean-floor spreading obtained by least-squares fitting (after LE PICHON, 1968).

| Method | Number of estimates | Latitude | Longitude | Standard deviation | Angular rate $(10^{-7} \text{ deg/y})$ |
|---|---|---|---|---|---|
| | *South Pacific* (separation of Antarctic and Pacific plates) | | | | |
| fracture zones | 6 | 70°S | 118°E | 4·5° | 10·8 |
| spreading rate | 11 | 68°S | 123°E | | 10·8 |
| | *Atlantic* (separation of America and Africa) | | | | |
| fracture zones | 18 | 58°N | 37°W | 2·9° | 3·7 |
| spreading rate | 9 | 69°N | 32°W | | 3·7 |
| | *North Pacific* (separation of America and Pacific) | | | | |
| fracture zones | 32 | 53°N | 47°W | 5·7° | 6·0 |
| | *Indian Ocean* (separation of Africa and India) | | | | |
| fracture zones | 5 | 26°N | 21°E | 0·6° | 4·0 |
| | *Arctic Ocean* (separation of America and Eurasia) | | | | |
| fracture zones | 4 | 78°N | 102°E | 9·1° | 2·8 |

Another implication of the global picture of plate movement is that the Indian Ocean and the mid-Atlantic ridge must be moving away from each other. This shows that the position of ocean ridges does not necessarily remain fixed on the Earth's surface; some or all of them must migrate.

The poles of rotation estimated for the spreading now taking place in the Pacific and Atlantic Oceans are close to each other and not far distant from the North Pole (Figure 7.15). This may not be fortuitous and may suggest that the Earth's rotation can influence the orientation of ocean-floor spreading now and in the past. On the other hand, the pole of rotation for the spreading Indian Ocean is in North Africa; Le Pichon showed that this is a geometrical necessity to produce compatibility with spreading in other oceans. In other words, the rates and orientations of the spreading in the expanding oceans are to some extent interdependent.

On examining spreading directions of the Atlantic in more detail, Le Pichon found that the poles of rotation for the North and South Atlantic differed significantly from each other. The north-west and south-west Atlantic semi-plates are converging slightly on each other as they move westwards, and the North and South American continents are moving towards each other. The surplus surface area is probably being destroyed in the complicated Caribbean region of island arcs, trenches and transform faults. A similar type of convergence between the South American and Antarctican plates may be related to the Scotia arc.

A similar incompatibility is found between the spreading in north and south parts of the Pacific Ocean when the fracture zones of the north-eastern Pacific were active 10 my and more ago. This would result in a compression zone towards the west interpreted as the Tuomotu ridge, and an extension zone towards the east interpreted as the Chile ridge. The conspicuous fracture zones of the north-eastern Pacific ceased to be active as transform faults 10 my ago, since when the direction of spreading in the northern Pacific has become more compatible with that in the south of the ocean.

Another kind of geometrical consequence of plate tectonics is that it may possibly explain curvature of some island arcs. FRANK (1968) has pointed out that if a flexible but inextensible spherical shell is bent inwards by an angle $\theta$, then the indented part forms an intersecting spherical surface having the same radius as the shell (Figure 7.21). The edge of the indented part is a circle whose diameter subtends an angle $\theta$ at the centre of the sphere. No other type of deformation can occur provided that the shell remains inextensible, as can be confirmed practically on an old ping-pong ball. The theorem applies equally validly to a plate which forms part of a spherical shell. Island arcs and trenches are believed to mark the position where oceanic lithosphere is bent downwards into the mantle at about 45° (p. 264). The theorem suggests that each arc-trench system should form a part of a circle which subtends an angle of about 45° at the Earth's centre. This would give a typical arc a radius on the Earth's surface of about 2500 km. This is in agreement with the observed curvature of some island arcs, but others do not agree well with Frank's hypothesis. Although Frank's argument may well be an oversimplification, no doubt the nearly universal convexity of island arcs is a consequence of spherical geometry.

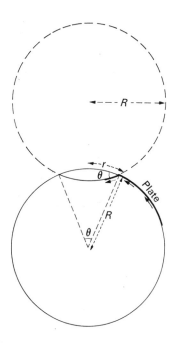

**Fig. 7.21** Diagram to show the geometry of an inward-bent inextensible plate on the surface of a sphere. The radius of the resulting arc measured along the spherical surface is given by $r = \frac{1}{2}R\theta$ where $R$ is the radius of the spherical surface and $\theta$ is the angle by which the plate is bent inwards, measured in radians.

A lot of detailed work on plate tectonics is in progress at the moment. The principles have been laid down, and now they are being applied with conspicuous success to the tectonics of more complicated regions such as the Mediterranean and the Caribbean. Le Pichon's analysis should be regarded as a first attempt which is likely to be modified in detail, although it has been a very successful beginning and has acted as a great stimulus to further work.

## 7.5 Conclusions

It has been shown that quite independent lines of evidence from many branches of geology, from palaeomagnetism, and from oceanic geophysics are consistent with each other in presenting an overwhelming case for continental drift. The new ideas of ocean-floor spreading and plate tectonics provide a geometrical framework within which continental drift can occur through known tectonic processes affecting the lithosphere, without causing any distortion to the vast majority of the Earth's surface. It will be shown in Chapter 8 that a modern assessment of the distribution and focal mechanism of earthquakes adds further support to the concept.

Ocean-floor spreading and plate tectonics do not only provide a geometrical explanation of continental drift. They also provide for the first time a unified explanation of the origin of the Earth's primary surface features:

(1) *Ocean ridges* are related to active formation of new crust, and their uplift is caused by the underlying mass deficiency in the mantle;

(2) *Rift valley* systems are incipient ocean ridges and lines of splitting of continents;

(3) *Ocean trenches* occur where oceanic crust is pushed/pulled down into the mantle;

(4) *Island arcs* occur on the continentward side of some trenches as a result of compression between the converging plates, and their arcuate shape is related to the bending down of oceanic crust at the associated trenches;

(5) *Young fold mountains* are produced by shortening of the continental crust as a result of compression between converging plates (the continental crust has too low a density for it to be carried down into the upper mantle);

(6) *Large strike-slip faults* are transform faults related to lateral movement of adjacent plates;

(7) *Active continental margins* occur where an oceanic and a continental block converge on each other;

(8) *Aseismic continental margins* occur where the adjacent parts of the oceanic and continental crusts form part of the same plate, and they mark the original line of split between the continents (and ancient rift system?).

Secondary tectonic features such as basins and geosynclines probably form as a consequence of the existence of a primary feature such as a mountain range or a continental margin.

The theory of plate tectonics has been developed without any need to refer to the underlying driving process in the mantle, which will be discussed in Chapter 9. However, it is worth mentioning at this stage that ocean-floor spreading at the present rate must have a profound effect on the upper mantle. Suppose that the average spreading rate is taken as 2 cm/y along 80 000 km of ocean ridge, and that the lithospheric plate which is recycled into the mantle at the trenches is 60 km thick. Then about $2 \times 10^8$ km$^3$ of lithosphere is recycled into the mantle every million years. The volume of the mantle down to 700 km depth is about $3000 \times 10^8$ km$^3$. This suggests that the whole of the topmost 700 km of the mantle could be recycled at the present rate in the order of 1500 my, and if the lower mantle is also involved the whole of the mantle could be recycled in less than the age of the Earth.

# 8     Fracture and flow in the crust and mantle

This chapter deals with the non-elastic response of the crust and the mantle to applied stress systems. The applied stresses which affect the Earth vary in duration from less than a second to over 1000 my. A serious difficulty is that our knowledge of the response to the longer duration stresses is particularly uncertain.

The occurrence of isostasy has shown that rocks at a depth of about 50–100 km and below must be able to flow and are much weaker than the surface layers above. The relatively strong surface layer of isostatic theory called the *lithosphere* is underlain by a weaker layer called the *asthenosphere* (p. 45). The boundary between them is likely to be gradational and probably occurs at shallow depth in the upper mantle. It will be shown in the chapter that the asthenosphere is probably about 400 km thick and that beneath it the rocks of the lower mantle may be much more resistant to flow.

## 8.1 Fracture and flow in solids

The type of deformation which a solid material undergoes is expressed in terms of the relationship between stress and strain. Stress is defined as force per unit area. The stress acting on any plane can be resolved into a normal stress or pressure acting at right angles to the surface and two components of shear stress acting within the plane. In any stressed medium there are three planes mutually perpendicular to each other in which the shear stress is zero. The three *principal pressures* ($\sigma_{max}$, $\sigma_{int}$, $\sigma_{min}$) act perpendicular to these planes. The state of stress at a point is completely specified by the magnitudes and directions of the three principal pressures. The shear stress is maximum in the two planes which bisect the directions of maximum and minimum principal pressure and is numerically equal to half the stress difference, i.e. $\tau_{max} = \frac{1}{2}(\sigma_{max} - \sigma_{min})$. In dealing with stresses within the Earth it is convenient to treat compression as positive and tension as negative (i.e. opposite to the usual convention).

Strain is defined as change in length per unit length. There are two types, linear strain in which the change in length is in the same direction as the initial length, and tangential strain in which they are at right angles. The state of strain can be completely specified by the directions and magnitudes of the three principal extensions.

Below the elastic limit solids deform according to Hooke's law with stress almost proportional to strain. Various types of non-elastic behaviour occur when the elastic limit is exceeded, as discussed below. A useful account is given by JAEGER (1962).

Solids may be classified as *brittle* or *ductile* substances depending on how they deform above the elastic limit. Brittle substances fracture without appreciable deformation; ductile substances deform appreciably by flow. Substances which are brittle under atmospheric conditions become ductile at high temperature and/or pressure. The brittle-ductile transition for the substance is a function of temperature and pressure.

The two main types of brittle fracture are *extension fracture* and *shear fracture*. Extension fracture occurs when one or more of the principal pressures is a tension; the plane of fracture occurs perpendicular to the direction of maximum tension. Shear fracture occurs under compression. In a homogeneous and isotropic substance the two complementary planes of fracture each subtend an angle of less than 45° to the maximum principal pressure (characteristically about 30°) and contain the axis of intermediate principal pressure (Figure 8.1.). According to the Coulomb-Navier hypothesis of

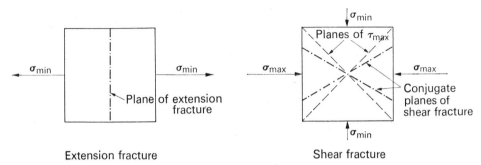

Extension fracture                                    Shear fracture

**Fig. 8.1**   Relationship between fracture planes and principal pressures for extension fracture and shear fracture.

shear fracture, the reason for this angle being less than 45° is the influence of internal friction on the attitude of fracture planes. The Griffith theory of fracture attributes brittle fracture of both kinds to local concentrations of stress caused by microscopic cracks within the substance. Observations on extension and shear fracture are in approximate agreement with the predictions of the theory.

In a *Newtonian viscous* substance the rate of strain $\dot{\epsilon}$ is proportional to the applied shear stress $\tau$, such that $\tau = \eta\dot{\epsilon}$ where $\eta$ is the coefficient of viscosity. Most liquids and many glasses exhibit Newtonian viscosity to a good approximation. Viscous flow is usually a thermally activated process, and the dependence of $\eta$ on absolute temperature $T$ is given by

$$\eta = \eta_0 e^{-E/kT}$$

where $E$ is the energy of activation and $k$ is Boltzmann's constant.

A *perfectly plastic* substance has a finite yield stress at which the body flows. The shear stress at yield cannot be exceeded whatever the rate of strain. Several yield criteria have been suggested, for example the Tresca criterion depends on the maximum shear stress reaching a critical value, whereas in the Von Mises criterion it is the elastic strain energy of distortion which reaches a critical value. Plastic flow occurs on two sets of slip planes perpendicular to each other. Many plastic substances show an increase in their yield stress with progressive plastic strain. This is known as *work hardening*. At low temperature the deformation of ductile metals is approximately plastic and generally shows work hardening. In metals, plastic deformation is produced by the movement of dislocations (lattice plane atomic mismatch) on slip planes accompanied by some movement of crystal boundaries.

Another type of deformation characteristic of metals and other crystalline materials at high temperature is *non-Newtonian viscous flow*. Above a finite yield stress the strain rate increases quasi-exponentially with increase of stress as shown in Figure 8.2. Substances which deform like this have a coefficient of viscosity which depends on strain rate. Non-Newtonian viscous flow is more closely approximated by perfect plasticity than by Newtonian viscosity.

Composite rheological models have been widely used to represent some types of non-elastic deformation. In these models, it is conventional to represent elastic deformation, viscous flow and finite yield stress by spring, dashpot and frictional contact respectively. A *Kelvin substance* (firmo-viscous)

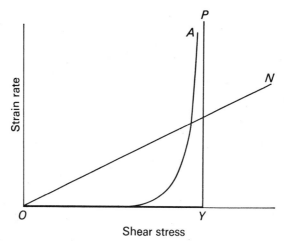

**Fig. 8.2** Strain rate plotted against stress for Newtonian viscosity (N), non-Newtonian (Andradean) viscosity (A) and ideal plasticity (OYP). Redrawn from OROWAN (1965), *Phil. Trans. R. Soc.,* **258A**, 285.

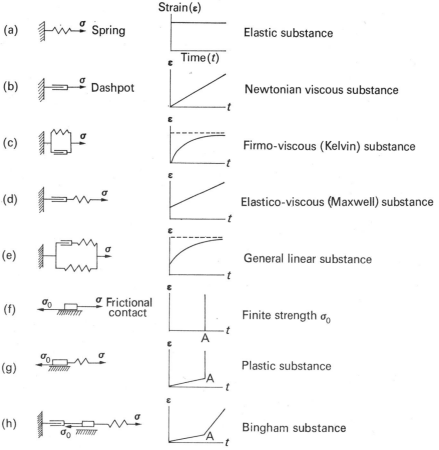

**Fig. 8.3** Rheological models for some common types of deformation mechanism, showing the corresponding strain-time relationships. For (a)–(e) the load is regarded as constant, but for (f)–(h) the strain increases linearly with time until the strength is reached at A and thereafter remains constant. Taken in part from JAEGER (1962(, *Elasticity, fracture and flow,* 2nd edition, pp. 100-104, Methuen.

shows an elastic deformation delayed by internal viscosity and is represented by a spring in parallel with a dashpot (Figure 8.3(c)). A *Maxwell substance* (elastico-viscous) is basically a Newtonian viscous substance showing an initial elastic strain and is represented by a dashpot in series with a spring (Figure 8.3(d)). A *Bingham body* is a plastic substance except that strain rate is directly proportional to stress above the yield stress and it is represented by a spring in series with a dashpot and a frictional contact (Figure 8.3(h)).

*Creep* is the name given to very slow non-elastic flow caused by a constant load. Creep mechanisms do not necessarily fit neatly into the simple models of flow described above. Several types of creep are known to occur in metals. Below about $0·2T_m$, where $T_m$ is the absolute melting temperature, strain is proportional to the logarithm of time; this is known as *logarithmic creep* or *α-creep*. It cannot produce large strains. It is produced by movement of dislocations and the progressive slowing down of the creep rate is believed to be caused by work hardening. Above about $0·5T_m$ the creep rate under a constant load becomes steady and independent of time; this is known as *hot creep* or *recovery creep* and it occurs when the temperature is high enough for dislocations to be renewed fast enough to inhibit work hardening. Hot creep follows the stress-strain relationships appropriate to a non-Newtonian viscous substance, with strain rate approximately proportional to the fourth power of the applied stress and with a finite creep strength. Between about $0·2T_m$ and $0·5T_m$ *transitional creep* or *β-creep* occurs. After a sufficiently long period of time transitional creep may become steady state creep.

At temperatures close to the melting point another type of creep is known to occur. This is *diffusion creep* or *Herring-Nabarro creep*. It is caused by migration of atoms in a stress gradient (GORDON, 1967). In strong contrast to recovery creep, the strain rate is proportional to the applied stress. Thus substances showing this sort of creep behave as if they possessed Newtonian viscosity.

## 8.2 Fracture and flow in the lithosphere

### The upper continental crust

Most rocks are brittle at the temperature and pressure prevailing in the uppermost part of the crust. At atmospheric temperature and pressure the compressive strength of granite is about 1·4 kbar and the tensile strength is about 40 bar. The strength of rocks increases with confining pressure but decreases with rise in temperature. Experiments suggest that the strength would be expected to increase with depth down to a few kilometres, but below about 5–10 km the temperature effect caused by the geothermal gradient would become dominant, causing strength to decrease with depth and brittle fracture to give way to ductile failure. An excellent review of experiments on rock deformation is provided by the Geological Society of America Memoir 79 (GRIGGS and HANDIN, 1960a).

In many places the basement rocks of continents are overlain by successions of sedimentary rocks. Sedimentary successions tend to be weakened by joints and bedding planes and by weak layers such as salt deposits and shales. Because of this the sedimentary successions tend to deform passively, adjusting as necessary to accommodate movements in the stronger basement rocks beneath. Our main concern is therefore to study the failure mechanism in the basement rocks.

Beneath a depth of about 100 m all three principal pressures are normally compressions. Consequently shear fracture would be expected to be the dominant failure mechanism of the upper crust. This is borne out by the prevalence of faulting. There are three main classes of faults observed by geologists, which have been interpreted by ANDERSON (1951) to depend on which of the three principal pressures is the vertical one (Figure 8.4). *Normal faults* have fracture planes dipping at about 60–70° and the downthrown side of the fault overlies the plane. According to Anderson normal faults are formed by shear fracture when $\sigma_{max}$ is vertical. In homogeneous rocks the fault trace is perpendicular to $\sigma_{min}$ and faulting results in local extension of the crust in this direction. Normal faulting, in common with the two other types, may occur on either of two conjugate fault planes but local inhomogeneities

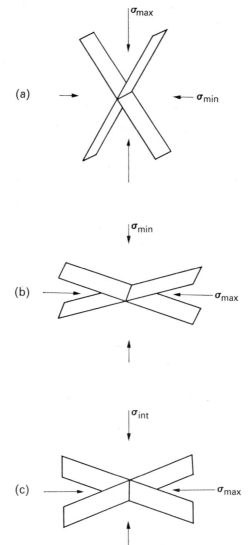

**Fig. 8.4** The subdivision of faults into three major classes by Anderson depends on (1) the Coulomb-Navier hypothesis of shear fracture, and (2) the three principal pressures near the Earth's surface being orientated horizontally and vertically. The three classes shown are:

(a) normal fault—$\sigma_{max}$ vertical;
(b) thrust fault—$\sigma_{min}$ vertical;
(c) strike-slip fault—$\sigma_{int}$ vertical.

Redrawn from ANDERSON (1951), *The dynamics of faulting*, pp. 14-16, Oliver and Boyd.

typically cause one of them to be dominant. *Thrust faults* are caused by fracture when $\sigma_{min}$ is vertical and in theory the fault plane should dip at 15–30°. Thrusts result in local shortening of the crust in the direction of $\sigma_{max}$. *Strike-slip faults* show horizontal movement on vertical fault planes and they occur when $\sigma_{int}$ is vertical.

Anderson's theory of faulting was one of the early successes of the application of mechanics to geological problems. There are, however, observed features of faulting which do not fit into the theory, although most of these may be explained by slight modifications. One particular difficulty lay in explaining the nearly horizontal thrust planes which show transport of many kilometres. However, HUBBERT and RUBEY (1959) have demonstrated that these may be explained in terms of the influence of pore-fluid which overcomes internal friction. The common occurrence of oblique-slip faulting was another difficulty. This may be explained without recourse to obliquely orientated stress systems by the presence of pre-existing planes of weakness (BOTT, 1959; JAEGER, 1960).

Dyke intrusion and the formation of certain types of joint are caused by extension fracture. Dyke swarms can produce local extension of the continental crust amounting to several kilometres and much larger extensions of the oceanic crust may be associated with ocean-floor spreading by the dyke intrusion mechanism. Dyke intrusion is believed to occur by the wedging effect of a sheet of magma under a higher hydrostatic pressure than the mean pressure of the intruded rocks (ANDERSON, 1951). This causes an actual tension to develop ahead of the advancing wedge of magma and this is relieved by extension fracture as shown in Figure 8.5. Both dykes and faults should form at right

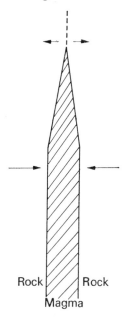

**Fig. 8.5**   The mechanism of dyke intrusion; a vertical extension fracture develops ahead of the upward rising wedge of magma which has a hydrostatic pressure exceeding the confining pressure of the intruded solid rocks. Redrawn from ANDERSON (1951), *The dynamics of faulting*, p. 24, Oliver and Boyd.

angles to the minimum principal pressure, which would be horizontal. Thus the study of dykes and normal faults can indicate past directions of tension in the continental crust.

The basement rocks of the upper crust are probably not normally hot enough to suffer significant deformation by creep in response to relatively small stress differences. Logarithmic and transitional types of creep would be expected to occur, but these would not produce large strains even over long geological periods. This is borne out by the quite large gravity anomalies which occur locally over Precambrian shields and are caused by igneous intrusions in the upper crust. These cause local stress differences of the order of 50–100 bar. Some of these intrusions have persisted over periods of more than 1000 my without 'creeping away'.

### The middle and lower continental crust

Experiments suggest that two important changes in the mechanical properties of rocks occur over the depth range of about 10–25 km. Firstly, the brittle-ductile transition occurs; GRIGGS, TURNER and HEARD (1960) have shown that no sudden fracture was observed in any rock apart from quartzite above 5 kbar and 500°C, which would be at an approximate depth of 20 km. Secondly, the compressive strength would be expected to decrease with depth because of the dominance of the temperature effect; for instance the compressive strength of dunite, pyroxenite and granite at 5 kbar confining pressure is 20 kbar at 25°C, 10 kbar at 500°C and 7 kbar at 800°C.

Rocks which have been deformed at high temperature and pressure are now visible at the Earth's surface in the ancient mountain-building belts. Deformation in these belts was accompanied by metamorphism. In the lower grades of metamorphism which correspond to shallower depths and lower temperatures deformation occured partly by recrystallization and partly by crushing and comminution of the rock (known as cataclasis), suggesting that this was the region of the brittle-ductile transition. At the higher grades, deformation took place mainly by recrystallization in which pore-fluid probably played an important part. The highest grade metamorphic rocks such as the granulite facies show less evidence of strong shear stress in their mineral fabrics. Thus observations on rocks deformed in orogenic belts are in full agreement with the predictions of experiments on deformation.

Mountain-building belts are exceptional regions of the crust where the release of strain energy is concentrated. We can be much less certain about the mechanism of deformation in the lower part of the normal continental crust. Experimental investigations on the creep of rocks at room pressure and at temperatures up to 750°C, made by MISRA and MURRELL (1965), are relevant. They found that the rocks studied, including microgranite and peridotite, show logarithmic creep below $0\cdot2T_m$, where $T_m$ is the absolute melting temperature, and transitional creep occurs at higher temperature. They did not reach a high enough temperature to observe steady state creep, but the important outcome of the investigation is that creep in rocks appears to follow a similar pattern of behaviour to metals. The results suggest that the lower part of the continental crust may be able to deform appreciably by transitional or steady state creep provided large enough stress differences exist and the temperature is high enough.

### The oceanic crust

The observed deformation of the oceanic crust contrasts strongly with that of the continental crust. Extension fracture associated with dyke intrusion near the crests of ocean ridges appears to be the most significant mechanism of major deformation, forming new oceanic crust at a rate of about 2–12 cm/y. Associated with these are the apparently large strike-slip displacements of up to several hundred kilometres along fracture zones resulting from transform faults. Both types of deformation produce much larger strain than is known to occur in the continental crust.

Away from the ocean ridges and fracture zones, the sediments of layer 1 of the oceanic crust show a remarkable lack of deformation.

## Stress systems affecting the lithosphere

The lithosphere is evidently much stronger than the deeper parts of the Earth. The fact that it has suffered widespread fracture and flow suggests that much larger stress differences occur in the crust than in the mantle beneath.

Local stress systems which affect the crust are caused by topographical features acting as loads, by lateral variations of density within the crust, and by upthrust at the base of the crust associated with changes in crustal thickness. Such stress systems play an important part in isostatic processes but are not the principal cause of tectonic activity. Local stress systems can be studied approximately by using two simple models:

(1) a load acting on the surface of a uniform elastic half-space;
(2) loading of an elastic crust underlain by a fluid substratum.

Both models are over-simplifications of natural systems but nevertheless they give an important insight into the strength of the crust and the mechanism of isostasy.

Consider a uniform elastic half-space subjected to a uniform two-dimensional surface load $\sigma$ extending from $x = +a$ to $x = -a$. The load causes a stress distribution as shown in Figure 8.6(a). The maximum stress difference is equal to $2\sigma/\pi$ which occurs along the semi-circular arc containing the ends of the load. Stress differences beneath the load exceed half this value down to a depth of 1·5 times the width of the load. Applying this to a topographical load 40 km wide, 500 m high with a density of 2·7 g/cm³, the maximum stress difference of 80 bar extends to 20 km depth. This is the sort of effect which a topographical uplift such as the northern Pennines would cause; the negative load caused by a low density granite batholith associated with a − 45 mgal gravity anomaly would cause stress differences of the same magnitude. JEFFREYS (1959) has shown that the stress difference below a large mountain range such as the Himalaya must locally exceed 1 kbar.

The stress differences shown in Figure 8.6(a) would not be large enough to cause failure of the crust. However, an interesting situation arises if a local stress system such as this one is superimposed on a large regional stress field. The result is that the local stress system modifies the regional stresses in the vicinity of the load. Figure 8.6(b) shows the above local stress system superimposed on a regional horizontal tension of 1 kbar. The resulting stress differences are slightly increased beneath the load but are reduced elsewhere. Such interaction of local and regional stress fields may be an important factor in locating the position where failure such as faulting first occurs (BOTT, 1965b).

The deformation of an elastic 'crust' underlain by a substratum of negligible strength must also be considered. If the load is narrower than the thickness of the crust, then the stress distribution does not differ greatly from that of the uniform half-space discussed above, although the maximum stress difference would be slightly greater. But if the load is wider than the thickness, the stress differences are much larger and are of the order of ten times the load. This suggests that the response of the crust to local loads depends critically on the rheological condition of the underlying mantle. Where the mantle is hot and weak (e.g. western U.S.A.) one would expect to find evidence of tectonic activity. Where the mantle is cool and relatively strong (e.g. beneath the Precambrian shields) one would expect the crust to be more resistant to failure, as is observed.

JEFFREYS (1959) has studied the deformation of an elastic crust subjected to a harmonic load $\sigma \cos 2\pi x/L$, where $L$ is the wavelength of the load. He treated the problem by the theory of bending of a thin elastic sheet. If the wavelength of the load is sufficiently large, then the sheet bends elastically to a state of approximate isostatic equilibrium. Jeffreys has shown that the vertical deformation $w$ of the mid-plane of the sheet is related to the deformation $w'$ which would occur if perfect isostatic

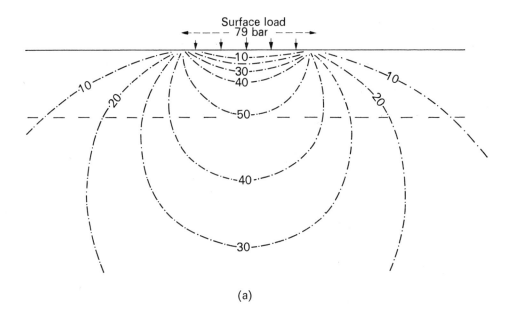

(a)

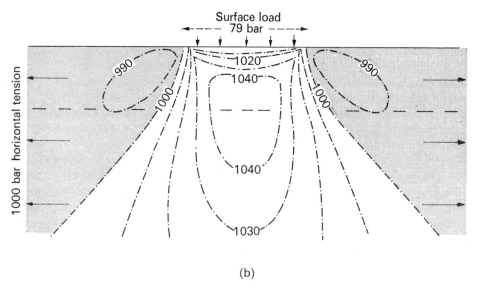

(b)

**Fig. 8.6**

(a) The stress-differences in bar produced by a uniform two-dimensional load of 79 bar acting on the surface of a homogenous elastic half-space, assuming plane strain. This load would be equivalent to a topographic elevation of about 300 m, or a positive gravity anomaly of 25-30 mgal caused by high density rocks in the uppermost part of the crust.

(b) The stress-differences in bar caused by the superimposition of the same surface load on a horizontal tension of 1000 bar.

Both distributions are independent of scale, but if the load is assumed to be 50 km wide, then the dashed line gives the approximate depth of the continental Moho. Redrawn from BOTT (1965b), *Submarine geology and geophysics*, p. 197, Butterworths.

equilibrium were attained by the following equation:

$$w/w' = g\rho/(aT^3/L^4 + g\rho)$$

where $a = 16\pi^4\mu(\lambda+\mu)/3(\lambda+2\mu)$, $\lambda$ and $\mu$ are Lamé's constants, $g$ is the value of gravity, $T$ is the thickness of the sheet and $\rho$ is the density of the fluid substratum. If $F = aT^3/L^4 g\rho < \frac{1}{2}$ then isostatic equilibrium is approximately attained, but if $F>2$ then the deformation is only a small fraction of the isostatic equilibrium value. Putting $T = 50$ km, $\lambda = \mu = 3\cdot3 \times 10^{11}$ dynes/cm$^2$, $\rho = 3\cdot3$ g/cm$^3$ and $g = 980$ cm/s$^2$, isostatic equilibrium is over 94% attained when the wavelength is over 900 km, but the deformation becomes less than 6% of the isostatic value when the wavelength is less than 225 km. These calculations suggest that large surface features such as mountain ranges, ocean ridges, continents and oceans, and large icecaps would approximately reach isostatic equilibrium by elastic bending of the lithosphere provided that the underlying mantle is weak enough. Features 200 km or less in width would only be able to attain isostatic equilibrium through failure of the lithosphere such as by faulting.

Another important situation occurs where a load acting on the surface of the crust is counterbalanced by an equal and opposite upthrust acting on the base of the crust. This is the appropriate model for isostatic equilibrium and is particularly important in relation to mountain ranges and to continental margins. It can be approximated by studying the stress system caused by equal and opposite pressures applied to both sides of a two-dimensional beam (Figure 8.7). The result is that the vertical pressure is increased between the load and the upthrust. This means that the crust beneath the load is effectively thrown into a state of horizontal tension relative to the adjacent crust.

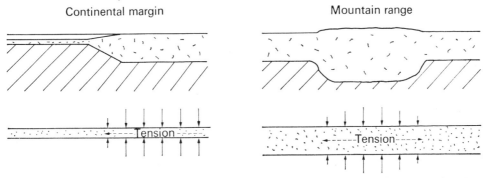

**Fig. 8.7** Production of regional tensile stress in the crust by the combined effect of a surface load and the upthrust of a root in isostatic equilibrium with the load, applied to continental margins and mountain ranges.

Local stress systems are unlikely to be the primary cause of most tectonic activity, apart from attainment of isostatic equilibrium. It is clear that the lithosphere has also been subjected to regional or global stress systems or both. Such stress systems could be caused by:

(1) drag of underlying convection currents on the lithosphere;
(2) the shouldering effect of dyke intrusion, particularly at ocean ridge crests;
(3) gravity gliding of a strong crust on a weaker substratum;
(4) expansion or contraction of the Earth.

These are illustrated in Figure 8.8. and are discussed in greater detail in Chapter 9.

## Can the crust buckle?

If a thin elastic plate is compressed along its length it may buckle before the compressive strength is reached. If the plate is viscous or plastic, resting on a fluid substratum of lower viscosity, then

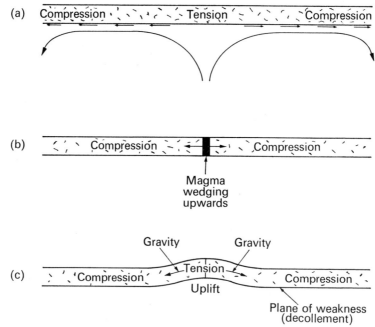

**Fig. 8.8**   Some possible causes of regional/global stress systems in the lithosphere:
(a) Drag of convection currents causes tension over rise and compression over sink;
(b) Magma intrusion causes compression;
(c) Gravity gliding causes tension above uplift and compression beyond (local stresses in the lithosphere would also be caused by the bending over the uplift.)

buckling may still occur. The problem arises as to whether the lithosphere can deform by buckling in response to horizontal compression. Such a mechanism has been suggested for the formation of geosynclines and trenches.

The problem of buckling of the lithosphere has been investigated from both theoretical and experimental standpoints by RAMBERG and STEPHANSSON (1964). They assumed that the lithosphere is an elastic or viscous plate overlying a fluid substratum. They found that the compressive stress needed to overcome the gravitational body forces caused by buckling greatly exceeds realistic estimates of the strength unless the lithosphere is unrealistically thin—250 m thick or less in the elastic model and 1·5 km or less in the viscous model. They point out that their conclusion would also apply to more complicated rheological models of the lithosphere and substratum.

Thus the buckling mechanism cannot cause deformation of the lithosphere or the crust as a whole. Failure must take place by fracture or flow long before the compressive stress is large enough to buckle such a thick layer.

## 8.3 The rheology of the mantle

*Creep processes in the mantle—theoretical considerations*

Apart from in the topmost few kilometres of the mantle which may belong approximately to the lithosphere, and in the deep earthquake zones (p. 264), creep is likely to be the most important deformation mechanism in the mantle.

By analogy with creep processes in metals, there are at least two types of creep which might give rise to large permanent strain within the mantle: *recovery creep*, involving movement of dislocations on slip planes; and *diffusion creep*. These two types of creep follow fundamentally different stress-strain relationships. Recovery creep produces non-Newtonian viscous flow when a finite strength has been exceeded whereas diffusion creep causes Newtonian viscous flow. *Recrystallization creep* produced through the influence of pore-fluids may also be very important in the mantle, but the stress-strain relationship for this type of creep is not well known. The predominant type of creep at a given position within the mantle depends on temperature, pressure, rate of strain, grain size and pore-fluids.

*Recovery creep* in rocks probably occurs at temperatures somewhat above $0 \cdot 5 T_m$ (MISRA and MURRELL, 1965). The temperature throughout most of the mantle probably exceeds this, especially in the upper mantle where the melting point is reached locally. Increase in temperature above $0 \cdot 5 T_m$, by analogy with metals, would be expected to bring about a rapid increase in creep rate and a decrease in the creep strength; nevertheless a finite creep strength would be expected even just below the melting point (OROWAN, 1965). Increase in confining pressure would be expected to reduce the creep rate and Misra and Murrell estimated that the creep rate of peridotite at 1200°C would be reduced by a factor of 100 by application of a confining pressure of 12 kbar. They extrapolated their observations on transitional creep in peridotite to estimate that the creep rate at 1200°C and 12 kbar confining pressure could be high enough to account for convective flow in the upper mantle. It is to be expected, therefore, that recovery creep does occur in the mantle wherever the shear stress exceeds the creep strength. It is more likely to occur in the upper mantle than deeper down because it is here that the temperature is closest to the melting point and the pressure is lowest.

*Diffusion creep* in metals is observed experimentally at temperatures above about $0 \cdot 85 T_m$. Theory suggests that it also occurs at lower temperatures but is either swamped by recovery creep or the rate is too slow to detect in experiments. Diffusion creep is believed to occur in all crystalline material and the question is whether the rate of creep is high enough to be of geophysical significance. It may be an important flow process in the mantle for the following reason. The strain rate of many natural processes within the Earth is lower than that of the experiments on metals by a factor of a million or more. The strain rate in diffusion creep depends linearly on the shear stress, but in recovery creep it depends on about the fourth power of the shear stress. Therefore reduction in the strain rate strongly enhances the relative importance of diffusion creep. For this reason, diffusion creep may be the predominant mechanism of steady flow in the mantle in response to small stresses of long duration.

A substance which deforms by diffusion creep behaves as a Newtonian viscous fluid. The coefficient of dynamic viscosity for grain radius $R$ is given by

$$\eta = kTR^2/10DV_a$$

where $k$ is Boltzmann's constant, $D$ is the diffusion coefficient and $V_a$ is the activation volume (MCKENZIE, 1968). $D$ is proportional to $e^{-G/kT}$ where $G$ is the free energy of activation. The effect of pressure on $\eta$ enters mainly through its influence on the free energy and depends on the activation volumes for vacancies and atomic jumps. GORDON (1965) has used a form of the above expression into which an error had crept to estimate the distribution of viscosity through the mantle assuming it to be made of the mineral periclase (MgO), which is possibly realistic for the lower mantle. The error has the effect of increasing the estimate of viscosity by a constant factor of 100. He obtained the free energy at zero pressure from experimental results and then extrapolated to high pressures using activation volumes estimated by analogy with close packed metals. Gordon was then able to predict the viscosity through the mantle for given assumptions of pressure and grain size. The resulting viscosity distribution (corrected for the error) shows a zone of low viscosity in the upper mantle and an increase in viscosity with depth below about 500 km depth, rising from $10^{19}$ P at 500 km to $10^{25}$ P at the core-mantle boundary.* The main criticism is that Gordon assumed a grain radius of $R = 0.05$ cm which is smaller than would be expected; the viscosity estimates would be 100 times bigger if the grain (i.e. crystal) dimension were 1 cm and would be in excellent agreement with other estimates. McKENZIE (1967) repeated the computations allowing for a higher activation energy in the upper mantle, and he produced the distribution shown in Figure 8.9.

Thus both recovery creep and diffusion creep are likely to occur in the mantle, although their relative importance is difficult to assess. Creep caused by recrystallization may also occur. It might be expected that recovery creep and recrystallization creep are dominant in the upper mantle and diffusion creep in the lower mantle.

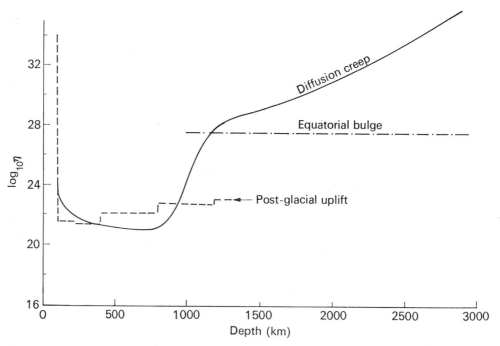

**Fig. 8.9**  Estimated viscosity distributions in the mantle based on theoretical computations for diffusion creep and on observations of the deformation of the mantle. $\eta$ is the dynamic viscosity in poise. Note that the diagram is based on the old (incorrect) formula for viscosity and assumes a grain radius of 0·05 cm. The diagram still applies after correction for the error (see McKENZIE, 1968) provided that the grain radius is taken as 0·5 cm, which is probably more realistic anyhow. Redrawn from McKENZIE (1967), *Geophys. J. R. astr. Soc.*, **14**, 303.

*The viscosity of the upper mantle*

An estimate of the apparent viscosity of the upper mantle can be made by studying the isostatic recovery of the Earth's surface after the application or removal of a surface load such as an icecap or a large lake (Figure 8.10). It is assumed that the lithosphere can be treated as an elastic layer about

* The c.g.s. unit of dynamic viscosity is 1 poise (P) = 1 g/cm s.

30–50 km thick and that the underlying asthenosphere behaves as a Newtonian viscous fluid. If the load is more than about 700 km in width, it is legitimate to assume that the lithosphere deforms passively by elastic bending and that the rate at which isostatic equilibrium is approached depends solely on the viscosity of the asthenosphere. If the load is known, then the rate of recovery at a given time after its removal (or application) provides an estimate of the viscosity. The calculation involves solution of the Stokes-Navier equation of viscous flow for a specified initial distribution of surface

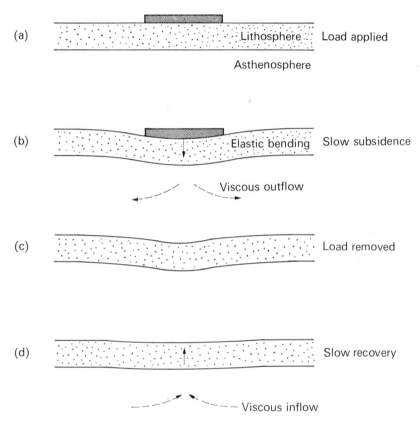

**Fig. 8.10**    The viscous response of the asthenosphere to the application and removal of a load such as an icecap on the surface of the elastic lithosphere.

pressure. This was done for a circular load assuming an infinite substratum of uniform viscosity by HASKELL (1935) and for two-dimensional harmonic loading by HEISKANEN and VENING MEINESZ (1958). McCONNELL (1965) repeated the calculations assuming the asthenosphere to be a layered viscous half-space, and TAKEUCHI and HASEGAWA (1965) extended the computational method to study the response of a layered viscous sphere to surface loading.

The best known application of the above method is to the uplift of Fennoscandia after melting of the Pleistocene icesheet. This icesheet was about 2·5 km thick and covered an area of 2500 × 1400 km² with the centre near the Gulf of Bothnia. It began to melt about 40 000 years ago and melting was practically complete about 10 000 years ago. The rate of uplift during the first half of the twentieth century is accurately known because precise levelling was carried out in Finland first between 1892 and 1910 and again between 1937 and 1953. The maximum rate of uplift of 1 cm/y occurs in the Gulf of Bothnia (Figure 8.11). Over a longer period, the pattern of uplift has been confirmed by the height of post-glacial raised beaches above the present sea-level, these being highest in the vicinity of the Gulf of Bothnia. Using the observed rate of uplift, HASKELL (1935) estimated the viscosity of the asthenosphere below Fennoscandia to be $3 \times 10^{21}$ P. Heiskanen and Vening Meinesz estimated it to be $10^{22}$ P. The preferred model of McConnell shows a viscosity of $3 \times 10^{21}$ P in the upper mantle increasing to about $10^{22}$ P at 400 km depth and $2 \times 10^{22}$ P at 1000 km.

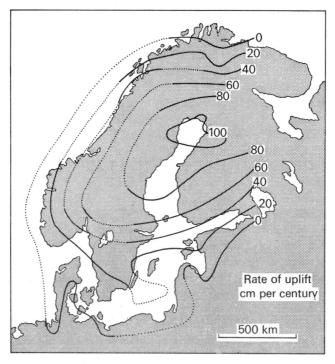

**Fig. 8.11** Contemporary rate of uplift of Fennoscandia. Redrawn from GUTENBERG (1959), *Physics of the Earth's interior*, p. 194, Academic Press.

Fennoscandia has already risen about 500 m since the last glacial period and probably has a further 200 m to rise. The rate of recovery towards equilibrium is conveniently measured by the *relaxation time*, which is the time required for the deviation from isostatic equilibrium to be reduced to $1/e$ of its initial value. The relaxation time is inversely proportional to the dimension of the load. It is about 5000 years for Fennoscandia.

The region of maximum glaciation in North America was centred on the southern part of Hudson Bay. This region is undergoing a similar uplift to Fennoscandia (GUTENBERG, 1954b). The North American and Fennoscandian icesheets had similar areal dimensions and thicknesses. The present-day uplift in southern Hudson Bay is about 1 cm/y. The North American uplift yields a viscosity estimate of $10^{22}$ P. The close agreement with the value for Fennoscandia is to be expected because both regions are Precambrian shields, which show little evidence of strong deformation since the beginning of the Cambrian.

A recently studied example of isostatic recovery is the uplift following the last drying up of Lake Bonneville in western Utah. This has been studied in careful detail by CRITTENDEN (1963a and b) and it is of exceptional interest because Utah lies in an active tectonic region with an anomalous upper mantle. Lake Bonneville was a Pleistocene lake which covered an area of about 50 000 km² west of Salt Lake City. It was filled periodically during the Pleistocene, the last period extending from 25 000 to 10 000 years ago. The geological history of the lake is well known. In particular, the ancient shoreline can be traced round the margins of the lake and on the hills which formed islands. This allows the depth of the lake and the deformation of the once level shoreline to be accurately worked out. The average depth of the last filling was about 145 m and the centre of the lake has subsequently risen about 64 m relative to the margins (Figure 8.12). This uplift is attributed to isostatic re-adjustment after removal of the load. Crittenden showed that the degree of isostatic equilibrium reached during the last period of loading and unloading was 75% or more.

The Lake Bonneville uplift gives an estimated value of $10^{21}$ P for the viscosity of the underlying asthenosphere. This is a factor of nearly ten smaller than the Fennoscandian and Canadian estimates. The difference is highly significant. It is highlighted by the observation that the relaxation time at Lake Bonneville is 4000 years, which is about the same as for Fennoscandia despite the radius of the load being ten times smaller. The most satisfactory explanation is that the active tectonic region of western U.S.A. is underlain by an upper mantle with much lower viscosity than occurs beneath the Precambrian shields, depending primarily on the different temperatures in the upper mantle.

Another important implication of the Lake Bonneville uplift is that the elastic lithosphere here must be less than 10 km thick, because otherwise it would prevent the movement. Crittenden concluded from this that the crust has also deformed by viscous flow. This does, however, lead to a difficulty. The stress differences caused by the Lake Bonneville load reached maximum values of about 20 bar, but the nearby Wasatch Mountains which are undergoing active tectonic deformation are associated with underlying stress differences from loading of at least 200 bar. Crittenden meets this difficulty by suggesting that active tectonic forces are holding the Wasatch Mountains out of perfect isostatic equilibrium.

Some caution is needed in interpreting the above estimates of viscosity. The observational data and interpretative techniques are as yet insufficient to demonstrate that the flow is Newtonian in preference to other types. A Newtonian model is assumed for the calculations and the results cannot therefore be treated as accurately valid unless the model is correct. It is possible that other types of creep behaviour in the upper mantle can give rise to the same observed phenomena. Despite these ambiguities, what the results do definitely show is that the strength and flow properties of the crust and upper mantle differ markedly between the shields and the active tectonic regions.

### The viscosity of the lower mantle

Before the era of artificial satellites it was believed that the Earth's equatorial bulge was exactly the size to be expected for a fluid rotating Earth in hydrostatic equilibrium. The great improvement in the estimates of the second and fourth zonal harmonics of the gravity potential resulting from measurements of satellite orbits shows that the observed flattening of $1/298\cdot25 \pm 0\cdot01\%$ exceeds the theoretical value by $0\cdot5\%$. This has considerable bearing on knowledge of the state of stress within the deep mantle, and if certain assumptions are made it enables us to estimate the viscosity of the lower mantle.

The theoretical value of the flattening which the Earth would have if in hydrostatic equilibrium

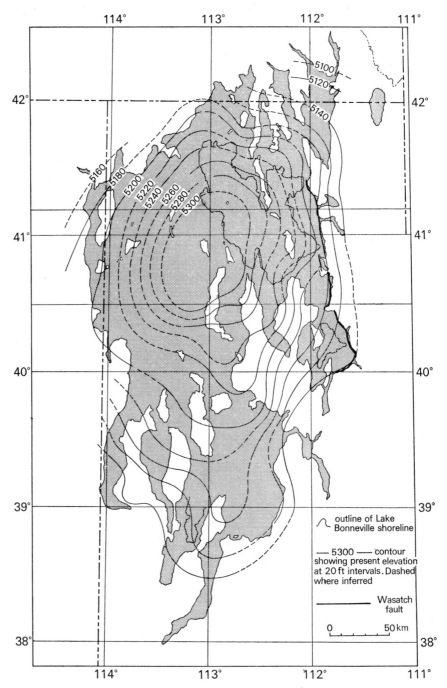

**Fig. 8.12** Deformation of the shoreline of Lake Bonneville since it last dried up. Redrawn from CRITTENDEN (1963a). *Prof. Pap. U.S. geol. Surv.*, **454-E**, p. E9.

can be calculated from the moment of inertia about the polar axis, $C$. From Chapter 1 (p. 4) we have

$$C/Ma^2 = \tfrac{1}{3}(f - \tfrac{1}{2}m)/H$$

where $M$ is the mass of the Earth,

$\quad a$ is the equatorial radius,

$\quad f$ is the observed flattening,

$\quad m$ is the ratio of centripetal acceleration to attraction at the equator,

and $\quad H$ is the precession constant.

The theoretical flattening $\epsilon$ can be calculated from an equation arising from the theory of the internal gravity field of the Earth (JEFFREYS, 1959, p. 143), based on solution of Clairaut's differential equation by Radau's approximation. The equation to the first order of accuracy is

$$C/Ma^2 = \tfrac{2}{3}\left(1 - \tfrac{2}{5}\sqrt{\tfrac{5}{2}(m/\epsilon) - 1}\right).$$

By using small second order corrections, it can be shown from this equation that the Earth would have a flattening of $1/299 \cdot 8$ if it were in hydrostatic equilibrium.

The discrepancy between the observed and the equilibrium values of the flattening shows that the Earth does deviate significantly from hydrostatic equilibrium. The present equatorial bulge is about 200 m larger than it should be. According to McKENZIE (1966), this represents excess gravitational energy of about $2 \times 10^{30}$ erg which is significantly larger than the gravitational energy associated with other harmonics of second or higher degree (although this conclusion has recently been challenged by Goldreich and Toomre (see below)). Consequently, it has been assumed that the deviation of the spheroid from the equilibrium figure is probably causally related to the Earth's rotation. This non-hydrostatic bulge almost certainly has its principle source in the mantle, because the core is fluid and the crust could not deviate from isostatic equilibrium to this extent.

Some possible explanations of the non-hydrostatic bulge are listed below. The last of these has received widest acceptance.

(1) Latitude dependent inhomogeneity in the chemical composition of the mantle could possibly cause high densities below the equator and low densities below the poles. The strength of the lower mantle would need to be about 160 bar to support the load (KAULA, 1963). It is difficult to see why such a density pattern should be related to the axis of rotation.

(2) A latitude dependent density variation could also be caused by systematic temperature differences beneath poles and equator, but the higher temperatures would need to be beneath the poles to explain the bulge. Such a density pattern could be carried by finite strength in the mantle as in (1), or alternatively it could be related to convection currents rising beneath the poles and sinking at the equator.

(3) It has been suggested by WANG (1966) that deviations from isostatic equilibrium on a global scale still remain after the unloading of the polar icecaps, but McKenzie has calculated that this effect is too small by an order of magnitude to explain the bulge.

(4) The most acceptable explanation, which is elaborated below, is that the viscosity of the lower mantle is high enough to delay attainment of the equilibrium figure as the Earth's rate of rotation progressively slows down through tidal friction.

MUNK and MACDONALD (1960) suggested that the non-hydrostatic bulge may represent a lag in attainment of hydrostatic equilibrium as the Earth's rate of rotation slows down. The slowing down occurs as rotational energy in the Earth-Moon system is lost through tidal friction (p. 20). The rate of deceleration is accurately known from astronomical observations and can be extrapolated back to the Palaeozoic using coral banding (p. 21). MACDONALD (1963) showed that the present-day figure of the Earth corresponds to the equilibrium figure about 10 my ago. He suggested that the viscosity of the lower mantle was causing a lag in the attainment of equilibrium, and he calculated a mean mantle viscosity of $10^{26}$ P to fit this explanation.

MCKENZIE (1966, 1967) has re-examined the problem, adopting a more realistic viscosity model for the mantle and taking into account the gravitational energy. He found that both the post-glacial uplift phenomena and the viscous lag in attainment of the equilibrium figure could be approximately explained by an outer shell 1000 km thick with a viscosity of $10^{22}$ P overlying a lower mantle of $10^{26}$ P which surrounds an inviscid core.

The existence of a fossil non-hydrostatic bulge of the equator has been questioned by GOLDREICH and TOOMRE (1969). They showed that the non-hydrostatic equatorial bulge is not in fact excessively large in comparison with other low degree harmonic components of the non-equilibrium shape of the Earth. In particular, they argued that previous workers have overestimated the gravitational energy associated with the non-equilibrium equatorial bulge in relation to other harmonics. They suggested that the pole of rotation follows the maximum principal non-hydrostatic moment of inertia, rather than vice versa. Thus the non-equilibrium bulge loses its unique significance and the above argument for a high viscosity of the lower mantle loses its force. Goldreich and Toomre considered that the pole of rotation of the Earth has wandered relative to the mantle during geological time, following the axis of maximum principal moment of inertia. They used this hypothesis to place an upper limit of about $5 \times 10^{24}$ P on the viscosity of the lower mantle.

McKenzie's interpretation of the viscosity distribution of the mantle as estimated from observations is shown in Figure 8.9. This shows good general agreement with theoretical predictions of Gordon and McKenzie based on diffusion creep. An important consequence of the suggested viscosity of $10^{2 6}$P below 1000 km depth is that convection in the lower mantle would be virtually inhibited. Alternatively, if Goldreich and Toomre are correct in their lower estimate, convection in the lower mantle may be possible.

*Has the mantle a finite creep strength?*

Finite strength implies that creep cannot occur until the stress difference (or some function of the stresses) exceeds a critical value. Earthquakes, involving sudden release of strain energy, occur down to a depth of 720 km, but those deeper than about 70 km are restricted to the narrow circum-Pacific belt and the Alpine-Himalayan belt. The deep earthquake belts must possess a finite strength over at least relatively short periods of time, but this cannot be taken to indicate that the normal rocks of the upper mantle possess a finite yield strength.

Gravity anomalies of wide horizontal extent show that stress differences do occur in the mantle. The non-hydrostatic equatorial bulge is believed to be supported by stress differences of about 160 bar in the lower mantle (p. 245). The higher harmonics of the anomalous gravity field revealed by satellite orbits, which represent the broad deviations of the geoid from the spheroid (p. 3), are caused by density anomalies in the mantle which according to KAULA (1963) give rise to stress differences of the order of 50–100 bar in the lower mantle. These stress differences could be borne by a finite strength in the mantle. But it is probably more realistic to suggest that they exist because of the high viscosity of the lower mantle which prevents them being dissipated by flow more rapidly than they have been produced. As density anomalies causing the higher harmonic gravity anomalies are of smaller areal extent, one would expect the relaxation time to be substantially longer than for the non-equilibrium bulge, possibly of the order of 100 my.

The only regions where there is a clear indication that the mantle may possess finite strength are (1) the deep earthquake belts, and (2) the relatively cool topmost mantle beneath Precambrian shields, where gravity anomalies in the overlying crust appear to have persisted without significant fracture or flow for periods of the order of 1000 my. But elsewhere the case for a finite creep strength for the mantle below about 70 km depth remains non-proven.

## 8.4 The anelasticity of the crust and mantle

### Introduction

Free elastic vibrations within the Earth do not continue indefinitely. The strain energy is progressively dissipated as heat through imperfections of elasticity often collectively referred to as solid friction. This effect is known as anelasticity. It occurs in all solids and is observed at indefinitely small strains. The anelastic properties of the Earth are of interest in connection with the damping of seismic waves, the Earth tides and the Chandler wobble.

Anelasticity causes free vibrations to decay in amplitude. The effect on forced vibrations such as Earth tides is to cause strain to lag behind the applied stress. Anelastic damping is conveniently measured by the quality factor $Q$ (KNOPOFF, 1964b). The most convenient definition for seismological purposes is that $2\pi Q^{-1}$ is equal to the fraction of the total strain energy dissipated per cycle. The logarithmic decrement, $ln(A_1/A_2)$ where $A_1$ and $A_2$ are successive maximum amplitudes, is approximately equal to $\pi Q^{-1}$ provided that $Q$ is fairly large. In dealing with forced vibrations, $Q^{-1}$ is the tangent of the phase angle between the applied stress and the strain. The alternative definitions of $Q$ are not exactly equivalent for seismic waves, but are close enough for our purposes.

Many experimental observations of anelasticity in metals and rocks have been made. The observed values of $Q$ for rocks are mostly within the range 50–1000 (BRADLEY and FORT, 1966). In both metals and rocks $Q$ appears to be nearly independent of frequency up to strains of about $10^{-6}$–$10^{-5}$, and it is not strongly dependent on temperature. $Q$ increases with confining pressure, suggesting that at low pressures pores play an important part in anelastic mechanisms. In fact the dominant anelastic process under experimental conditions is grain boundary sliding, causing frictional losses. This process would be inhibited by pressure below a few kilometres depth, suggesting that the experimental results are not relevant to the anelasticity of the mantle.

GORDON and NELSON (1966) suggested that the anelasticity of the mantle is produced by the following processes, in order of probable decreasing importance: (1) grain boundary damping by viscous processes; (2) stress-induced ordering of defects in crystal structures of various types; and (3) damping caused by vibration of dislocations. These would be expected to give rise to values of $Q$ which are dependent on frequency. Another possible mechanism which they suggested may occur in the upper mantle is damping through flow of partially melted material through intergranular spaces. Other known anelastic processes, such as losses caused by ferromagnetic interactions and thermal anelasticity caused by heating and cooling of adjacent compressed and rarefied regions and the consequent flow of heat, are unlikely to contribute significantly to damping in the mantle.

Attempts have been made to interpret anelastic effects in terms of simple rheological models, such as firmo-viscosity (Kelvin substance) or logarithmic creep, with only limited success.

### Damping of seismic waves

Observations on the decay with time of seismic waves and free oscillations can be used to estimate the anelastic properties of the crust and mantle within the period range 1 second to 1 hour. The method is to determine the amplitude of a seismic wave at different distances from the source, measured in terms of wavelength. A correction needs to be applied for attenuation caused by geometrical spreading, and the remaining decay is attributed to anelastic effects and yields an estimate of $Q$. The decay of free oscillations can be measured by analyzing strain seismometer or gravimeter records at successive periods of time after excitation. Four partly interrelated values of $Q$ can be measured, $Q_\alpha$ for compression waves, $Q_\beta$ for shear waves, $Q_R$ for Rayleigh waves and spheroidal free oscillations and $Q_L$ for Love waves and torsional free oscillations.

The straightforward use of body waves to determine the $Q_\alpha$ and $Q_\beta$ distributions with depth in the mantle is not practicable because of the large variations in amplitude produced by low velocity layers and second order discontinuities. A more successful method applicable to $S$ waves is to study nearly normally incident waves which have been repeatedly reflected between the core-mantle boundary and the Earth's surface (Figure 8.13). For $S$ waves the reflection coefficient at both boundaries can be taken

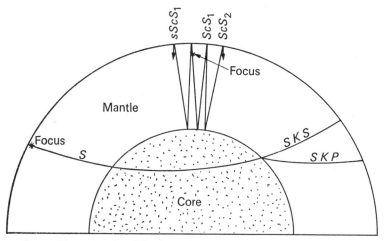

**Fig. 8.13**   Ray paths used for estimating $Q$ in the mantle from body wave amplitudes.

as unity, and the geometrical spreading factor can be accurately calculated. By comparing the amplitudes of successive reflections, such as $ScS_1$ and $ScS_2$ or $ScS_2$ and $ScS_3$, the average attenuation in the mantle can be estimated. By using deep focus earthquakes, and comparing the amplitudes of $ScS$ and the later arrival $sScS$ which is first reflected from the surface, the average attenuation between the depth of the focus and the Earth's surface can be estimated. A more sophisticated approach is to carry out a spectral analysis on the reflected $S$ pulse, thereby obtaining estimates of $Q_\beta$ for the different frequency components in the pulse (KOVACH and ANDERSON, 1964).

Using the above methods, Kovach and Anderson estimated that the average $Q_\beta$ for the whole mantle is 600; the estimate for the mantle above 600 km is 200 and for the lower mantle is 2200. This suggests the existence of a region of high attenuation in the upper mantle. Recent studies show that $Q_\beta$ may decrease with increasing period of the $S$ wave (SATO and ESPINOSA, 1967).

A more satisfactory method of estimating $Q$ in the mantle is to use surface waves and free oscillations (ANDERSON and ARCHAMBEAU, 1964; ANDERSON, 1967c). A single seismograph station can be used to measure the attenuation of trains of surface waves which pass round the Earth more than once. The decay in amplitude can be determined for waves of different period in the wave-train. Accurate corrections can be applied for geometrical spreading and for other effects such as the ellipticity of the

Earth. This enables $Q$ to be determined for each selected wavelength. The observations can be extended to the long period end of the spectrum by observing the decay of free oscillations. The resulting values of $Q_R$ and $Q_L$ depend on period (Figure 8.14).

Figure 8.14 shows qualitatively the result of the low $Q$ zone in the upper mantle. The longer period waves which significantly penetrate the lower mantle are less strongly attenuated, yielding higher values of $Q$. Anderson and Archambeau have shown that the $Q_L$ results can be inverted to give an estimate of the $Q_\beta$ distribution in the mantle provided that $Q$ is independent of frequency (Figure 8.15). Once $Q_\beta$ is known, $Q_\alpha$ can then be obtained with lesser accuracy by inverting the $Q_R$ data.

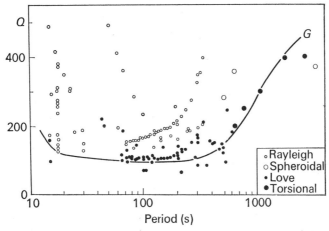

**Fig. 8.14**  Estimates of $Q_R$ and $Q_L$ as a function of period from the attenuation of surface waves and free oscillations. The curve G represents the theoretical $Q_L$ values for the $Q_\beta$ model G, Figure 8.15. Redrawn from ANDERSON and ARCHAMBEAU (1964), *J. geophys. Res.*, **69**, 2079.

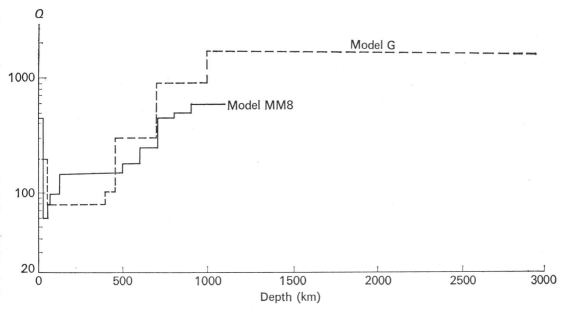

**Fig. 8.15**  Two shear wave $Q_\beta$ models for the mantle based on the attenuation of surface waves and free oscillations. Redrawn from ANDERSON and O'CONNELL (1967), *Geophys. J. R. astr. Soc.*, **14**, 288.

The general picture which emerges is as follows. PRESS (1964) has used the attenuation of the $L_g$ phase (p. 40) to show that $Q_\beta$ in the crust is probably about 450. Love wave attenuation shows that $Q_\beta$ has a minimum value of about 80 between depths of about 50 and 150–350 km. Much higher values of the order of 2000 are believed to be characteristic of the lower mantle. This is in broad agreement with the results of body wave attenuation studies.

Another important result is that the ratio $Q_\alpha/Q_\beta$ established from surface wave and free oscillation attenuation is about 2·25. This high ratio is confirmed by preliminary studies on the ratio of amplitudes of $SKS$ and $SKP$ (Figure 8.13), which suggests that the ratio for the whole mantle is at least 2·5 (KOVACH, 1967). This ratio would be about 1·1–1·5 if the anelastic effects equally influenced the rigidity and bulk moduli. The high observed value of the ratio would be expected if practically all the attenuation is caused by anelastic influence on the rigidity modulus.

The low $Q$ (high attenuation) zone in the upper mantle is usually interpreted in terms of proximity to the melting point. This is supported by the high observed ratio of $Q_\alpha$ to $Q_\beta$. However, other interpretations may be possible. GORDON (1967) has shown that the low $Q$ zone does not necessarily mean a low strength or viscosity, because anelasticity is produced by quite different processes from creep although both are temperature dependent. Thus the anelasticity distribution should be interpreted with caution.

*Damping of the Chandler wobble*

The Chandler wobble affecting the Earth's axis of rotation appears to be a randomly excited damped oscillation. It was shown earlier how the well-established period of oscillation of 1·20 years gives an estimate of the rigidity of the core (p. 158). What concerns us here is the damping of the wobble, which provides an estimate of $Q$ for oscillations of about one year period.

Estimates of the decay time of the Chandler wobble vary from about 2·2 to 22·4 years. MUNK and MACDONALD (1960) concluded that it is between 10 and 30 years. The corresponding values of $Q$ lie between 30 and 80. For the most likely $Q$ value of 30–40, a uniform Maxwell (elastico-viscous) Earth would have a viscosity of about $10^{20}$ P but for a Kelvin (firmo-viscous) Earth it would be $10^{17}$ P. If it is assumed that the mantle possesses Newtonian viscosity, then the Chandler wobble places a lower limit of $10^{20}$ P on the viscosity because otherwise much stronger damping would occur.

The mechanics of the damping is not known. It could be related to processes in the core, mantle or oceans. If, as seems likely, it is mainly caused by viscous processes in the mantle, a problem arises. This is because: (1) the wobble suffers much stronger damping than seismic waves or free oscillations; and (2) such strong damping implying low viscosity is inconsistent with the perseverance of the Earth's non-equilibrium bulge over periods of several million years.

One way out of the above dilemma has been suggested by MCKENZIE (1967). The above models assume a homogeneous mantle, which is not altogether true. He considered that the wobble is damped by the less viscous upper part of the mantle and that the lag in attainment of hydrostatic equilibrium of several million years is caused by the much more viscous lower mantle.

*Can the poles wander?*

This question is of interest in connection with the interpretation of palaeomagnetic and palaeo-climatological data. It has been pointed out (e.g. GOLDREICH and TOOMRE, 1969) that the apparent motion of continents relative to the pole over the last 300–500 years, as deduced from palaeomagnetic studies, is generally much larger than is needed to account for relative movement of the continents. This is taken to suggest that polar wandering has occurred as well as continental drift. IRVING and

ROBERTSON (1969) have also pointed out that a comparison of the history of ocean-floor spreading with palaeomagnetic results leaves some inconsistencies which can be explained if the pole has wandered. Let us recast their argument. Data on ocean-floor spreading interpreted according to plate tectonics gives us the relative motion of the major plates which formed the lithosphere during the Tertiary. It turns out that most of the relative motion is in an east-west direction (p. 225). However, palaeo-magnetic studies show that the north Pacific and west Atlantic Oceans have moved northwards relative to the pole by an average of 16° since the Cretaceous. This may require polar wandering in addition to the relative movement between plates. Thus there is some indication from recent magnetic studies of various kinds that the pole may have wandered significantly during geological time.

There are two basic types of polar wandering: (1) the pole of rotation may have wandered in relation to the Earth as a whole; or (2) the pole of rotation may have remained fixed relative to the lower mantle, but the lithosphere has bodily slipped over the underlying part of the Earth causing apparent polar wandering. The possibility that the pole of rotation can wander relative to the Earth as a whole was discussed and rejected by Sir George Darwin during the last century. The question has been re-opened by GOLD (1955).

A rotating body is in stable equilibrium when the maximum principal moment of inertia coincides with the axis of rotation. This is the condition for minimum kinetic energy of rotation. Because of the equatorial bulge, the Earth rotates about its maximum principal axis of inertia and therefore the axis of rotation would be expected to remain stable both in space and relative to the fixed surface features. However, tectonic and meteorological activity cause small changes in the pattern of surface loading such as icecaps and newly formed mountain belts and these produce small changes in the directions of the principal axes of inertia amounting to a small fraction of a degree. The result would be that the pole would wander to a new position. For example, Gold shows that if the whole of South America were raised by 3 m, the change in moment of inertia would cause the pole to migrate by about $(1/1000)°$.

If the Earth behaves as a rigid body, the polar wandering caused by tectonic and glacial events would be negligible. Gold argued that if the Earth is capable of yielding by flow, the equatorial bulge (which is caused by rotation) will migrate towards the new equator $(1/1000)°$ away, and the deviation between the axis of rotation and the principal moment of inertia will be re-established. Thus the load will cause a progressive migration of the pole until the load itself has settled on the equator. The rate of migration depends entirely on how fast the equilibrium figure of the Earth can be re-established. Gold suggested that the decay time of about 10 years appropriate to the Chandler wobble would also apply to the re-adjustment of the equatorial bulge. Taking the 3 m uplift of South America as an example, the pole would take about 10 years to migrate $(1/1000)°$ or $10^5$ years to migrate 10° of latitude. This argument suggested to Gold that quite large polar wandering would be expected to occur over periods of geological time.

GOLDREICH and TOOMRE (1969) have extended Gold's argument by showing that the redistribution of mass within the Earth associated with the movement of continents (and presumably associated plates) would itself cause the pole of rotation to follow the axis of maximum non-hydrostatic moment of inertia provided that the viscosity of the mantle is less than about $5 \times 10^{24}$ P. They show that the movement of $n$ continents would result in migration of the pole of rotation at a faster rate than that of the individual continents by a factor of $n^{\frac{1}{2}}$.

The crucial uncertainty is the rheological condition of the lower mantle. MUNK and MACDONALD (1960) have shown that a small finite strength of about 10 bar in the mantle would inhibit polar wandering. An even more serious difficulty arises from estimates of high viscosity for the lower mantle. If the delay in attainment of an equilibrium bulge of the equator is as high as 10 my, then it would take

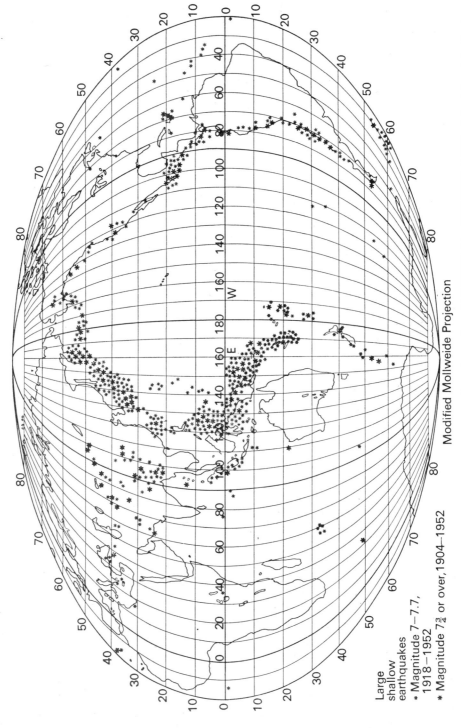

Large
shallow
earthquakes
* Magnitude 7–7.7,
  1918–1952
* Magnitude 7¾ or over, 1904–1952

Modified Mollweide Projection

**Fig. 8.16**  World map of large shallow earthquakes, 1904-1952. Redrawn from GUTENBERG and RICHTER (1954), *Seismicity of the Earth and associated phenomena*, p. 14, Princeton University Press.

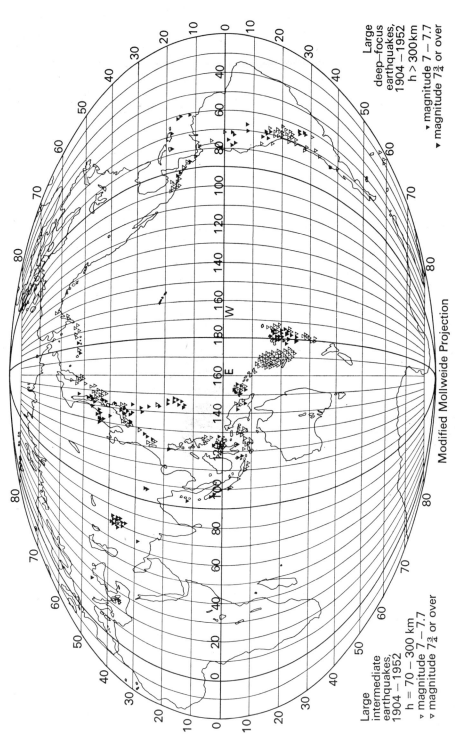

Modified Mollweide Projection

**Fig. 8.17** World map of large intermediate and deep earthquakes, 1904-1952. Redrawn from GUTENBERG and RICHTER (1954), *Seismicity of the Earth and associated phenomena*, p. 15, Princeton University Press.

Large
intermediate
earthquakes,
1904 – 1952

h = 70 – 300 km
▽ magnitude 7 – 7.7
▼ magnitude 7¾ or over

Large
deep-focus
earthquakes,
1904 – 1952
h > 300 km

▼ magnitude 7 – 7.7
▼ magnitude 7¾ or over

$10^{11}$ years for the pole to migrate $10°$ of latitude. On the other hand, if the lower mantle has a much lower viscosity making it possible for the non-equilibrium bulge to be dissipated in a much shorter time as argued by GOLDREICH and TOOMRE (1969), then wandering of the pole relative to the Earth as a whole is possible and has probably occurred. Even if the lower mantle does have a high viscosity, apparent polar wandering by slippage of the lithosphere as a whole remains a possibility.

## 8.5 Earthquakes

### Introduction

Earthquakes are produced by sudden release of strain energy in relatively localized regions of the crust and upper mantle. The study of the mechanism and distribution of earthquakes gives important insight into the process of strain release in the lithosphere, and in certain restricted seismic belts down to a depth of about 700 km in the mantle.

Earthquakes are classified according to depth of focus as follows:

| | |
|---|---|
| 0–70 km | shallow focus |
| 70–300 km | intermediate focus |
| below 300 km | deep focus |

No earthquakes have been detected below 720 km depth. Over 75% of the energy is released in shallow focus events and only about 3% in deep focus events.

The world-wide distribution of large earthquakes is shown in Figures 8.16 and 8.17. 75% of shallow earthquakes, 90% of intermediate and nearly all deep events occur beneath the circum-Pacific belt of island arcs, deep trenches and mountain ranges. The foci tend to cluster near a plane which dips at about $45°$ beneath the continents adjacent to the Pacific and which underlies the belt of active volcanoes. Most of the remaining large earthquakes occur in the Alpine-Himalayan belt.

A significant belt of small shallow focus earthquakes follows the crest of the ocean ridge system and extends along the East African rift belt. These are not large enough to show up on Figure 8.16, but they do appear on the worldwide plot of all epicentres located by the U.S. Coast and Geodetic Survey during the period 1961–1967 (Figure 8.18).

Small earthquakes do occur elsewhere but the total energy release from them is insignificant. Some of these (but not all) are associated with volcanic activity.

### Magnitude and energy release

The old method of measuring the size of an earthquake was by use of an intensity scale based on the surface effects of the shock. By making field investigations of the area affected, an isoseismal map showing regions of equal intensity could be produced. Unfortunately the maximum intensity is not a good measure of the energy released because it is strongly affected by the depth of focus and by local conditions. A much better approach is to use a quantity related to the strain energy release known as the *magnitude* which is determined by the amplitude of seismic waves at a distance from the earthquake.

RICHTER (1935) introduced the concept of magnitude as a measure of the size of shallow focus earthquakes of southern California. He defined magnitude in terms of the maximum trace amplitude observed at an epicentral distance of 100 km on a standard type of short-period torsion seismometer; this quantity is now known as local magnitude ($M_L$). GUTENBERG and RICHTER (1936) extended the concept to determination of magnitude ($M$) of shallow focus events at greater distances by measurement of the maximum ground motion amplitude in microns of surface (Rayleigh) waves at 20 second period. Subsequently there has been considerable discussion about the best relationship to use, although they are all of the form

$$M = \log A + af_1(\Delta, h) + b_1 \text{ (at } 20 \pm 2 \text{ second period)}$$

or

$$M = \log(A/T) + af_1(\Delta, h) + b_2$$

where $A$ = maximum amplitude,
$T$ = period,
$\Delta$ = epicentral distance,
$h$ = focal depth

and $a$ and $b$ are constants, empirically determined.

A formula recently suggested for shallow focus earthquakes is

$$M = \log(A/T) + 1 \cdot 66 \log \Delta + 3 \cdot 3 \text{ (see BÅTH, 1966).}$$

Because of the less effective generation of surface waves in deep focus events, GUTENBERG (1945) introduced another magnitude scale ($m$) based on the maximum amplitude of body waves. The empirical relationship for body wave magnitude is of the foim

$$m = \log(A/T) + af(\Delta, h) + b$$

where $f$ is as before an empirically determined function of epicentral distance and focal depth which allows for geometrical spreading and damping. For shallow focus tectonic earthquakes $m$ and $M$ are related with a very broad scatter by an empirical formula such as that given by BÅTH (1966):

$$m = 0 \cdot 56M + 2 \cdot 9.$$

However, this relationship does not apply to surface explosions for which the ratio of surface wave to body wave amplitudes is smaller than would be predicted, or for mid-ocean ridge earthquakes for which the ratio is larger than predicted.

The elastic wave energy released by an earthquake can be approximately calculated from the magnitude. The relationship between magnitude and energy release $E$ in erg has been subject to discussion and change, and the preferred formula given by Båth is

$$\log E = (12 \cdot 24 \pm 1 \cdot 35) + (1 \cdot 44 \pm 0 \cdot 20)M.$$

RICHTER (1958), using a slightly different formula, estimated the average annual release of energy to be about $10^{25}$ erg, which is notably smaller than earlier estimates. Most of the annual release of energy comes from a few really large shocks. It should be remembered that the total energy released is only partly converted to elastic wave energy. Much of the energy must be converted directly to heat by friction.

A useful method for depicting strain release as a function of time was introduced by Benioff. The square root of energy released by a sequence of earthquakes is estimated from the magnitudes and is plotted against time as a cumulative curve called an 'elastic-strain rebound characteristic'. The method can be used to study global or regional strain release for different types of earthquake, or can be applied to a series of events such as an aftershock sequence. Figure 8.19 shows the strain release curve for all large shallow earthquakes with magnitude of 8·0 or greater, 1904–1955 (BENIOFF, 1955). This suggests that strain release has in the past occurred in cycles, *Aa*, *Bb* etc., but that the cycles have decreased in prominence. Figure 8.20 shows the application of the method to the aftershock sequence of the Kern County earthquake of California, July 21, 1952.

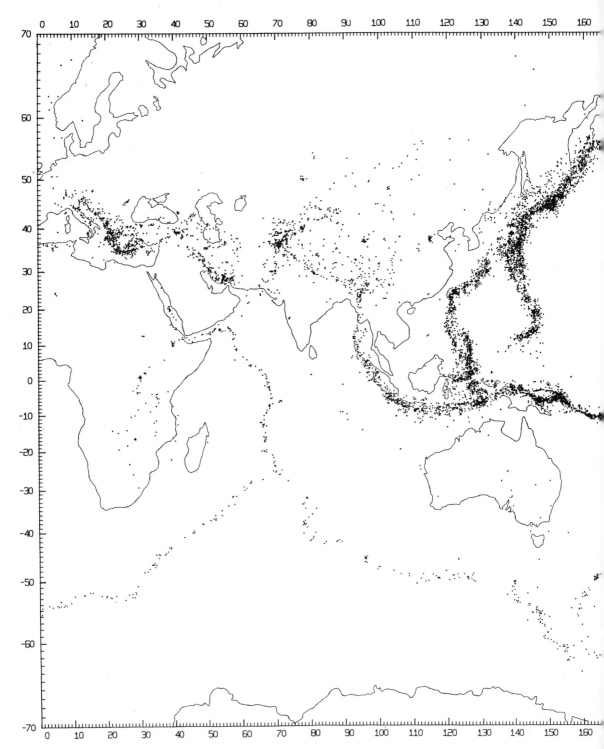

**Fig. 8.18** World-wide distribution of earthquake epicentres (0-700 km depth) for the period 1961-1967, as compiled from the U.S. Coast and Geodetic Survey records. After BARAZANGI and DORMAN (1969), *Bull. seism. Soc. Am.,* **59**, in pocket.

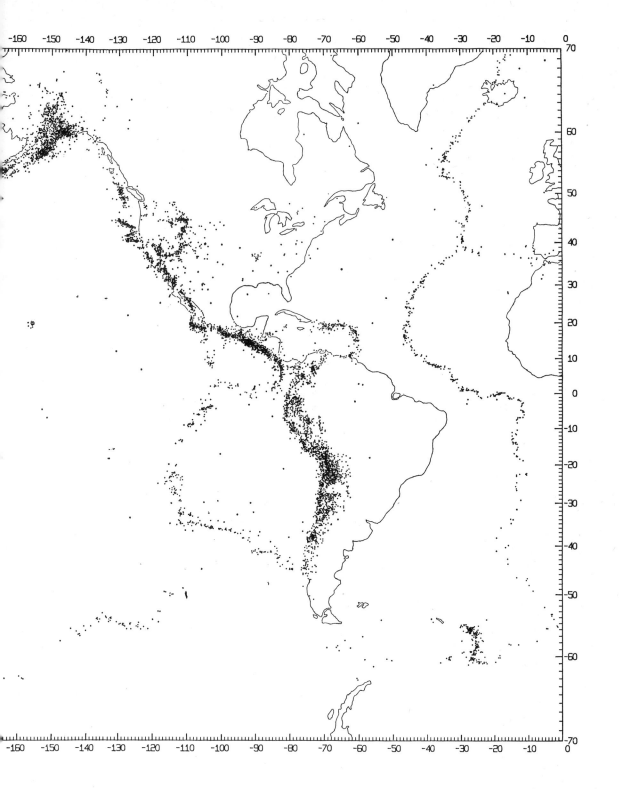

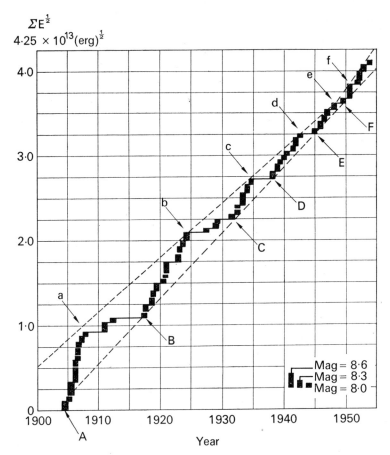

**Fig. 8.19**    Elastic-strain rebound characteristic for shallow earthquakes, $M \geqslant 8\cdot0$, 1904–1954. $E^{\frac{1}{2}}$ is the square root of the energy released, estimated from the empirical formula $\log_{10}E^{\frac{1}{2}}=4\cdot5+0\cdot9M$ (which differs from Båth's more recent formula). Redrawn from BENIOFF (1955), *Crust of the Earth, Geol. Soc. Am. spec. Pap.,* **62**, 71.

*Focal mechanism*

The most widely accepted explanation of shallow earthquakes is Reid's elastic-rebound theory which was formulated after the San Francisco earthquake of 1906 when a right-lateral movement of up to 21 ft occurred on the San Andreas fault. The theory attributes earthquakes to the progressive accumulation of strain energy in tectonic regions and the sudden release of this energy by faulting when the fracture strength is exceeded (Figure 8.21). The common observation that the largest earthquakes are accompanied by visible faulting adds support to the theory.

The modern method of studying strain release at the focus is to observe the pattern of ground motion at distance from the earthquake. It was pioneered by BYERLY (1926) and has been further developed by groups in Canada, U.S.A., Japan, Russia and elsewhere. STAUDER (1962) has given a comprehensive review of the method and an excellent short account is given by BENIOFF (1964). A

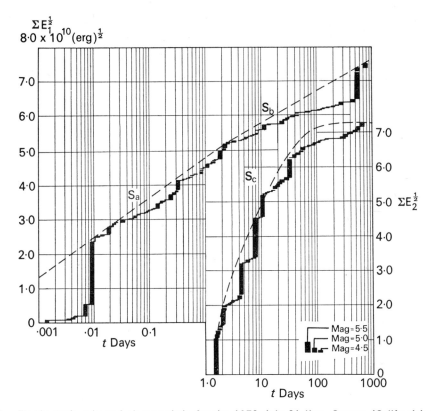

**Fig. 8.20**  Elastic-strain rebound characteristic for the 1952 July 21 Kern County (California) earthquake aftershock sequence. Shocks from the two sides of the fault are plotted separately. Redrawn from BENIOFF (1955), *Crust of the Earth, Geol. Soc. Am. spec. Pap.*, **62**, 64.

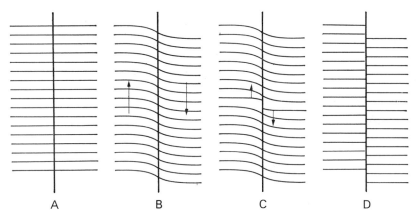

**Fig. 8.21**  Sketch illustrating Reid's elastic-rebound earthquake source mechanism. Redrawn from BENIOFF (1964), *Science, N.Y.*, **143**, 1400.

great improvement in the quality of focal mechanism studies has been made by Sykes and his co-workers (e.g. SYKES, 1967). Sykes has made use of the World-Wide Standardized Seismograph Network which was established by the U.S. Coast and Geodetic Survey in 1962. This considerably improved the quality, consistency and geographical distribution of records. Sykes has also found that records from long-period seismographs give the most consistent results for $P$ and $PKS$ arrivals. The outcome is that Sykes can obtain reliable results on focal mechanism for earthquakes of magnitude 5·5 or greater, thereby considerably increasing the scope and accuracy of the method.

The usual method for studying the focal mechanism of an earthquake is to observe the polarity of the first motion of $P$ and $S$ waves at seismological stations spread over the Earth's surface. For each station, it is noted whether the first $P$ pulse arriving is a compression or a dilatation, and the sense of

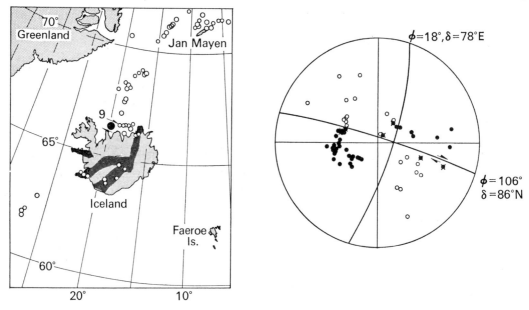

**Fig. 8.22**   Focal mechanism solution for the Icelandic earthquake of March 28, 1963, shown as event 9 on the map. The initial directions of the observed $P$ arrivals are plotted on the lower hemisphere of the projection, compressions as solid circles and dilatations as open circles. The two great circles shown on the projection are the nodal surfaces, either of which could be interpreted as the fault plane. The movement must be perpendicular to their intersection and must therefore be strike-slip; the movement would be in the right-lateral sense if the $\phi = 106°$ plane were the fault plane or left-lateral if the $\phi = 18°$ plane were the fault plane. Local geology suggests that the $\phi = 106°$ plane is the actual fault. Redrawn from SYKES (1967), *J. geophys. Res.*, **72**, 2143.

the first $S$ arrival is determined. The initial direction at source of the ray reaching the station is computed from a knowledge of the travel-path. The results for all stations are then plotted on a projection which shows the initial directions of the rays and their polarity. Figure 8.22 is an example for an Icelandic earthquake of 1963; this shows that the first pulse is compressional in the ENE and WSW quadrants and dilatational in the other quadrants.

The observed pattern for an earthquake is then compared with the theoretical patterns for different types of source mechanism. The two most commonly used models are a Type I source which is a single couple, and a Type II source which consists of two couples at right angles acting in the same

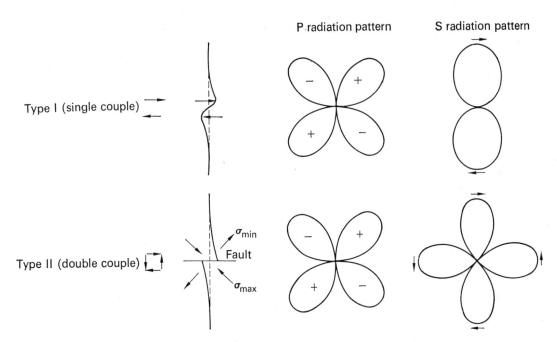

P radiation pattern        S radiation pattern

Type I (single couple)

Type II (double couple)

**Fig. 8.23**   *P* and *S* radiation patterns for Type I and Type II sources. Taken in part from BENIOFF (1964), *Science, N.Y.,* **143**, 1401 and 1402.

plane (Figure 8.23). The initial *P* motion is identical for both types of source. In two diagonally opposite sectors the initial pulse is compressional and the initial ground motion is away from the source; in the other two sectors the initial pulse is dilatational and the ground motion is towards the source. The two types of source can be distinguished by using the *S* radiation pattern. A Type I source gives rise to two shear wave lobes while a Type II source shows four lobes. More complicated source models, such as a moving source representing a fracture spreading along a fault plane, have also been studied.

It was once believed that the mechanism of elastic-rebound could be represented best by a Type I source. However, the observed initial motion of *S* waves showed that the four lobe pattern of a Type II source was characteristic of most earthquakes. At first this cast some doubt on the hypothesis of elastic rebound; but it was subsequently shown that faulting is better represented by a double couple, because fracture results in relief of the maximum shearing stress on two conjugate planes at right angles to each other and not on just one of them. The problem, now, is to explain the few remaining Type I sources which have been observed. Aki has suggested that these may represent dip-slip faults for which the radiation pattern is modified by the fault plane intersecting the Earth's surface (BENIOFF, 1964).

The main value of focal mechanism studies is to determine the attitude of the fault planes and the direction of movement associated with earthquakes. The *P* wave radiation pattern (e.g. Figure 8.22) shows two nodal planes at right angles to each other which separate the compressional and dilatational quadrants. The fault plane is interpreted as being parallel to one of these nodal planes. The *S* radiation pattern for a Type I source shows which of the two planes should be chosen, but for a Type II source the ambiguity remains (Figure 8.23) and it becomes necessary to use knowledge of the local geology to select the more likely fault plane. Once the fault plane is known, the *P* radiation pattern determines the sense of the fault movement uniquely. In Figure 8.22 the two nodal planes are (1) strike 18°, dip 78°E; and (2) strike 106°, dip 86°N. Local geology suggests that (2) is the more likely fault plane; it can now be deduced that the movement is almost horizontal and right-lateral.

*The physical mechanism at the focus*

Earthquakes with foci shallower than about 20 km probably occur as the result of faulting. The main problem of earthquake mechanism is to explain the focal processes causing deeper events. The theory of faulting appears inadequate to explain these because brittle fracture would not be expected to occur at the high confining pressures below 15–30 km depth. The shear fracture theory of faulting requires that the shear stress on the fracture plane exceeds the sum of the cohesive strength and the internal friction. The internal friction is proportional to the confining pressure, and thus the frictional stresses which need to be overcome become excessive at such depths. The only realistic conclusion is that some other mechanism causes earthquakes in the upper mantle, where they occur. Several suggestions have been made in recent years, some of which are discussed below.

OROWAN (1960) has suggested that creep instability may explain intermediate and deep focus earthquakes. Creep instability has been observed to occur during flow of the hot creep type. This is a consequence of the quasi-exponential increase of strain rate with stress characteristic of non-Newtonian viscosity and it would not occur with Newtonian viscous flow caused by diffusion creep. Creep instability causes a local concentration of stress which is suddenly released. Creep instability can occur in homogeneous material but its occurrence may be encouraged by the presence of geometrical irregularities such as pockets of local fusion.

GRIGGS and HANDIN (1960b) have developed a shear-melting hypothesis for the origin of deep earthquakes. This depends on the weakening associated with a phase change, particularly fusion. If the dissipation of strain energy can be concentrated in a sufficiently thin layer, then it may cause melting at the tip as a crack propagates. Griggs and Handin showed that a local 'flaw' is needed to initiate the process and they suggested that local pockets of magma could provide such a flaw.

The presence of local pockets of magma or other fluids such as water may assist the earthquake mechanism in other ways. For instance, HUBBERT and RUBEY (1959) showed that the presence of pore-fluid can vastly reduce the effect of internal friction allowing shear fracture to occur at much higher pressures than the experiments on dry rocks indicate. Another mechanism was suggested by RALEIGH and PATERSON (1965) who have shown experimentally that serpentine at 500°C and 3·5–5 kbar can suffer shear fracture of the brittle type when decomposing to olivine, talc and water. The Raleigh-Paterson mechanism is probably restricted to the depth range of 20–60 km but analogous processes may occur deeper and the crack may propagate into the crust. Among other suggested mechanisms involving magma, ROBSON, BARR and LUNA (1968) have attributed some earthquakes to extension fracture occurring during segregation of the magma into a crack.

The physical processes causing intermediate and deep earthquakes are still in doubt. However, if they are related to the downsinking process associated with ocean-floor spreading, they would be associated with exceptional conditions including very low temperature.

## 8.6 Earthquakes and ocean-floor spreading

The theory of ocean-floor spreading (Chapter 7) is one of the most spectactular contributions of modern geophysics. It interprets the relatively strong lithosphere in terms of about six main plates, each moving relative to the others at velocities of the order of 2–20 cm/y. If the theory is correct, the Earth's main earthquake activity would be expected to be concentrated where the plates come into contact with each other. This is borne out by the world map of epicentres (Figure 8.18) which shows that the circum-Pacific and Alpine-Himalayan compression features correspond to broad epicentre belts including large magnitude events and that the ocean ridge extension features correspond to a narrower belt of smaller magnitude events. Most of the known earthquakes are concentrated in these belts. The lack of activity in the extensive areas between the belts is equally spectacular. But we need to examine the seismological evidence relating to ocean-floor spreading in more detail.

According to the theory, the three main types of boundary between 'plates' of lithosphere are:

(1) extension fracture and normal faulting associated with ocean ridges;
(2) transform faulting at fracture zones;
(3) thrust faulting in the compression belts.

Each type of boundary ought to be associated with a characteristic type of seismicity and focal mechanism in agreement with the appropriate type of faulting. ISACKS, OLIVER and SYKES (1968) have recently reviewed the seismological evidence with this in view. They find that the evidence is in remarkably good agreement with the predictions of the ocean-floor spreading hypothesis, thereby adding new support to the theory and new meaning to studies of earthquake activity.

*Ocean ridges and transform faults*

A belt of shallow focus earthquakes follows the crest of the ocean ridge system along its entire length and extends along the East African rift system. Most of the recorded earthquakes are in the magnitude range 4·0–6·0, so that the total release of energy is small compared with the circum-Pacific belt. The epicentres occur in two main settings as follows (SYKES, 1967): (1) near the crest of the ridge, where both isolated events and earthquake swarms indicative of volcanic activity occur; and (2) along fracture zones. FRANCIS (1968) has shown that the total energy released by the fracture zone events is probably at least 100 times as great as that released by the crestal events, and that the magnitude-frequency distributions differ between the two types of event. This suggests a fundamental difference in focal mechanism which is borne out by other evidence outlined below.

Sykes has demonstrated that the fracture zone epicentres tend to cluster between the offset sections of the crest but are relatively rare beyond these positions (Figure 8.24). This is exactly what would be expected if the fracture zones are transform faults rather than conventional strike-slip faults, for which the activity would be expected to extend beyond the ridges on both sides.

Sykes has also investigated the focal mechanism of about 17 events from the ridges. He found that the crestal events and those from the East African rift system are caused predominantly by normal faulting, with the inferred extension direction perpendicular to the axis of the ridge. On the other hand, the fracture zone events indicate strike-slip movement on steeply dipping planes, with one of the two possible fault planes corresponding to the fracture zone (e.g. Figure 8.22); moreover, Sykes found that the sense of the strike-slip movement indicates transform faulting, but is in the opposite direction to the movement on a strike-slip fault needed to displace the ridge crests (Figure 8.24). Later work along the same lines by Sykes and his co-workers fully agrees with these conclusions.

Thus the seismicity and focal mechanism studies support the idea that the ocean ridges are extension features where new crust is formed and that the fracture zones are transform faults related to ocean-floor spreading rather than wrench faults offsetting the ridge crests.

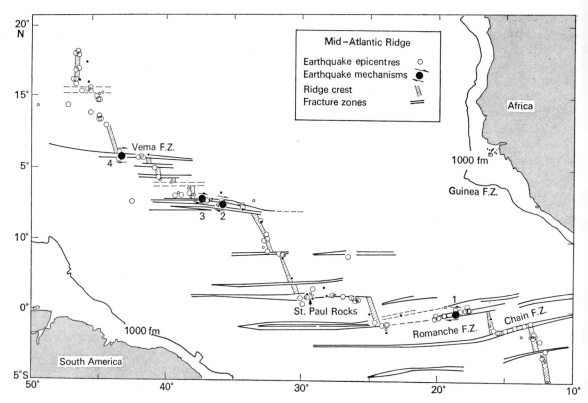

**Fig. 8.24** Epicentres of earthquakes (1955–1965) along the equatorial section of the mid-Atlantic ridge, showing four focal mechanism solutions for fracture zone events. Note that the epicentres cluster (1) along the crest, and (2) on the fracture zones between adjacent portions of crest, but not beyond. The focal mechanism solutions all indicate strike-slip movement in the sense expected for transform faults. After Sykes (1967), *J. geophys. Res.*, **72**, 2137.

## The circum-Pacific belt

According to ideas on ocean-floor spreading, the circum-Pacific belt is a compression feature which marks the junction between the outward spreading Pacific crust and the inward moving continental blocks which are progressively overriding the Pacific and reducing its areal extent. It is the belt where about 85% of the global release of strain energy is concentrated.

It has been known for some time that the earthquake foci tend to cluster on a plane which reaches the surface beneath the continentward flanks of the trenches and dips at about 45° away from the Pacific Ocean (GUTENBERG and RICHTER, 1954; BENIOFF, 1955). Recently SYKES (1966) has used the greatly improved seismograph network to re-study the pattern of earthquake foci associated with parts of the circum-Pacific belt. In particular, he has studied in detail the distribution of earthquakes in the Tonga-Kermadec trench region. His results confirm the existence of the steeply dipping belt of foci (Figures 8.25 and 8.26) and show that the belt is less than 100 km wide and in some places is probably not more than 20 km wide. At the north end of the Tonga trench, where the island arc and volcanic belts bend at right angles towards the west, the earthquake belt bends in harmony, confirming that the same underlying tectonic process causes both earthquakes and island arc.

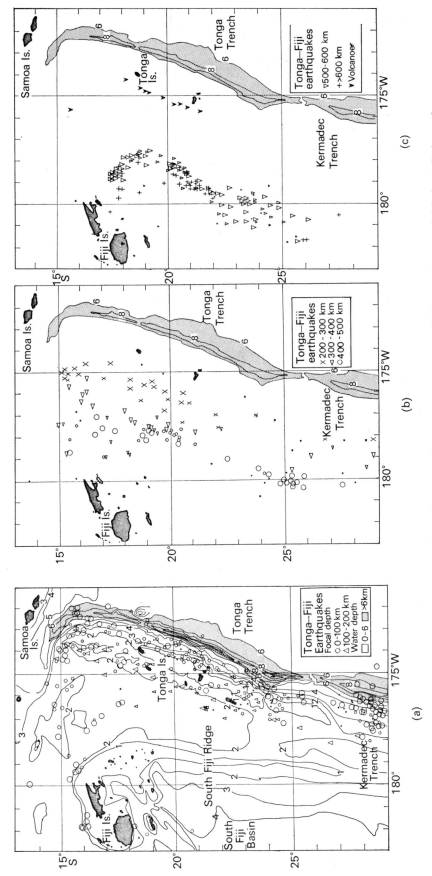

**Fig. 8.25** Tonga-Fiji earthquake epicentres: (a) shallow, showing water depth in km; (b) intermediate; and (c) deep. Redrawn from SYKES (1966), *J. geophys. Res.*, **71**, 2984–2986.

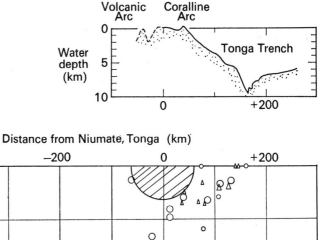

**Fig. 8.26**   Vertical section orientated perpendicular to the Tonga arc, showing earthquake foci during 1965. Circles represent earthquakes projected from within 0–150 km north of the section, and triangles from 0–150 km south. Topography exaggerated by about 13:1 is shown above. The inset shows enlargement of the region of deep focus events. Redrawn from ISACKS, OLIVER and SYKES (1968), *J. geophys. Res.*, **73**, 5871.

Insight into the rheological structure of the Tonga deep earthquake belt has come from the study of body wave amplitudes (especially *S*) from deep events which have reached the surface along different paths (OLIVER and ISACKS, 1967). The waves reaching the Tonga station traverse the length of the earthquake belt and these show much larger amplitude and higher frequency content than waves which reach Fiji and Rarotonga stations through more typical upper mantle material (Figure 8.27). This suggested to Oliver and Isacks that the deep earthquake zone is also a belt of high *Q* of about 1000, in contrast to the more normal *Q* of 150 averaged over the paths to Fiji and Rarotonga. The seismic zone appears to lie near the upper margin of the high $Q$ belt, which is about 100 km thick. They interpreted the high $Q$ belt as the Pacific lithosphere which is disappearing into the mantle as a result of ocean-floor spreading, and which is considerably cooler than the adjacent asthenosphere.

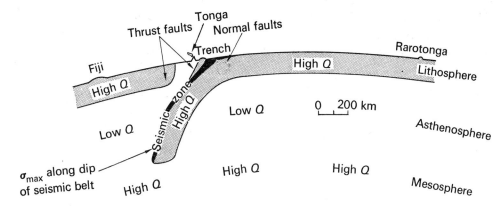

**Fig. 8.27** Hypothetical section through Fiji, Tonga and Rarotonga based on attenuation of seismic waves. It should be noted that the boundaries between high Q and low Q regions are not well determined and should be taken as a first approximation. The prevalent types of earthquake mechanism are shown. Adapted from OLIVER and ISACKS (1967), *J. Geophys. Res.*, **72**, 4272.

Early focal mechanism studies on the circum-Pacific belt suggested that strike-slip faulting is the main mechanism of faulting in this belt. This led to the suggestion that the surrounding continents are rotating clockwise in relation to the Pacific Ocean (BENIOFF, 1959), which was supported in part by the observed movements on strike-slip faults such as the San Andreas fault. ISACKS, SYKES and OLIVER (1969) have now shown that much of the old data is unreliable and that dip-slip faulting predominates in shallow, intermediate and deep focus events.

They found that the shallow focus earthquakes beneath the landward flanks of the trenches and beneath the island arcs are characteristically caused by thrust faults with the axis of horizontal compression orientated perpendicular to the axis of the trench. A few events indicated strike-slip faulting associated with transform faults. There is some indication that normal faulting predominates beneath the axis and oceanward flanks of the trenches. STAUDER (1968) has found this pattern to be particularly well displayed for shallow earthquakes associated with the Aleutian trench and island arc. Twelve events from beneath the oceanward flank and axis of the trench all display maximum horizontal extension perpendicular to the axis of the trench. The more numerous events from beneath the landward flank and the arc are mainly compressional and indicate underthrusting of the ocean-floor beneath the arc. Stauder concluded that the shallow events are in full agreement with the ocean-floor spreading hypothesis, with underthrusting on the landward side and extension caused by bending of the oceanic lithosphere on the oceanward side.

The intermediate and deep focus events appear to be Type II sources involving shear fracture (despite an earlier suggestion of recent years that they may be caused by impulsive forces acting downwards). However, the principal stress directions are not horizontal and vertical as in the shallow events but are aligned parallel and perpendicular to the dip of the earthquake belt. In the uncontorted parts of the deep earthquake belt, ISACKS and MOLNAR (1969) found that either the maximum or the minimum compression is orientated parallel to the dip of the belt, while the intermediate principal direction is horizontal and is parallel to the strike of the belt.

These focal mechanism studies show that the intermediate and deep focus earthquakes are not caused by shearing at the upper boundary of the sinking plate of lithosphere. Rather, they are caused by release of stress *within* the sinking plate of lithosphere, which must be assumed to be stronger than the adjacent asthenosphere because of its lower temperature. As the plate sinks, it warms up and gradually weakens; thus the earthquake belts associated with the more rapidly sinking plates would be expected to extend to the greater depths, which is borne out by observations.

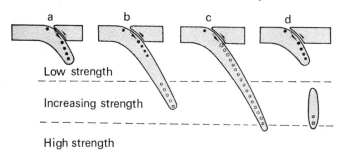

Fig. 8.28   Models to illustrate four possible types of stress distribution associated with sinking slabs of lithosphere. Open circles represent downdip compression and solid circles represent downdip extension.

(a)  Slab of lithosphere sinking into the asthenosphere without encountering resistance, producing extension at all depths;

(b)  Slab starts to meet resistance to sinking near the bottom of the asthenosphere, producing extension near the top and compression near the base;

(c)  Slab meets strong resistance to sinking at the base of the asthenosphere, producing compression at all depths;

(d)  Bottom part of slab breaks off.

Horizontal extension affects the upper part of the slab, in all four models, where the slab bends downwards, and horizontal compression affects it where the downgoing slab underthrusts the adjacent plate of lithosphere. Redrawn from ISACKS and MOLNAR (1969), *Nature, Lond.,* **223,** 1123.

ISACKS and MOLNAR (1969) have suggested that the downward gravitational pull of the cool and consequently dense sinking plate would produce a tension orientated parallel to the dip of the plate, provided that the downward movement was unimpeded. On the other hand, if the plate abutted against the resistant lower boundary of the asthenosphere, then at least the lower part would be thrown into compression parallel to the dip. These two possible situations and variants on them are illustrated diagrammatically in Figure 8.28. Isacks and Molnar have been able to find examples of the predicted situations. They found that down-dip extension predominates, with gaps in the belt in some regions, in the Middle American, New Hebridean and Chilean parts of the circum-Pacific belt. In contrast, the Tonga and north Honshu parts of the belt are continuous in depth and indicate compression in the dip direction.

In conclusion, several aspects of seismology of the circum-Pacific earthquake belt are in excellent agreement with the hypothesis of ocean-floor spreading and add strong support to the validity of the hypothesis.

# 9 The origin of the Earth's surface features

## 9.1 Introduction

Large horizontal movements of the lithosphere affect the Earth's surface and cause primary tectonic activity including continental drift (Chapter 7). These movements occur because the relatively strong lithosphere is subjected to stresses which locally exceed its strength. The associated strain energy is dissipated in mobile belts principally as heat, elastic wave energy and gravitational energy of root formation. The release of elastic energy by earthquakes alone is about $10^{25}$ erg/y (p. 255). Thus strain energy must be continually pumped into the lithosphere at a rate exceeding $10^{25}$ erg/y, by a process which originates in the mantle beneath. It is this underlying process which is the subject of this chapter.

The main known source of available energy within the Earth is the internal heat supply which escapes at a rate of $10^{28}$ erg/y. The simplest explanation of the fundamental tectonic process is that the lithosphere is strained as a by-product of heat escaping from the Earth. The Earth acts as a heat engine, converting a small fraction of the escaping heat into strain energy concentrated into the lithosphere. Most heat engines are highly inefficient, partly because of fundamental limitations on efficiency described by the second law of thermodynamics, and partly because of the practical difficulties of obtaining maximum efficiency. If the terrestrial heat engine were only 1% efficient, with 99% of the input energy escaping as heat flow, then $10^{26}$ erg/y would still be available for straining the lithosphere. This is ten times more energy than is released annually by earthquakes. The idea is feasible, and the remaining problems are (1) to find whether this is indeed the process by which tectonic energy is released, and if so (2) to determine how the terrestrial heat engine may work.

## 9.2 The contraction hypothesis

The contraction hypothesis was suggested as a mechanism of mountain-building by DE BEAUMONT (1852) and by DANA (1847). The mathematical treatment of the contraction of a cooling Earth was developed by DAVISON (1887) and by DARWIN (1887). Until quite recently the hypothesis had widespread support from many eminent geologists and geophysicists (e.g. JEFFREYS, 1959) but as a result of recent discoveries about continental drift and ocean-floor spreading the contraction hypothesis is now generally regarded as superseded.

The basic idea is that the Earth contracts because it is believed to be cooling. This may occur through (1) normal thermal contraction, (2) outgassing of volatiles from the Earth, and (3) outward migration of phase boundaries as the temperature drops. Jeffreys has suggested that the available shortening of the Earth's circumference over geological time may be of the order of 200–600 km. In the region where cooling is most rapid, contraction causes tangential tension which may be relieved by flow if the strength is exceeded. The overlying shell is thrown into a state of tangential compression as gravity causes it to collapse inwards on top of the shrinking sphere beneath. DAVISON (1887) showed that there would be a 'level of no strain' between the regions of compression above and tension below. This used to be identified with the steepening of the circum-Pacific deep earthquake zone at about 300 km depth.

According to the contraction hypothesis, tectonic activity is caused by the release of compressive stress in the outer shell, which contains the lithosphere. The lithosphere is too thick to buckle but it can yield by fracture if the compressive strength is exceeded. The strain can be most effectively relieved by two arcuate fracture systems of global dimension, roughly at right angles to each other. These form the two belts of Tertiary mountain-building.

It has been increasingly recognized that the contraction hypothesis is inadequate as a fundamental theory of tectonic activity. Some of the main arguments against it in order of increasing cogency are as follows:

(1) The idea that the Earth is cooling cannot be proved or disproved and is merely a hangover from Lord Kelvin's interpretation of terrestrial heat flow (p. 185); modern interpretations rather suggest that the Earth may be in thermal balance or even heating up slightly.

(2) It is doubtful whether the theory can account for all the apparent crustal shortening in mountain ranges formed since the start of the Cambrian, let alone through the whole of geological time; thus the contraction hypothesis is scarcely adequate to explain the one type of major surface feature for which is was originally postulated.

(3) Contraction implies that the lithosphere is under continual and general compression, which conflicts with the widespread occurrence of tension features such as normal faults.

(4) The hypothesis has no explanation to offer for other surface features such as ocean ridges, rift valleys, plateau uplifts; it is quite inadequate to have any bearing at all on ocean-floor spreading and continental drift, and if mountain-building is related to the spreading ocean-floor there is no longer any need for a contracting Earth. It has become irrelevant to tectonic problems.

## 9.3 The expanding Earth hypothesis

The expanding Earth hypothesis was originally suggested by HILGENBERG (1933) and HALM (1935). It has recently been adopted by some geologists and geophysicists as an explanation of continental drift, ocean ridge formation and the progressive withdrawal of the seas from the continental regions. The idea is that the sialic shell was once continuous and has been fragmented into the present continents as the Earth's surface area has increased about threefold and the radius increased from an initial value of 3500–4000 km. From a different standpoint, some physicists following DIRAC (1938) believe that the universal gravitational constant $G$ is decreasing with time, which would inevitably cause expansion of the Earth as pressure became less.

EGYED (1957) has been one of the main proponents of the expanding Earth hypothesis. Following Ramsey (p. 159), he suggested that the outer and inner core are formed of successively higher pressure modifications of ultrabasic mantle material. He supposed that the high pressure forms are metastable and are inherited from the formation of the Earth, when it consisted entirely of the highest pressure form. With the passage of time, the metastable phases have gradually reverted in two stages to the stable low density phase which forms the mantle. The mean density of the Earth has consequently decreased from about 17 g/cm$^3$ to its present value of 5·5 g/cm$^3$. More recently, EGYED (1960) has attributed expansion to the decrease of $G$ with time.

Egyed worked out the rate of expansion in four ways. (1) Using the present area of continental crust as an estimate of the original surface area of the Earth, he found that the radius has increased by 0·06 cm/y over the Earth's life. (2) Assuming the initial density to be 17 g/cm$^3$, he estimated the increase in radius as 0·046 cm/y. (3) Using palaeogeographic maps of Strahow and Termier to trace the progressive withdrawal of the seas from the continents since the Cambrian, he estimated the increase in the area occupied by the oceans since then; this yielded estimates of 0·066 cm/y (Strahow) and 0·04 cm/y (Termier) for the rate of increase of the radius. (4) The slowing down of the Earth's rate of rotation, normally attributed to tidal friction, was interpreted as the result of an increasing moment of inertia as the Earth expanded which suggested that the radius is increasing at 0·05 cm/y. Egyed argued that the close agreement between these four independent estimates of the rate of expansion is strong support for the hypothesis. Egyed's concept of expansion is shown in Figure 9.1.

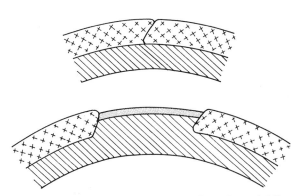

**Fig. 9.1**   The formation of ocean basins and the mechanism of continental drift according to the expanding Earth hypothesis. Redrawn from EGYED (1957), *Geol. Rdsch.,* **46**, 110.

More radical forms of the expansion hypothesis have been suggested by CAREY (1958) and by HILGENBERG (1962), both of whom believe that most of the expansion has taken place over the last 200–300 my. Carey adopted the expanding Earth hypothesis because he found that the continents fitted together much better on a globe of much smaller radius. This argument for expansion is weakened because he fitted the continental coastlines, whereas the edges of the continental shelves ought to have been used. Carey interpreted the post-Palaeozoic separation of the continents as the result of a rapid expansion of the Earth, the Carboniferous radius being 0·7 times the present value. The variation of radius with time for three possible models of an expanding Earth is shown in Figure 9.2.

If the Earth expands either through reversion of high density unstable phases to lower density stable forms as Egyed supposed or by thermal expansion, then thermal and chemical energy are converted to gravitational potential energy as the radius increases. Expansion would also occur as a result of reduction in pressure if $G$ is inversely proportional to the age of the universe as supposed by DIRAC (1938). DICKE (1962) believed that $G$ may be decreasing by about 3 parts in $10^{11}$ per year and he showed that this would cause an increase in the Earth's radius of about 0·002 cm/y. In contrast with the other possible causes of expansion, the gravitational potential energy would be progressively reduced. According to JORDAN (1962), energy of an unspecified form must be fed into the Earth from outside to balance this loss.

How can the expanding Earth hypothesis provide tectonic energy? As the Earth expands (whatever the cause), the soft interior deforms by flow and the strong lithospheric shell is stretched. The strain energy of a spherical shell, such as a football, increases as it is blown up. However, the reverse is true for the Earth because it is initially highly compressed. Stretching of the outer shell of the Earth causes the horizontal compression to be reduced slightly. This *reduces* the strain energy as it is blown up from inside. The energy which is released by normal faulting is gravitational energy; the

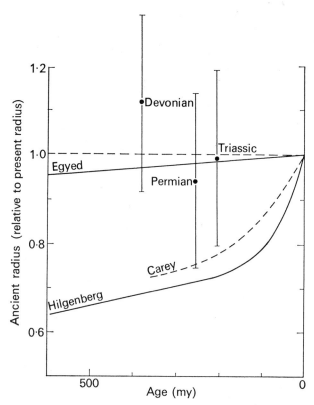

**Fig. 9.2**   Changes in the Earth's radius since the beginning of the Cambrian according to three versions of the expanding Earth hypothesis, compared with estimates obtained palaeomagnetically. The limits shown on the palaeomagnetic determinations are estimates of the standard deviation of the means. Redrawn from IRVING (1964), *Paleomagnetism*, p. 290, John Wiley.

appropriate mechanism depends on the wedge subsidence idea, which has been applied to rift valley formation (p. 62), and modifications of it. Thus the role of the tension in the lithosphere is to initiate tensile faulting, and the source of tectonic energy according to the expansion hypothesis must be gravitational energy rather than strain energy. It is also possible that magmatic activity is produced by lowering of the melting point as pressure is reduced, providing a possible further source of available energy for tectonism.

   Two types of geophysical observation have been used with partial success to test the expansion hypothesis. The older method is to determine the ancient radius by palaeomagnetic measurements. The palaeolatitudes at a given time in the past are determined at two points as far apart as possible on a continental mass which has kept its original dimensions. The results give an estimate of the ancient angle subtended by the two points at the Earth's centre; combining this with the known distance apart gives us an estimate of the radius at that time. WARD (1963) used this method to estimate the radii in the Devonian, Permian and Trias to be 1·12, 0·94 and 0·99 times the present radius respectively. None of these differs significantly from the present radius. These estimates disagree with the rapid rates of expansion supposed by Carey and Hilgenberg but are not precise enough to discount the slower rates of expansion supposed by Egyed and Dicke (Figure 9.2).

The other method depends on estimates of the past moment of inertia of the Earth (p. 22). RUNCORN (1964) used fossil rugose coral growth ring estimates of the length of day and month in the Devonian. He then showed that the Devonian moment of inertia lay between $0.994 \pm 0.003$ and $0.999 \pm 0.003$ times the present value, the two values depending on extreme estimates of the mechanism of tidal dissipation (p. 20). The Devonian moment of inertia would be about 0.94 times the present value for a rate of expansion of 0.05 cm/y as suggested by Egyed. Thus Runcorn's result appears to rule out Egyed's expansion hypothesis as well as the much more drastic forms of Carey and Hilgenberg. However, the slower rate of expansion of 0.002 cm/y supposed by Dicke is not inconsistent with Runcorn's estimated moment of inertia in the Devonian.

Is it feasible that ocean-floor spreading should occur as a result of expansion of the Earth? As new oceanic crust formed, there would be no need for destruction of surface because the total area of the ocean would be correspondingly increased. This would require an increase in radius of the order of 2 cm/y, with the Earth's surface area increasing threefold over the last 200–300 my. There is a more sophisticated objection pointed out by LE PICHON (1968). Most of the spreading now occurs along lines of latitude; if it were caused by expansion, then the Earth would be increasing in equatorial circumference but not in polar circumference. This could only occur if the Earth's equatorial bulge was increasing at a rate of about 2 cm/y, which is ostensibly not occurring.

To conclude, the rapid expansion hypotheses of Carey and Hilgenberg are inconsistent with what evidence we have, but a slower rate of expansion as advocated by Dicke cannot be ruled out. Expansion of the Earth cannot be the driving mechanism behind ocean-floor spreading, continental drift or the associated tectonic activity. Thus a slow expansion may be occurring but its tectonic effects would be swamped by ocean-floor spreading and associated processes. The expansion hypothesis appears to have no obvious relevance to the origin of the Earth's major surface features.

## 9.4 The convection hypothesis

### Introduction

The hypothesis that sub-crustal convection currents cause mountain-building was originally suggested in 1839 by Hopkins and was later used by FISHER (1881). The idea that convection currents in the solid mantle cause both continental drift and mountain belts was introduced by HOLMES (1928, 1931, 1933) and others at about the same time. Holmes envisaged a convection current upwelling beneath a super-continent such as Gondwanaland, causing it to split and dragging the fragments apart as shown in Figure 9.3. The circum-Pacific and the Alpine-Himalayan belts were interpreted as compression features formed near the sinking surrents. Holmes recognized that mantle convection would produce tension near the upwelling currents and compression near the sinking currents and between them. Since then, the ocean ridge system has been interpreted as the tension feature associated with the upwelling currents. This early version of the convection hypothesis is a remarkably accurate forecast of the modern concept. The main difference is that Holmes thought that sialic continents had floated over the simatic oceanic crust whereas we now believe that new oceanic crust is formed during drift.

There has been renewed interest in the convection hypothesis ever since palaeomagnetic results started to support the continental drift hypothesis. This is because some form of convection in the mantle is the only plausible mechanism yet suggested for drift. The demonstration that the ocean-floors are spreading away from the ridges at up to 6 cm/y adds further support to the convection hypothesis, but raises the new problem of how the convection currents are coupled to the moving plates of lithosphere.

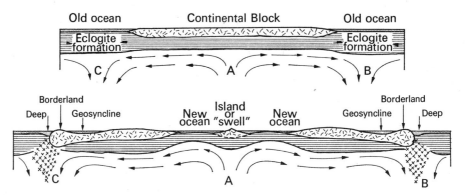

**Fig. 9.3**   The convection hypothesis according to Holmes. The upper diagram shows 'sub-continental circulation in the substratum with eclogite formation from the basaltic rocks of the intermediate layer above B and C where the sub-continental currents meet sub-oceanic currents and turn down'.
   The lower diagram shows 'distention of the continental block on each side of A with formation of new ocean floors from rising basaltic magma. The front parts of the advancing continental blocks are thickened into mountainous borderlands with oceanic deeps in the adjoining ocean floor due to the accumulation of eclogite at B and C.' Redrawn from HOLMES (1933), *J. Wash. Acad. Sci., 23*, 188.

Free convection occurs in fluids when the density distribution deviates from stable equilibrium. The resulting buoyancy forces cause flow to occur until equilibrium has been established. Irregularities in the density distribution can be produced both by thermal and by chemical disequilibrium. If the density anomalies are produced as rapidly as they are dissipated, then some form of regular convection pattern is produced; this is characteristic of thermal convection in a Newtonian viscous fluid. It is possible that episodic thermal convection may occur in fluids possessing more complicated rheological properties. The dominant type of convection believed to occur in the mantle is thermal, although penetrative convection following the production of a fluid phase is undoubtedly the mechanism which causes magma to rise to the surface.

In discussing the convection hypothesis, our object is to find out as much as possible about the pattern of mantle convection and how it can be coupled to the rigid plates which form the lithosphere. Two threads run through the argument. Firstly, there is the search for a pattern of convection consistent with the observed facts of plate tectonics; secondly, there is the theoretical knowledge about convection which places further limitations on the allowable patterns. As thermal convection is primarily a mechanism for rapid upward transfer of heat, it is important to examine the relationship between heat flow and postulated convection patterns.

### Feasibility of thermal convection in the mantle

As a first step, it is convenient to assume that the mantle is a Newtonian viscous fluid. Although this is unlikely to be true, the approach does give some indication of the feasibility of mantle convection and of the influence of sphericity and rotation on the convection pattern.

Free thermal convection may occur in a fluid when it is heated from below. The raised temperature at the bottom causes a reduction in density through thermal expansion; buoyancy forces cause the hot light fluid to rise to the surface where it loses heat and consequently increases in density and sinks to the bottom again. BÉNARD (1900) showed experimentally that convection in a fluid layer starts when the upward heat flow by conduction exceeds a critical limit. A regular pattern of convection cells then became established in his experiments, hexagonal in plan view and rectangular in cross-section. Hot fluid rises at the centre of each cell and cold fluid sinks near the margin. On substantially increasing the amount of heat transferred by convection, the regular pattern disappears and the flow eventually becomes turbulent.

The condition for the onset of convection was derived by RAYLEIGH (1916) in explanation of Bénard's observations. Rayleigh assumed a homogeneous and incompressible Newtonian fluid possessing a uniform kinematic viscosity $v$*. The upper and lower boundaries were assumed to be perfect conductors of heat and to be free in the sense that they exert no shear stress on the fluid. Rayleigh showed that convection starts when the dimensionless number $R = \alpha\beta g d^4/\kappa v$ exceeds $27\pi^4/4$, where $\alpha$ is the coefficient of thermal expansion, $\beta$ is the temperature gradient, $g$ is gravity, $d$ is the thickness of the layer and $\kappa$ is the thermal diffusivity. $R$ is now called the Rayleigh number. The horizontal dimension of a cell is about $2\sqrt{2}d$.

JEFFREYS (1930) and KNOPOFF (1964a) have shown that Rayleigh's result applies to a compressible fluid provided that $\beta$ is regarded as the difference between the actual temperature gradient and the adiabatic gradient. This is called the *superadiabatic gradient*. The onset of convection requires a somewhat higher Rayleigh number if one or both boundaries are rigid or if the convection occurs in a spherical shell. For instance, Knopoff showed that the critical Rayleigh number for a spherical shell with outer radius twice the inner one and with both boundaries rigid is 2380.

The extent to which rotation affects convection depends on the magnitude of the dimensionless Taylor number $T = (2wd^2/v)^2$ where $w$ is the angular velocity of rotation. The rotational influence becomes significant when $T$ is unity or exceeds it. Applying the formula to the mantle by putting $w = 7\cdot27 \times 10^{-5}$ rad/s, $d = 3 \times 10^8$ cm and $v = 10^{21}$ S, we find that $T$ is about $10^{-16}$ which indicates that the Earth's rotation is unlikely to affect the pattern of Newtonian convection in the mantle. However, rotation would be expected to affect strongly the convection pattern in the much less viscous outer core (p. 172).

Convection as described above is primarily a mechanism of heat transfer. The efficiency is measured by the Nusselt number which is the ratio of total heat transferred to heat transferred by thermal

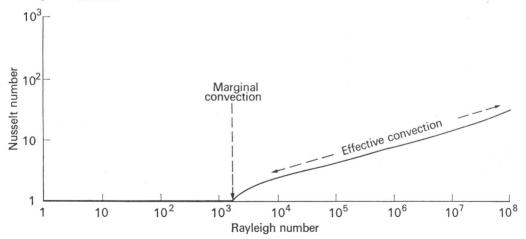

**Fig. 9.4**   Nusselt number as a function of Rayleigh number for free convection in a viscous fluid. Adapted from ELDER (1965), *Terrestrial heat flow*, p. 234, American Geophysical Union.

* Kinematic viscosity is the ratio of dynamic viscosity to density. The c.g.s. unit is 1 stokes (S) = 1 cm²/s.

conduction. The experimentally determined relationship between Nusselt number and Rayleigh number for Newtonian fluid convection is shown in Figure 9.4. At marginal stability the Nusselt number is unity which means that a negligible amount of heat is transferred by convection. Effective upward transfer of heat by mantle convection would require the Nusselt number to be about ten or more; the corresponding value of the Rayleigh number is about $10^6$ to $10^7$ (Figure 9.4).

Table 9.1 shows the superadiabatic gradient in the mantle which would cause (1) marginal convection, and (2) effective transfer of heat, assuming Rayleigh-Bénard convection in a Newtonian fluid. The estimate of viscosity for the upper mantle is based on post-glacial uplift (p. 242) and that for the lower

**Table 9.1** The superadiabatic temperature gradient needed for convection in the mantle, computed from the Rayleigh numbers assuming $\alpha = 2 \times 10^{-5}/°C$, $g = 10^3$ cm/s² and $\kappa = 0.01$ cm²/s.

| | | Kinematic viscosity (S) | Marginal convection ($R = 2000$) | Effective convection ($R = 10^6$) |
|---|---|---|---|---|
| upper mantle ($d = 4 \times 10^7$ cm) | | $10^{21}$ | $4 \cdot 0 \times 10^{-2}°$C/km | $2 \cdot 0 \times 10°$C/km |
| lower mantle ($d = 2 \times 10^8$ cm) | | $10^{27}$ | $6 \cdot 2 \times 10$ | $3 \cdot 1 \times 10^4$ |
| whole mantle* ($d = 2 \cdot 9 \times 10^8$ cm) | (a) | $10^{21}$ | $1 \cdot 4 \times 10^{-5}$ | $7 \cdot 0 \times 10^{-3}$ |
| | (b) | $10^{27}$ | $1 \cdot 4 \times 10$ | $7 \cdot 0 \times 10^3$ |

\* (a) and (b) both assume uniform viscosity for the whole mantle.

mantle on the lag in attainment of the equilibrium bulge (p. 246). Increase in the thermal diffusivity $\kappa$ because of radiative heat transfer would cause a corresponding increase in the temperature gradient required for the onset of convection.

Table 9.1 shows that fairly effective thermal convection might be expected to occur in the upper mantle; the effectiveness might be further increased by non-Newtonian viscosity (p. 229). On the other hand, if the estimate of a high viscosity for the lower mantle is correct, convection in the lower mantle or through the whole mantle would be completely inhibited by the impermissibly high superadiabatic gradient required for the onset. This is the main argument for suggesting that convection is restricted to the upper mantle. It has also been suggested that the phase changes in the transition zone would prevent mantle-wide convection, but this is incorrect: a convection current could pass through the transition zone provided that there is a sufficiently large temperature difference across it to overcome the steepening of the adiabatic gradient which is associated with the zone (VERHOOGEN, 1965; BOTT, 1967c). However, the suggested iron enrichment in the lower mantle relative to the upper mantle (p. 143) would inhibit mantle-wide convection.

It can be tested whether convection currents in the upper mantle occur by laminar or turbulent flow by computing Reynolds number $Re = vd/v$, where $v$ is the velocity of flow. Putting $v = 10$ cm/y $\doteq 3 \times 10^{-7}$ cm/s, $d = 400$ km $= 4 \times 10^7$ cm and $v = 10^{21}$ S in the expression, we find that $Re \doteq 10^{-20}$. This very small value of $Re$ indicates that the flow is laminar and that turbulence is highly unlikely to occur in any form of mantle convection. However, if the Rayleigh number becomes as high as about $10^8$ unsteady flow may be expected to occur (KNOPOFF, 1964a).

To summarize, the results of applying the theory of convection in Newtonian fluids to the mantle are as follows:

(1) Convection might be expected to occur in the upper mantle and to be effective as a mechanism of heat transfer;

(2) Convection in the lower mantle or through the whole depth range of the mantle may be inhibited by the high viscosity supposed for the lower mantle (but see p. 246);

(3) The Earth's rotation appears to be unlikely to influence the pattern of convection (but see p. 225);

(4) Turbulent flow is unlikely to occur.

It must be emphasized that convection in the mantle, or part of it, is likely to be a very complex process and quite unlike the ideal Rayleigh-Bénard type described above. Factors may include: (1) a complicated rheological structure, not necessarily of Newtonian viscous type; (2) distribution of heat sources within and above the convection cell as well as below it; (3) influence of the strong lithosphere above and its mechanism of deformation on the convection; and (4) the fact that convection is probably far from marginal. These and other factors may make the problem quite inaccessible to analytical solution, even if we knew the relevant physical properties.

## 9.5 Convection in the upper mantle

*The broad pattern of convection*

It was shown above that convection currents in the mantle are probably restricted to the uppermost 500 km or thereabouts. The next stage is to discover the broad pattern of upper mantle convection currents as viewed from the Earth's surface looking downwards.

The convection hypothesis was put forward primarily as an explanation of continental drift and of the Earth's major surface features rather than as a mechanism of heat transfer in the mantle. An acceptable pattern of convection must therefore be capable of accounting for the present movements of lithospheric plates associated with ocean-floor spreading and for the location of the major tension and compression features at the Earth's surface. Using this as a guideline, the present-day pattern of convection should be as follows. The convection currents upwell beneath the ocean ridges or near them and possibly also beneath the rift systems which are continental extensions of the ocean ridges. The horizontal flow of the cooling currents is away from the ridges and towards the continents. The position of the sinking currents is less well defined but would be expected to be beneath continental margins or the continents, and particularly beneath the circum-Pacific belt and the Alpine-Himalayan belt.

An exact correspondence between the upwelling currents and the ocean ridge crests would not necessarily be expected. The crest of the ridge system is affected by some sharp changes in direction and by offsets interpreted as transform faults; these irregularities have been inherited from the initial split of the continents and have been perpetuated by the mechanism of plate tectonics. One would expect the underlying convection current system, if cellular, to have a much more regular pattern of horizontal flow, which may include a component of flow along the ridge as well as perpendicular to it.

It was shown in Chapter 7 that the southern mid-Atlantic ridge and the Indian Ocean ridges are migrating away from each other at a rate of about 3 cm/y, and have been doing so since Africa and South America split apart from each other. At least one of these ridges must be migrating laterally. This suggests that the underlying pattern of convection currents can also migrate laterally in sympathy with the movement of the lithospheric plates above.

The horizontal dimension of a convection cell would probably need to be of the order of an ocean width to have relevance to ocean-floor spreading and continental drift. Rayleigh-Bénard cells in the upper mantle with a width of about 1000 km would be expected to cause a more complicated pattern of surface features and it is difficult to see how a unified pattern of stresses sufficient to cause ocean-floor spreading could be produced by them. Mathematical expression of this argument is obtained by analysing the Earth's topography and major surface features in terms of spherical harmonics (PREY, 1922; COODE, 1966). The surface topography, representing the distribution of continents and oceans, peaks at the third, fourth and fifth harmonics. The mid-ocean ridges, mountain ranges and trenches peak at the fifth degree harmonic. The underlying convection cells would be expected to have a similar harmonic content. The fifth harmonic has a wavelength of about 8000 km, which is about the width of an ocean.

The postulated convection cells in the upper mantle would need to be about ten times as long as they are thick, looking rather like longitudinal sections of a sausage. This is quite unlike the standard Rayleigh-Bénard convection cell. A possible explanation suggested by TOZER (1965a) is that the presence of radioactive heat sources within the convection cell causes the ratio of cell length to thickness to increase. Whatever the explanation, it warns us that the simple Rayleigh theory of Newtonian convection is an oversimplification.

One of the main complications is that the heat has to escape through the relatively strong lithospheric plate. During this process strain energy is fed into the lithosphere, which is relieved by tectonic activity. This is the heart of the terrestrial heat engine. Two possible processes by which convected heat escapes through the lithosphere, causing it to be stressed, are outlined in the following sections.

*Cellular convection in the upper mantle*

We first explore the idea that there are tabular convection cells in the upper mantle, with a horizontal dimension equal to several times the vertical extent (Figure 9.5(a)). These would be expected to exert a viscous drag on the overlying lithosphere, causing tension above the uprising currents and compression above the sinking currents and between them. Magma would form by partial fusion of the uprising convection current as pressure was reduced, and this would rise towards the crest of the ocean ridge system above by penetrative convection. Heat would escape by thermal conduction through the overlying lithosphere and by discharge of magma at and near the surface of the ocean ridges. Two main problems need to be investigated: (1) how does this model fit in with the observed pattern of oceanic and continental heat flow? and (2) how is strain energy fed into the lithosphere?

(1) *The heat flow problem*: In dealing with the escape of heat from the crust and topmost mantle, it is quite adequate to neglect the Earth's curvature and to assume a plane two-dimensional model. Furthermore, the horizontal component of heat flow is much smaller than the vertical component and can therefore also be neglected without serious error. The problem is to search for models of convection which are consistent with the observed pattern of heat flow; in doing this, it will be found that certain forms of the convection hypothesis are ruled out.

The upper part of a convection current must cool by thermal conduction of heat to the Earth's surface. The original versions of the ocean-floor spreading hypothesis (HESS, 1962; DIETZ, 1961) implied that the convection current itself reaches the ocean-floor. The lithosphere travels horizontally at the same rate as the underlying convection current. The pattern of heat loss associated with this version of the convection hypothesis is shown in Figure 9.6 for spreading rates of 1 and 6 cm/y. The initial temperature beneath the ocean ridge crest (distance 0 km) is assumed to be 1100°C at all depths, this being the approximate temperature of basalt magma. The observed heat flow pattern differs from the computed pattern in several important respects. In particular, the observed pattern is much more uniform than that shown in the models. Although high heat flow is observed over ocean ridges, the average heat flow out to a spreading distance of 32 my is much less than that of the models, and beyond this distance the observed heat flow is greater than that of the models. Furthermore, the overall pattern of heat flow does not depend on the spreading rate in the way shown in the models.

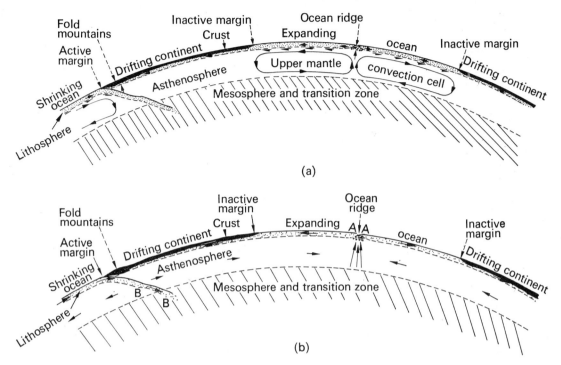

**Fig. 9.5** Two concepts of convection in the upper mantle:
(a) Long cellular convection cells exerting a drag on the overlying lithosphere, which may at the same time be subjected to further straining by the mechanism appropriate to (b).
(b) The Orowan-Elsasser type of convection driven by magma wedging and gravity gliding at A (Figure 8.8) and by the downward gravitational pull of the cool dense lithosphere at B.

A much more uniform pattern of oceanic heat flow would be produced by cooling of the convection current through a stationary overlying layer. This is illustrated by the model of heat loss through a 60 km thick layer shown in Figure 9.7. The corresponding temperature-depth profiles at different distances from the crest of the ridge are shown in Figure 9.8. These models show that an almost uniform pattern of oceanic heat flow can be produced by cooling of a convection current through an overlying lithosphere about 60 km thick provided that the current flows sufficiently fast. A velocity of about 20 cm/y would be appropriate if thermal conductivity is of the order of 0·008 cal/cm s °C, but a slower velocity of about 4 cm/y would be acceptable if radiative heat transfer became dominant in the convection cell, enabling the cooling to penetrate deeper. This model does not account for the high and variable heat flow associated with ocean ridges, but this can be explained without difficulty by upward migration of magma (p. 199).

A more realistic model would be a combination of the models in Figures 9.6 and 9.7, with a fast convection current overlain by a slower spreading lithosphere. This model would, however, produce a higher heat flow than observed over the ocean ridges. It might be regarded as satisfactory if some other mechanism of heat-removal were effective out to a spreading distance of about 32 my age.

What would happen to the pattern of heat flow if a convection current of the type shown in Figure 9.7 were to flow beneath the continental margin? This is illustrated in Figure 9.9. In the model shown, the combined effect of radioactivity in the continental crust and heat loss from the convection current

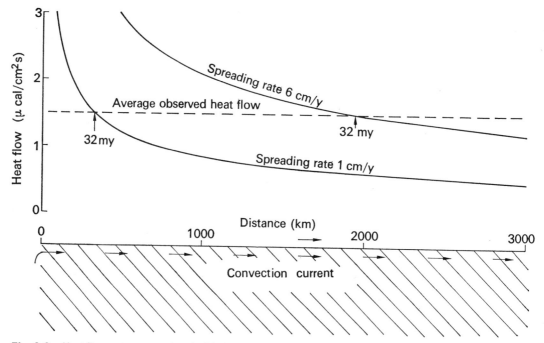

**Fig. 9.6** Heat flow pattern associated with the cooling of a convection current reaching the Earth's surface and emerging at the origin. It is assumed that the convection current flows horizontally and that the velocity of the cooling upper part is constant with depth, at rates of 6 cm/y or 1 cm/y. The initial temperature at all depths beneath the origin is taken as 1100°C, the diffusivity as 0·01 cm²/s and the thermal conductivity as 0·0077 cal/cm s°C. Radioactive heat sources have been assumed absent or negligible.

makes the continental heat flow about 20% higher than the oceanic. The heat flow in the model increases abruptly across the continental margin onto the continent. The theoretical predictions of this model may not agree with observations. The realistic conclusion is that the convection currents would be expected to sink near the continental margin, only locally passing beneath the continents.

To summarize this short discussion of heat flow in relation to cellular convection in the upper mantle, the pattern of oceanic heat flow could be explained provided that some of the heat beneath ocean ridges is removed by a process other than thermal conduction which is not yet understood.

(2) *How the lithosphere is strained*: Convection currents flowing in the asthenosphere would be expected to exert a viscous drag on the overlying lithosphere. Maximum tension would be produced over the upwelling current at ocean ridges and maximum compression above the downsinking currents and between them. In order to estimate the magnitude of the stresses produced in the lithosphere, consider a two-dimensional system of convection cells of horizontal dimension $L$ overlain by a lithosphere of thickness $T$ (Figure 9.10). The shearing stress exerted on the base of the lithosphere is $\eta\dot{\epsilon}$ where $\eta$ is the coefficient of viscosity and $\dot{\epsilon}$ is the strain rate. This would cause a horizontal tension of $\eta\dot{\epsilon}L/2T$ over the upwelling current and an equal compression over the downsinking current, upon which a constant stress might be superimposed. Let us suppose that the convection current flows at 24 cm/y at 100 km depth and at 4 cm/y at 60 km depth; then the strain rate in the boundary zone is about $1\cdot6\times10^{-13}$/s. If $\eta = 10^{21}$ P as indicated by post-glacial uplift (p. 242), the shearing stress on the base of the lithosphere becomes 160 bar. Putting $L = 3000$ km and $T = 60$ km, the above expression shows that the maximum values of tension and compression produced by the drag

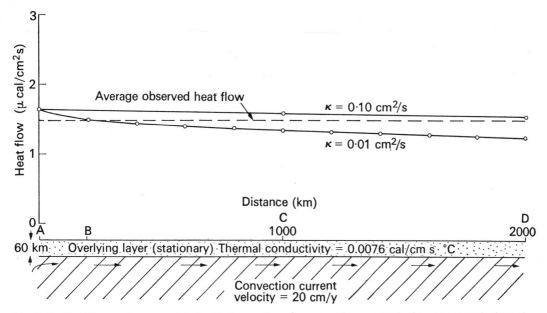

**Fig. 9.7** Heat flow pattern associated with the cooling of a convection current in the upper mantle through a stationary overlying layer 60 km thick, calculated by computer for diffusivity values in the convection cell of (1) 0·01 cm²/s (no significant radiative heat transfer), and (2) 0·1 cm²/s (radiative heat transfer dominant). Radioactive heat sources in the cooling part of the convection cell and in the overlying layer have been neglected. The assumed initial temperature-depth distribution at A and computed distributions below points B, C and D are shown in Figure 9.8. Redrawn partly after BOTT (1967c), *Geophys. J. R. astr. Soc.,* **14**, 420.

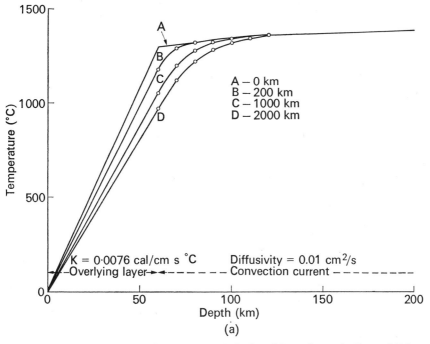

**Fig. 9.8** (a)   The temperature-depth profiles at points A, B, C and D as shown in Figure 9.7 for an assumed diffusivity of 0·01 cm²/s within the convection cell, appropriate to no significant radiative heat transfer. Curve A shows the temperature distribution below A which has been used as the 'initial conditions' for the computations shown in Figures 9.7 and 9.9. Partly after BOTT (1967c), *Geophys. J. R. astr. Soc.,* **14**, 421.

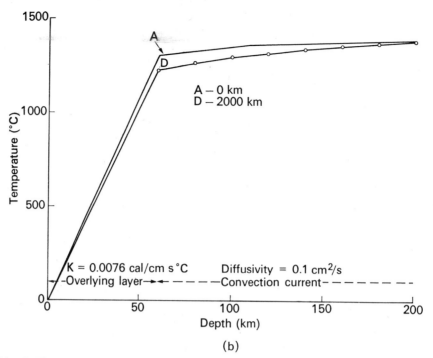

Fig. 9.8 (b)    As Figure 9.8 (a) for points A and D except that diffusivity within the convection cell is assumed to be 0·1 cm²/s appropriate to dominant radiative heat transfer. Partly after BOTT (1967c), *Geophys. J. R. astr. Soc.*, **14**, 421.

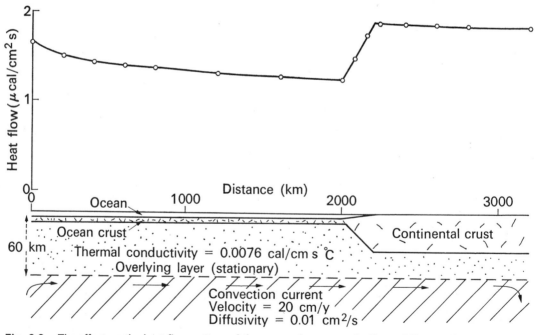

Fig. 9.9    The effect on the heat flow pattern of the convection current in Figure 9.7 extending beneath the continental crust. Uniform distribution of radioactive heat sources in the continental crust is assumed to contribute 0·88 $\mu$cal/cm² s to the continental heat flow, but radioactive heat sources have been assumed absent elsewhere in the cooling part of the convection cell and the overlying stationary layer. Redrawn from BOTT (1967c), *Geophys. J. R. astr. Soc.*, **14**, 423.

are 4 kbar. This would be more than adequate to explain tectonic activity. Although the model is oversimplified, it does show that substantial horizontal pressures, involving both tension and compression, can be produced in the lithosphere by quite a weak drag exerted over a large horizontal distance. However, if the horizontal dimension of the convection cells was only about 400 km such large stresses could not be produced in this way.

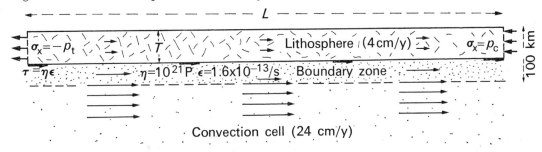

**Fig. 9.10**  The viscous drag of a horizontally flowing convection current on the base of the lithosphere. As the horizontal forces acting on the block of lithosphere must be in equilibrium, $\tau L = (p_c + p_t)T$. A simplified boundary zone 40 km thick with a uniform velocity gradient has been assumed, the velocity above it being the spreading rate of 4 cm/y and the velocity below being the convection cell maximum velocity of 24 cm/y.

## The Orowan-Elsasser type of upper mantle convection

A radically different concept of upper mantle convection has developed from the original ideas of OROWAN (1965) and ELSASSER (1969). In place of the elongated convection cells in the asthenosphere, in the new concept it is the lithosphere of the oceans which forms the fast-moving cooling part of the system as it spreads laterally from ocean ridges. The reverse current is a relatively slow migration of material back towards the ridges in the upper mantle (Figure 9.5(b)). Thus the asthenosphere is nearly stationary and the overlying plates move.

As in cellular convection, the driving force behind the movement is basically the release of thermal energy assisted by chemical differentiation following from partial fusion. However, the plates of lithosphere move in response to forces applied at or near their ends rather than by drag exerted on the base. The stresses can originate in three ways, all of which may be important (p. 237 and Figure 8.8):

(1) Beneath the ocean ridge crests, forcible emplacement of magma under pressure would exert boundary forces on the adjacent plates of lithosphere, tending to force them apart. Let us investigate the magnitude of this effect by supposing that there is a vertical column of magma 40 km high in the underlying upper mantle. The density of the magma would be about 2·65 g/cm³ and that of the solid mantle about 3·25, producing a density contrast of $-0·6$ g/cm³. If the magma pressure is equal to the confining pressure in the mantle at the base of the column, then it will have an excess pressure of $\Delta\rho gh = 2·4$ kbar at the top of the column. This excess pressure would apply a horizontal compression of up to 2·4 kbar to the lithospheric plates as the magma was injected upwards as a dyke. The underlying magma would not necessarily need to be a discrete column, but could be in the form of a pore-fluid provided that there was continuity between top and bottom.

(2) Regional stresses in the lithosphere may be caused by its tendency to slide down the flanks of uplifted regions such as ocean ridges. This mechanism, which is akin to gravity gliding tectonics, would only work if the lithosphere is decoupled from the underlying substratum by a plane or zone of weakness such as the asthenosphere would be expected to form. If the litho-

sphere were free to move it would glide down the slopes; otherwise, tension would be developed over the crestal region of the uplift and compression on the lower part of the flanks and beyond them. If the uplift amounted to 3 km, the difference in horizontal pressure between the crest and the adjacent ocean basin would be up to 0·7 kbar, depending on the amount of stress carried by the substratum.

(3) The cold, sinking plates of lithosphere beneath the circum-Pacific belt would be expected to have a relatively high density because of their low temperature. This would tend to make the plate sink of its own accord, dragging the surface part of the plate down after it. Once this process had started, it would create a convective environment favourable to continuing the process. The tensile stress produced in this way would be a maximum where the plate bends downwards at the oceanward side of the trench. This process may assist the rapid spreading of the Pacific Ocean, but it can hardly be relevant to the movement of lithospheric plates which are overriding the Pacific Ocean.

The above processes causing the lithosphere to be stressed depend on the ability of the lithosphere to transmit stress over long horizontal distances without buckling (p. 238).

Although the Orowan-Elsasser scheme of convection overcomes any difficulty in envisaging a Newtonian viscous upper mantle, it does come up against the following difficulties: (1) It is not obvious how the hypothesis can explain the relatively uniform heat flow of ocean basins. The expected pattern of heat flow would be similar to that of Figure 9.6 for the appropriate spreading rate, except that the fall-off of heat flow as a function of distance from the ridge would be even greater than in the Figure because of loss of heat to the relatively cool mantle beneath the moving lithosphere. (2) It is not obvious how the fundamental difference in the properties of the sub-oceanic and sub-continental upper mantle can be produced by this mechanism.

To conclude, the problem of how the upper mantle convection works is not yet clear. The writer thinks that the relatively uniform heat flow of ocean basins and the fundamental difference in upper mantle properties beneath oceans and continents are best explained by long cellular convection cells which are dominantly restricted to the sub-oceanic mantle. However, the Orowan-Elsasser mechanism for stressing the lithosphere may play an important part in the ocean-floor spreading process, supplementing the stresses produced by drag on the base of the lithosphere. Furthermore, it is clear that some control over the position of the convection cells is applied by the geometry of plate tectonics, so that the convection cells migrate as oceans widen.

### The changing pattern of convection

The present episode of continental drift and ocean-floor spreading started about 300 my ago with the opening up of the Indian and Atlantic Oceans. It cannot continue at the present rate for more than about 200 my because the Pacific would have closed up by then. The life-span of the present direction of ocean-floor spreading appears to be about 500 my. We have seen that there must be a constant migration of the convection pattern as new oceans are increased in size. However, a much more radical change in the plan and direction of convection must have occurred near the beginning of the Mesozoic.

What happened before the present episode of ocean-floor spreading started? Presumably most of the oceanic crust produced by earlier periods of spreading has now been destroyed. However, the tectonic history of the continents suggests that major Earth movements go back to the earliest recorded Precambrian rocks 3000 my ago, and palaeomagnetic results show that the continents were moving relative to each other in the Precambrian. The simplest interpretation is that the present episode of ocean-floor spreading is the last in a series of discrete episodes which have occurred over the Earth's geological history, possibly since the original formation of the crust. As the process of spreading causes the continental crust to be shortened across young fold mountain belts, it is apparent that the process has caused a progressive increase in the area of the oceans, with a corresponding reduction of the continental area. It is thus possible that the continents with much thinner crust occupied most or all of the Earth's surface when they were originally formed.

Support for four major tectonic 'events' during the Earth's geological history comes from a survey of radioactive age dates initially made by GASTIL (1960). Gastil plotted the ages on a histogram and found that four peaks occur at ages of 350, 1100, 1800 and 2700 my. A histogram based on more age determinations than were available to Gastil is shown in Figure 9.11. The peaks represent periods of maximum tectonic activity. RUNCORN (1962) suggested that these peaks may represent periods when the pattern of convection currents is changed.

Runcorn suggested that the convection pattern might change as the core grew in size (Figure 9.12). Following Urey, he supposed that the core formed slowly over geological time by steady separation of the iron phase from the mantle. This would cause the radius of the core-mantle boundary to increase from zero to its present value. According to theoretical calculations of Newtonian convection in a spherical shell made by Chandhrasekhar, the convection pattern would progressively change from an $n = 1$ pattern at the start to an $n = 5$ pattern at present, as the mantle became progressively thinner. This suggested that there would be four major readjustments of the convection pattern during the life of the Earth. Runcorn identified these periods of changing pattern with the four radioactive dating peaks of Gastil (Figure 9.13).

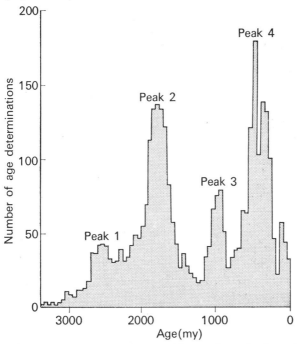

**Fig. 9.11** Frequency histogram of igneous and metamorphic age determinations plotted against geological time. Redrawn from DEARNLEY (1966), *Physics Chem. Earth*, **7**, 6, Pergamon Press.

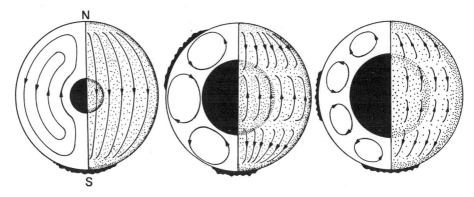

**Fig. 9.12**  Mantle-wide convection with different sizes of the core. *Left,* n=1 ; *middle,* n=3 ;*right,* n=4. After RUNCORN (1962), *Continental drift,* p. 35, Academic Press.

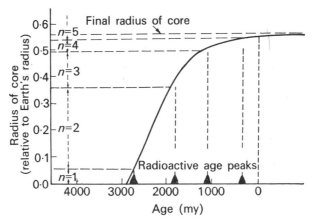

**Fig. 9.13**  Growth of the Earth's core compared with Gastil's radioactive age determination peaks. Redrawn from RUNCORN (1962), *Continental drift,* p. 36, Academic Press (reprinted from *Nature, Lond.,* **195**, 1249).

Runcorn's argument can be criticized because (1) the core probably formed relatively rapidly (p. 190), and (2) mantle-wide convection on which his hypothesis depends now seems unlikely to occur. The radioactive dating peaks may still, however, mark changes in the pattern of ocean-floor spreading. Presumably each episode of spreading terminates when the 'Pacific type' ocean at that time has shrunk as far as is possible. There may then be a period of quiescence when temperatures in the mantle would build up, and eventually a new episode of spreading would be initiated by the upwelling of a new convection current beneath one of the continental masses, causing a split which would form the embryo of a new expanding ocean.

A suggestion as to how this might occur is shown in Figure 9.14. An earlier episode of spreading is shown in (a). This terminates when the shrinking ocean has vanished or has reached the state when further reduction in area is impossible. When this stage has been reached, the other two oceans in Figure 9.14(b) can expand no longer. A new episode of spreading will start once one of the two remaining oceans develops active continental margins where oceanic crust can sink into the mantle; this ocean then becomes a newly formed shrinking ocean. Expansion of the other ocean can then

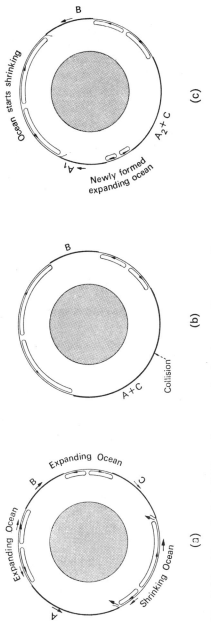

**Fig. 9.14**   Stages in the initiation of a new episode of ocean-floor spreading:
(a) The earlier episode of ocean-floor spreading, showing two expanding oceans and one shrinking one;
(b) The earlier episode terminates as the shrinking ocean vanishes and the continents on either side come into collision;
(c) A new convection cell develops beneath A, causing a split with the formation of a new expanding ocean. One of the expanding oceans now becomes a shrinking ocean as a new pattern of movement is started.
The process is shown for cellular sub-oceanic convection cells in the upper mantle, but it would  equally well apply to other types of convection such as the Orowan-Elsasser mechanism.

continue and a new ocean may be initiated by the start of a convection cell beneath one of the continental masses, as shown in Figure 9.14(c).

In Figure 9.14 it is assumed that the convection in the upper mantle is cellular and that the cells are mainly restricted to the sub-oceanic mantle as suggested by the global heat flow pattern (p. 280). This implies that the convection cells change in horizontal extent as the oceans change in size, and that some of them migrate in position. It also implies that the temperatures at depths of about 300–600 km beneath the continents will progressively build-up because of lack of convective heat transfer. It is this build-up of temperature which may ultimately start the new convection cell forming beneath a continental mass, causing splitting and continental drift of the fragments (BOTT, 1964). Figure 9.14 illustrates the change in pattern associated with one particular form of the cellular convection hypothesis. A similar type of changeover would occur for other types of cellular convection hypothesis or for the Orowan-Elsasser type of convection.

## 9.6 General conclusions

Although knowledge of the structure of the Earth's interior has been growing at an increasing rate throughout the twentieth century, it has only been during the last few years that a satisfying theory of the origin of the Earth's surface features has started to emerge. It has been a long-standing puzzle to understand the fundamental process of energy release within the Earth which has made the surface of our planet so strikingly different from those of the other inner planets and of the Moon. After many false starts, it has been the discovery of ocean-floor spreading which has led us to a new and unified understanding of this process.

The mechanism causing ocean-floor spreading has been attributed to the escape of heat from the Earth's interior. The Earth is acting as an inefficient heat-engine and a small fraction of the escaping heat amounting to $1/1000$–$1/100$ is converted into strain energy in the lithosphere which is released by tectonic activity concentrated along mobile belts. The only known process by which the lithosphere can be continually strained in this way is by some sort of convection in the mantle. Modern knowledge of the rheology of the mantle has suggested that the convection is probably restricted to the upper mantle. There still remains much difference of opinion about the exact form the convection takes and how it causes the lithosphere to be strained. Some possibilities have been discussed in this chapter.

What is the next stage? Undoubtedly our knowledge of the structure of the Earth's interior will increase and we shall have further insight into the mechanism of convection as strain measurements are increasingly made on the Earth's surface. We can also expect to see a new lease of life given to our understanding of geological processes. The geology of the continental surface has been worked out in exceptional detail over the last 150 years, but a fundamental understanding of many of the well-known observations still eludes us. The new framework of the ocean-floor spreading theory should give us fresh insight into many of these problems of igneous, structural and metamorphic geology. Some geological features will be related directly to ocean-floor spreading and plate tectonics. Other features such as sedimentary basins may become understood as secondary structures, in that their formation depends on the prior existence of a primary feature such as a mountain range or a continental margin.

# References/Author Index

In the References/Author Index, italic numbers in square brackets refer to pages on which figures appear, and bold numbers to pages on which tables appear.

ADAMS, J. L. [20]

AGGER, H. E. and CARPENTER, E. W. (1964). A crustal study in the vicinity of the Eskdalemuir seismological array station. *Geophys. J. R. astr. Soc.*, **9**, 69–83. [*44*, 45, 114]

AIRY, G. B. (1855). On the computation of the effect of the attraction of mountain-masses, as disturbing the apparent astronomical latitude of stations in geodetic surveys. *Phil. Trans. R. Soc.*, **145**, 101–104. [47]

AKI, K. [261]

AKI, K. and PRESS, F. (1961). Upper mantle structure under oceans and continents from Rayleigh waves. *Geophys. J. R. astr. Soc.*, **5**, 292–305. [115]

AKIMOTO, S. and FUJISAWA, H. (1965). Demonstration of the electrical conductivity jump produced by the olivine-spinel transition. *J. geophys. Res.*, **70**, 443–449. [125, *125*]

AKIMOTO, S. and FUJISAWA, H. (1968). Olivine-spinel solid solution equilibria in the system $Mg_2SiO_4$–$Fe_2SiO_4$. *J. geophys. Res.*, **73**, 1467–1479. [141, 149]

ALFVÉN, H. (1967). On the origin of the solar system. *Q. Jl R. astr. Soc.*, **8**, 215–226 [17]

ALLEY, C. O. and others (1970). Laser ranging retro-reflector: continuing measurements and expected results. *Science, N.Y.*, **167**, 458–460. [26]

ANDERSON, O. L. (1965). Lattice dynamics in geophysics. *Trans. N.Y. Acad. Sci.*, ser. II, **27**, 298–308. [185]

ANDERSON, D. L. (1967a). Latest information from seismic observations. In *The Earth's mantle*, pp. 355–420, edited by Gaskell, T. F., Academic Press, London and New York. [108]

ANDERSON, D. L. (1967b). Phase changes in the upper mantle. *Science, N.Y.*, **157**, 1165–1173. [141, *142*, 142, 150]

ANDERSON, D. L. (1967c). The anelasticity of the mantle. *Geophys. J. R. astr. Soc.*, **14**, 135–164. [248]

ANDERSON, D. L. (1968). Chemical inhomogeneity of the mantle. *Earth & planet. Sci. Lett. (Neth.)*, **5**, 89–94. [142]

ANDERSON, D. L. and ARCHAMBEAU, C. B. (1964). The anelasticity of the Earth. *J. geophys. Res.*, **69**, 2071–2084. [248, *249*]

ANDERSON, D. L. and O'CONNELL, R. (1967). Viscosity of the Earth. *Geophys. J. R. astr. Soc.*, **14**, 287–295. [*249*]

ANDERSON, D. L. and TOKSÖZ, M. N. (1963). Surface waves on a spherical Earth 1. Upper mantle structure from Love waves. *J. geophys. Res.*, **68**, 3483–3500. [111]

ANDERSON, E. M. (1951). *The dynamics of faulting and dyke formation with applications to Britain*, Second edition, Oliver and Boyd, Edinburgh and London, 206 pp. [231, *232*, 233, *233*]

BACKUS, G. (1958). A class of self-sustaining dissipative spherical dynamos. *Ann. Phys.*, **4**, 372–447. [171]

BAKER, H. B. (1911). The origin of the Moon. *Detroit Free Press*, 23 April, 1911. [201]

BANKS, R. J. (1969). Geomagnetic variations and the electrical conductivity of the upper mantle. *Geophys. J. R. astr. Soc.*, **17**, 457–487. [120]

BANKS, R. J. and BULLARD, E. C. (1966). The annual and 27 day magnetic variations. *Earth & planet. Sci. Lett. (Neth.)*, **1**, 118–120. [120]

BARAZANGI, M. and DORMAN, J. (1969). World seismicity maps compiled from ESSA, Coast and Geodetic Survey, epicenter data, 1961–1967. *Bull. seism. Soc. Am.*, **59**, 369–380. [*256–257*]

BÅTH, M. (1966). Earthquake energy and magnitude. *Physics Chem. Earth*, **7**, 115–165. [255]

BECK, A. E. (1965). Techniques of measuring heat flow on land. In *Terrestrial heat flow*, pp. 24–57, edited by Lee, W. H. K. *Geophys. Monogr.* No. 8, American Geophysical Union, Washington, D.C. [176]

BÉNARD, H. (1900). Les tourbillons cellulaires dans une nappe liquide. *Revue gén. Sci. pur. appl.*, **11**, 1261–1271 and 1309–1328. [274]

BENIOFF, H. (1955). Seismic evidence for crustal structure and tectonic activity. *Spec. Pap. geol. Soc. Am.*, **62**, 61–74. [255, *258*, *259*, 264]

BENIOFF, H. (1959). Circum-Pacific tectonics. In *The mechanics of faulting, with special reference to the fault-plane work (a symposium)*, edited by Hodgson, J. H. *Publs Dom. Obs.*, **20**, 395–402. [267]

BENIOFF, H. (1964). Earthquake source mechanisms. *Science, N.Y.*, **143**, 1399–1406. [258, *259*, *261*, 261]

BERNAL, J. D. (1936). Geophysical discussion. *Observatory*, **59**, 268. [141]

BERRY, M. J. and WEST, G. F. (1966). Reflected and head wave amplitudes in a medium of several layers. In *The*

*Earth beneath the continents*, pp. 464–481, edited by Steinhart, J. S. and Smith, T. J., *Geophys. Monogr.* No. 10, American Geophysical Union, Washington, D.C.                                                           [*35*]

BIRCH, F.                                                                                                      [64]

BIRCH, F. (1952). Elasticity and constitution of the Earth's interior. *J. geophys. Res.*, **57**, 227–286.      [140]

BIRCH, F. (1958). Interpretation of the seismic structure of the crust in the light of experimental studies of wave velocities in rocks. In *Contributions in geophysics in honor of Beno Gutenberg*, pp. 158–170, edited by Benioff, H., Ewing, M., Howell, B. F., Jr. and Press, F., Pergamon Press, London, New York, Paris and Los Angeles.
                                                                                                               [64]

BIRCH, F. (1960). The velocity of compressional waves in rocks to 10 kilobars, part 1. *J. geophys. Res.*, **65**, 1083–1102.                                                                                                       [64]

BIRCH, F. (1961a). The velocity of compressional waves in rocks to 10 kilobars, part 2. *J. geophys. Res.*, **66**, 2199–2224.                                                                                             [64, 65, *66*]

BIRCH, F. (1961b). Composition of the Earth's mantle. *Geophys. J. R. astr. Soc.*, **4**, 295–311.       [145, *160*]

BIRCH, F. (1964). Density and composition of mantle and core. *J. geophys. Res.*, **69**, 4377–4388.          [112]

BIRCH, F. (1965). Speculations on the Earth's thermal history. *Bull. geol. Soc. Am.*, **76**, 133–153.       [190]

BIRCH, F. and CLARK, H. (1940). The thermal conductivity of rocks and its dependence upon temperature and composition. *Am. J. Sci.*, **238**, 529–558 and 613–635.                                                          [185]

BLACKETT, P. M. S.                                                                                            [165]

BLACKETT, P. M. S. (1947). The magnetic field of massive rotating bodies. *Nature, Lond.*, **159**, 658–666.   [170]

BLUNDELL, D. J. and PARKS, R. (1969). A study of the crustal structure beneath the Irish Sea. *Geophys. J. R. astr. Soc.*, **17**, 45–62.                                                                                          [45]

BOLDIZSÁR, T. (1964). Heat flow in the Hungarian basin. *Nature, Lond.*, **202**, 1278–1280.                  [199]

BOLT, B. A.                                                                                                   [144]

BOLT, B. A. (1962). Gutenberg's early *PKP* observations. *Nature, Lond.*, **196**, 122–124.           [*154*, 154]

BOLT, B. A. (1964). The velocity of seismic waves near the Earth's center. *Bull. seism. Soc. Am.*, **54**, 191–208.
                                                                                                         [154, *155*]

BOSCOVITCH, R. J.                                                                                             [45]

BOTT, M. H. P. (1954). Interpretation of the gravity field of the Eastern Alps. *Geol. Mag.*, **91**, 377–383.  [55]

BOTT, M. H. P. (1959). The mechanics of oblique slip faulting. *Geol. Mag.*, **96**, 109–117.                 [233]

BOTT, M. H. P. (1961). The granitic layer. *Geophys. J. R. astr. Soc.*, **5**, 207–216.                   [54, 68]

BOTT, M. H. P. (1964). Convection in the Earth's mantle and the mechanism of continental drift. *Nature, Lond.*, **202**, 583–584.                                                                                              [288]

BOTT, M. H. P. (1965a). Formation of oceanic ridges. *Nature, Lond.*, **207**, 840–843.                       [88]

BOTT, M. H. P. (1965b). The deep structure of the northern Irish Sea—a problem of crustal dynamics. In *Submarine geology and geophysics*, pp. 179–204, edited by Whittard, W. F. and Bradshaw, R., Colston Papers No. 17, Butterworths, London.                                                                            [235, *236*]

BOTT, M. H. P. (1967a). Geophysical investigations of the northern Pennine basement rocks. *Proc. Yorks. geol. Soc.*, **36**, 139–168.                                                                                            [*54*]

BOTT, M. H. P. (1967b). Solution of the linear inverse problem in magnetic interpretation with application to oceanic magnetic anomalies. *Geophys. J. R. astr. Soc.*, **13**, 313–323.                                     [*80*, 82]

BOTT, M. H. P. (1967c). Terrestrial heat flow and the mantle convection hypothesis. *Geophys. J. R. astr. Soc.*, **14**, 413–428.                                                                                       [276, *281*, *282*]

BOTT, M. H. P. and SCOTT, P. (1964). Recent geophysical studies in south-west England. In *Present views of some aspects of the geology of Cornwall and Devon*, pp. 25–44, edited by Hosking, K. F. G. and Shrimpton, G. J., Royal Geological Society of Cornwall.                                                                        [*52*, *53*]

BOTT, M. H. P., HOLDER, A. P., LONG, R. E. and LUCAS, A. L. (1970). Crustal structure beneath the granites of south-west England. In *Mechanism of igneous intrusion*, pp. 93–102, edited by Newall, G. and Rast, N., *Geol. J. Spec.* Issue No. 2.                                                                                   [*35*, *36*, *37*]

BOUGUER, P.                                                                                                    [45]

BRADLEY, J. J. and FORT, A. N., JR. (1966). Internal friction in rocks. In *Handbook of physical constants*, Revised edition, edited by Clark, S. P., Jr., *Mem. geol. Soc. Am.*, **97**, 175–193.                               [247]

BROOKS, J. A. (1962). Seismic wave velocities in the New Guinea-Solomon Islands region. In *The crust of the Pacific basin*, pp. 2–10, edited by Macdonald, G. A. and Kuno, H., *Geophys. Monogr.* No. 6, American Geophysical Union, Washington, D.C.                                                                                [106]

BRUNE, J. and DORMAN, J. (1963). Seismic waves and earth structure in the Canadian Shield. *Bull. seism. Soc. Am.*, **53**, 167–209.                                                                                     [107, 116]

BULLARD, Sir EDWARD                                                                                 [144, 170, 178]

BULLARD, E. C. (1936). Gravity measurements in East Africa. *Phil. Trans. R. Soc.*, **235A**, 445–531.         [61]

BULLARD, E. C. (1967). Electromagnetic induction in the Earth. *Q. Jl R. astr. Soc.*, **8**, 143–160.  [117, *118*, *119*, 124]

BULLARD, E. C. and GELLMAN, H. (1954). Homogeneous dynamos and terrestrial magnetism. *Phil. Trans. R. Soc.*, **247A**, 213–278. [*170*, 171, *172*, *173*]

BULLARD, E. C. and GRIGGS, D. T. (1961). The nature of the Mohorovičić discontinuity. *Geophys. J. R. astr. Soc.*, **6**, 118–123. [130]

BULLARD, E. C., EVERETT, J. E. and SMITH, A. G. (1965). The fit of the continents around the Atlantic. In *A symposium on continental drift*, *Phil. Trans. R. Soc.*, **258A**, 41–51. [202, *203*]

BULLEN, K. E. [94, 158]

BULLEN, K. E. (1936). The variation of density and the ellipticities of strata of equal density within the Earth. *Mon. Not. R. astr. Soc. geophys. Suppl.*, **3**, 395–401. [140]

BULLEN, K. E. (1958). Solidity of the inner core. In *Contributions in geophysics in honor of Beno Gutenberg*, pp. 113–120, edited by Benioff, H., Ewing, M., Howell, B. F., Jr. and Press, F., Pergamon Press, London, New York, Paris and Los Angeles. [158]

BULLEN, K. E. (1963). *An introduction to the theory of seismology*, Third edition, Cambridge University Press, London and New York, 381 pp. [**9**, 94, 95, 144, 152]

BYERLY, P. (1926). The Montana earthquake of June 28, 1925, G.M.C.T. *Bull. seism. Soc. Am.*, **16**, 209–265. [258]

BYERLY, P. (1956). Subcontinental structure in the light of seismological evidence. *Adv. Geophys.*, **3**, 105–152. [41]

CAGNIARD, L. (1953). Basic theory of the magneto-telluric method of geophysical prospecting. *Geophysics*, **18**, 605–635. [120]

CANER, B. and CANNON, W. H. (1965). Geomagnetic depth-sounding and correlation with other geophysical data in western North America. *Nature, Lond.*, **207**, 927–929. [122, *123*]

CANN, J. R. (1968). Geological processes at mid-ocean ridge crests. *Geophys. J. R. astr. Soc.*, **15**, 331–341. [83]

CANTWELL, T. and MADDEN, T. R. (1960). Preliminary report on crustal magnetotelluric measurements. *J. geophys. Res.*, **65**, 4202–4205. [121]

CAPON, J., GREENFIELD, R. J. and LACOSS, R. T. (1969). Long-period signal processing results for the large aperture seismic array. *Geophysics*, **34**, 305–329. [*99*, *100*]

CAREY, S. W. (1958). The tectonic approach to continental drift. In *Continental drift, a symposium*, pp. 177–355, edited by Carey, S. W., University of Tasmania, Hobart. [202, 271]

CARPENTER, E. W., MARSHALL, P. D. and DOUGLAS, A. (1967). The amplitude-distance curve for short period teleseismic P-waves. *Geophys. J. R. astr. Soc.*, **13**, 61–70 [108]

CASSINI, J. [1]

ČERVENÝ, V. (1966). On dynamic properties of reflected and head waves in the n-layered Earth's crust. *Geophys. J. R. astr. Soc.*, **11**, 139–147. [33]

CHAPMAN, S. (1919). The solar and lunar diurnal variations of terrestrial magnetism. *Phil. Trans. R. Soc.*, **218A**, 1–118. [120]

CHAPMAN, S. and BARTELS, J. (1940). *Geomagnetism*, volumes I and II, Oxford University Press, London and New York, 1049 pp. [164]

CHINNERY, M. A. and TOKSÖZ, M. N. (1967). P-wave velocities in the mantle below 700 km. *Bull. seism. Soc. Am.*, **57**, 199–226. [108]

CLARK, S. P., JR. (1957). Absorption spectra of some silicates in the visible and near infrared. *Am. Miner.*, **42**, 732–742. [192]

CLARK, S. P., JR. (editor) (1966). *Handbook of physical constants*, Revised edition, *Mem. geol. Soc. Am.*, **97**, 587 pp. [**64**, *65*, 159, 161]

CLARK, S. P., JR. and RINGWOOD, A. E. (1964). Density distribution and constitution of the mantle. *Rev. Geophys.*, **2**, 35–88. [*126*, 127, 138, *138*, 145, *147*]

CLAYTON, D. D. (1964). Cosmoradiogenic chronologies of nucleosynthesis. *Astrophys. J.*, **139**, 637–663. [16]

CLEARY, J. and HALES, A. L. (1966). An analysis of the travel times of P waves to North American stations, in the distance range 32° to 100°. *Bull. seism. Soc. Am.*, **56**, 467–489. [114]

CLOSS, H. and LABROUSTE, Y. (editors) (1963). *Recherches séismologiques dans les Alpes occidentales au moyen des grandes explosions en 1956, 1958 et 1960. Séismologie*, sér. XXII, **2**, Centre National de la Recherche Scientifique, 241 pp. [58, 114]

COLLINSON, D. W., CREER, K. M. and RUNCORN, S. K. (1967). *Methods in palaeomagnetism*, Elsevier, Amsterdam, London and New York, 609 pp. [165]

CONRAD, V. (1925). Laufzeitkurven des Tauernbebens vom 28. November, 1923. *Mitt. Erdb-Kommn Wien*, No. 59, 1–23. [29, 41]

COODE, A. M. (1966). An analysis of major tectonic features. *Geophys. J. R. astr. Soc.*, **12**, 55–66. [278]

COOK, A. H. (1967). Gravitational considerations. In *The Earth's mantle*, pp. 63–87, edited by Gaskell, T. F., Academic Press, London and New York. [4, 149]

CORON, S. (1963). Aperçu gravimétrique sur les Alpes occidentales. In *Recherches séismologiques dans les Alpes occidentales au moyen des grandes explosions en 1956, 1958 et 1960*, pp. 31–37, edited by Closs, H. and Labrouste, Y. *Séismologie*, sér. XXII, **2**, Centre National de la Recherche Scientifique. [*56, 58*]

COWLING, T. G. (1934). The magnetic field of sunspots. *Mon. Not. R. astr. Soc.*, **94**, 39–48. [*170*]

COX, A. and DOELL, R. R. (1960). Review of palaeomagnetism. *Bull. geol. Soc. Am.*, **71**, 645–768. [*168*]

COX, A., DOELL, R. R. and DALRYMPLE, G. B. (1964). Reversals of the Earth's magnetic field. *Science, N.Y.*, **144**, 1537–1543. [*213*]

CREER, K. M. (1957). Palaeomagnetic investigations in Great Britain V. The remanent magnetization of unstable Keuper marls. *Phil. Trans. R. Soc.*, **250A**, 130–143. [*166*]

CREER, K. M. (1965). Palaeomagnetic data from the Gondwanic continents. In *A symposium on continental drift*, *Phil. Trans. R. Soc.*, **258A**, 27–40. [*206, 207, 208*]

CRITTENDEN, M. D., JR. (1963*a*). New data on the isostatic deformation of Lake Bonneville. *Prof. Pap. U.S. geol. Surv.*, **454E**, 1–31. [*243, 244*]

CRITTENDEN, M. D., JR. (1963*b*). Effective viscosity of the Earth derived from isostatic loading of Pleistocene Lake Bonneville. *J. geophys. Res.*, **68**, 5517–5530. [*243*]

DALY, R. A. [*30, 64*]

DANA, J. D. (1847). Geological results of the Earth's contraction in consequence of cooling. *Am. J. Sci.*, Ser. 2, **3**, 176–188. [*269*]

DARWIN, SIR GEORGE [*26, 251*]

DARWIN, G. H. (1887). Note on Mr. Davison's paper on the straining of the Earth's crust in cooling. *Phil. Trans. R. Soc.*, **178A**, 242–249. [*269*]

DAVIES, D. (1968). A comprehensive test ban. *Sci. J., Lond.*, November 1968, 78–84. [*7*]

DAVIES, D. and MCKENZIE, D. P. (1969). Seismic travel-time residuals and plates. *Geophys. J. R. astr. Soc.*, **18**, 51–63. [*106*]

DAVISON, C. (1887). On the distribution of strain in the Earth's crust resulting from secular cooling; with special reference to the growth of continents and the formation of mountain chains. *Phil. Trans. R. Soc.*, **178A**, 231–242. [*269*]

DEARNLEY, R. (1966). Orogenic fold-belts and a hypothesis of Earth evolution. *Physics Chem. Earth*, **7**, 1–114. [*285*]

DESCARTES, R. [*16*]

DE BEAUMONT, E. (1852). *Notice sur les systèmes de montagnes*, 3 volumes, P. Bertrand, Paris. [*269*]

DICKE, R. H. [*16*]

DICKE, R. H. (1962). The Earth and cosmology. *Science, N.Y.*, **138**, 653–664. [*271*]

DICKE, R. H. (1966). The secular acceleration of the Earth's rotation and cosmology. In *The Earth-Moon system*, pp. 98–164, edited by Marsden, B. G. and Cameron, A. G. W., Plenum Press, New York. [*21*]

DIETZ, R. S. [*208*]

DIETZ, R. S. (1961). Continent and ocean basin evolution by spreading of the sea floor. *Nature, Lond.*, **190**, 854–857. [*82, 278*]

DIRAC, P. A. M. (1938). A new basis for cosmology. *Proc. R. Soc.*, **165A**, 199–208. [*270, 271*]

DIX, C. H. (1965). Reflection seismic crustal studies. *Geophysics*, **30**, 1068–1084. [*37*]

DOBRIN, M. B. (1960). *Introduction to geophysical prospecting*, Second edition, McGraw-Hill Book Company, New York, Toronto and London, 446 pp. [*31*]

DOHR, G. and FUCHS, K. (1967). Statistical evaluation of deep crustal reflections in Germany. *Geophysics*, **32**, 951–967. [*37*]

DORMAN, J., EWING, J. and ALSOP, L. E. (1965). Oscillations of the Earth: new core-mantle boundary model based on low-order free vibrations. *Proc. natn. Acad. Sci. U.S.A.*, **54**, 364–368. [*146*]

DORMAN, J., EWING, M. and OLIVER, J. (1960). Study of shear-velocity distribution in the upper mantle by mantle Rayleigh waves. *Bull. seism. Soc. Am.*, **50**, 87–115. [*111, 115*]

DOUGLAS, A. and CORBISHLEY, D. J. (1968). Measurement of dT/d$\Delta$. *Nature, Lond.*, **217**, 1243–1244. [*110*]

DRAKE, C. L. and GIRDLER, R. W. (1964). A geophysical study of the Red Sea. *Geophys. J. R. astr. Soc.*, **8**, 473–495. [*62*]

DRAKE, C. L., EWING, M. and SUTTON, G. H. (1959). Continental margins and geosynclines: the east coast of North America north of Cape Hatteras. *Physics Chem. Earth*, **3**, 110–198. [*89*]

DU TOIT, A. L. [*206*]

DU TOIT, A. L. (1937). *Our wandering continents, an hypothesis of continental drifting*, Oliver and Boyd, Edinburgh and London, 366 pp. [*201, 202*]

DUTTON, C. E. [*45*]

EADE, K. E., FAHRIG, W. F. and MAXWELL, J. A. (1966). Composition of crystalline shield rocks and fractionating effects of regional metamorphism. *Nature, Lond.*, **211**, 1245–1249. [**67**]

EATON, J. P. (1963). Crustal structure from San Francisco, California, to Eureka, Nevada, from seismic-refraction measurements. *J. geophys. Res.*, **68**, 5789–5806. [*34, 44*]

EATON, J. P. and MURATA, K. J. (1960). How volcanoes grow. *Science, N.Y.*, **132**, 925–938.    [127]

EGYED, L. (1957). A new dynamic conception of the internal constitution of the Earth. *Geol. Rdsch.*, **46**, 101–121.    [270, *271*]

EGYED, L. (1960). Dirac's cosmology and the origin of the solar system. *Nature, Lond.*, **186**, 621–622.    [270]

ELDER, J. W. (1965). Physical processes in geothermal areas. In *Terrestrial heat flow*, pp. 211–239, edited by Lee, W. H. K., *Geophys. Monogr.* No. 8, American Geophysical Union, Washington, D.C. [*193*, 194, 197, *275*]

ELSASSER, W. M.    [170]

ELSASSER, W. M. (1969). Convection and stress propagation in the upper mantle. In *The application of modern physics to the Earth and planetary interiors*, pp. 223–246, edited by Runcorn, S. K., Wiley-Interscience, London, New York, Sydney and Toronto.    [283]

EWING, J. and EWING, M. (1967). Sediment distribution on the mid-ocean ridges with respect to spreading of the sea floor. *Science, N.Y.*, **156**, 1590–1592.    [217, 220, *221*]

EWING, J. and ZAUNERE, R. (1964). Seismic profiling with a pneumatic sound source. *J. geophys. Res.*, **69**, 4913–4915.    [74]

EWING, J., WORZEL, J. L., EWING, M. and WINDISCH, C. (1966). Ages of horizon A and the oldest Atlantic sediments. *Science, N.Y.*, **154**, 1125–1132.    [75, 77]

EWING, M.    [72, 84]

EWING, M. (1965). The sediments of the Argentine basin. *Q. Jl R. astr. Soc.*, **6**, 10–27.    [74, *74*, 76, 77]

EWING, M. and others (1969). *Initial reports of the Deep Sea Drilling Project*, volume I., Washington, D.C. (U.S. Government Printing Office), 672 pp.    [78]

EWING, M. and PRESS, F. (1959). Determination of crustal structure from phase velocity of Rayleigh waves Part III: the United States. *Bull. geol. Soc. Am.*, **70**, 229–244.    [31]


FAURE, G. and HURLEY, P. M. (1963). The isotopic composition of strontium in oceanic and continental basalts: application to the origin of igneous rocks. *J. Petrology*, **4**, 31–50.    [135, *136*]

FISH, R. A., GOLES, G. G. and ANDERS, E. (1960). The record in the meteorites. III. On the development of meteorites in asteroidal bodies. *Astrophys. J.*, **132**, 243–258.    [188]

FISHER, SIR RONALD    [165]

FISHER, O. (1881). *Physics of the earth's crust*. Macmillan and Co., London    [273]

FRANCIS, T. J. G. (1968). The detailed seismicity of mid-oceanic ridges. *Earth & planet. Sci. Lett. (Neth.)*, **4**, 39–46    [263]

FRANK, F. C. (1968). Curvature of island arcs. *Nature, Lond.*, **220**, 363.    [226]

FUCHS, K., MÜLLER, S., PETERSCHMIDTT, E., ROTHE, J. P., STEIN, A. and STROBACH, K. (1963). Essais d'interprétation séismique. VI.A. Essai d'interprétation N° 1. In *Recherches séismologique dans les Alpes occidentales au moyen des grandes explosions en 1956, 1958 et 1960*, pp. 118–171, edited by Closs, H. and Labrouste, Y. *Séismologie*, sér. XXII, **2**, Centre National de la Recherche Scientifique.    [*57, 58*]


GAST, P. W. (1960). Limitations on the composition of the upper mantle. *J. geophys. Res.*, **65**, 1287–1297. [11, 135]

GASTIL, G. (1960). The distribution of mineral dates in time and space. *Am. J. Sci.*, **258**, 1–35.    [285]

GELLIBRAND, H.    [164]

GERSTENKORN, H. (1955). Über Gezeitenreibung beim Zweikörperproblem. *Z. Astrophys.*, **36**, 245–274.    [22, 26]

GERSTENKORN, H. (1969). The earliest past history of the Earth-Moon system. *Icarus*, **11**, 189–207.    [26]

GERSTENKORN, H. (1970). The early history of the Moon. In *The Moon and Planets*, NATO Advanced Study Institute 9th–16th April, 1970, held at University of Newcastle-upon-Tyne (abstract).    [26]

GILBERT, W.    [169]

GIRDLER, R. W. (1964). Geophysical studies of rift valleys. *Physics Chem. Earth*, **5**, 121–156.    [*59, 60*, 62]

GOLD, T. (1955). Instability of the Earth's axis of rotation. *Nature, Lond.*, **175**, 526–529.    [251]

GOLD, T. (1963). Problems requiring solution. In *Origin of the solar system*, pp. 171–174, edited by Jastrow, R. and Cameron, A. G. W., Academic Press, New York and London.    [16]

GOLDREICH, P. (1966). History of the lunar orbit. *Rev. Geophys.*, **4**, 411–439.    [26]

GOLDREICH, P. and TOOMRE, A. (1969). Some remarks on polar wandering. *J. geophys. Res.*, **74**, 2555–2567.    [246, 250, 251, 254]

GORDON, R. B. (1965). Diffusion creep in the Earth's mantle. *J. geophys. Res.*, **70**, 2413–2418.    [239]

GORDON, R. B. (1967). Thermally activated processes in the Earth: creep and seismic attenuation. *Geophys. J. R. astr. Soc.*, **14**, 33–43.    [231, 247]

GORDON, R. B. and NELSON, C. W. (1966). Anelastic properties of the Earth. *Rev. Geophys.*, **4**, 457–474.    [247]

GORSHKOV, G. S. (1958). On some theoretical problems of volcanology. *Bull. volcan.*, Ser. II, **19**, 103–114.    [127]

GREEN, D. H. and RINGWOOD, A. E. (1963). Mineral assemblages in a model mantle composition. *J. geophys. Res.*, **68**, 937–945.    [**134**, 134, 137]

GREEN, D. H. and RINGWOOD, A. E. (1967). An experimental investigation of the gabbro to eclogite transformation and its petrological applications. *Geochim. cosmochim. Acta*, **31**, 767–833.                                                       [68]

GRIGGS, D. and HANDIN, J. (editors) (1960a). *Rock deformation (a symposium)*, *Mem. geol. Soc. Am.*, **79**, 382 pp. [231]

GRIGGS, D. and HANDIN, J. (1960b). Observations on fracture and a hypothesis of earthquakes. In *Rock deformation (a symposium)*, pp. 347–364, edited by Griggs, D. and Handin, J., *Mem. geol. Soc. Am.*, **79**.                            [262]

GRIGGS, D. T., TURNER, F. J. and HEARD, H. C. (1960). Deformation of rocks at 500° to 800°C. In *Rock deformation, (a symposium)*, pp. 39–104, edited by Griggs, D. and Handin, J., *Mem. geol. Soc. Am.*, **79**.                              [234]

GUNN, R. (1947). Quantitative aspects of juxtaposed ocean deeps, mountain chains and volcanic ranges. *Geophysics*, **12**, 238–255.                                                                                                          [93]

GUTENBERG, B.                                                                                                                 [8]

GUTENBERG, B. (1924). Der Aufbau der Erdkruste auf Grund geophysikalischer Betrachtungen. *Z. Geophys.*, **1**, 94–108.                                                                                                                        [72]

GUTENBERG, B. (1945). Amplitudes of *P*, *PP*, and *S* and magnitude of shallow earthquakes. *Bull. seism. Soc. Am.* **35**, 57–69.                                                                                                           [255]

GUTENBERG, B. (1953). Wave velocities at depths between 50 and 600 kilometers. *Bull. seism. Soc. Am.*, **43**, 223–232.                                                                                                                       [106]

GUTENBERG, B. (1954a). Effects of low velocity layers. *Geofis. pura appl.*, **28**, 1–10.                                    [40]

GUTENBERG, B. (1954b). Postglacial uplift in the Great Lakes region. *Arch. Met. Geophys. Bioklim.*, **7A**, 243–251.          [242]

GUTENBERG, B. (1959). *Physics of the Earth's interior*, Academic Press, New York and London, 240 pp.
                                                                    [8, 9, *9*, 14, 41, 112, 152, *153, 154, 242*]

GUTENBERG, B. and RICHTER, C. F. (1936). On seismic waves (third paper). *Beitr. Geophys.*, **47**, 73–131.     [254]

GUTENBERG, B. and RICHTER, C. F. (1954). *Seismicity of the Earth and associated phenomena*, Second edition, Princeton University Press, Princeton, New Jersey, 310 pp.                                                      [*252, 253, 264*]

HAGEDOORN, J. G. (1959). The plus-minus method of interpreting seismic refraction sections. *Geophys. Prospect.*, **7**, 158–182.                                                                                                            [31]

HALES, A. L. and DOYLE, H. A. (1967). *P* and *S* travel time anomalies and their interpretation. *Geophys. J. R. astr. Soc.*, **13**, 403–415.                                                                                          [117, 137, 139]

HALES, A. L. and SACKS, I. S. (1959). Evidence for an intermediate layer from crustal structure studies in the eastern Transvaal. *Geophys. J. R. astr. Soc.*, **2**, 15–33.                                                              [42, *42*]

HALM, J. K. E. (1935). An astronomical aspect of the evolution of the Earth. *J. astr. Soc. S. Afr.*, **4**, 1–28.     [270]

HAMILTON, E. I. (1965). *Applied geochronology*, Academic Press, London and New York, 267 pp.     [**12**, 13, 14]

HARLAND, W. B., SMITH, A. G. and WILCOCK, B. (editors) (1964). *The Phanerozoic time-scale*, *Q. Jl geol. Soc. Lond.*, **120s**, 458 pp.                                                                                                  [13, **15**]

HARRIS, P. G. and ROWELL, J. A. (1960). Some geochemical aspects of the Mohorovičić discontinuity. *J. geophys. Res.*, **65**, 2443–2459.                                                                                                     [130]

HARTMANN, W. K. (1967). On the nature of the infrared nebula in Orion. *Astrophys. J.*, **149**, L87–L90.     [16]

HASKELL, N, A. (1935). The motion of a viscous fluid under a surface load, I. *Physics*, **6**, 265–269.     [241, 242]

HASKELL, N. A. (1953). The dispersion of surface waves on multilayered media. *Bull. seism. Soc. Am.*, **43**, 17–34.
                                                                                                                           [103]

HAYES, D. E. (1966). A geophysical investigation of the Peru-Chile trench. *Marine Geol.*, **4**, 309–351.   [90, 91, *92*]

HAYFORD, J. F.                                                                                                                [47]

HAYS, J. D. and OPDYKE, N. D. (1967). Antarctic radiolaria, magnetic reversals, and climatic change. *Science, N.Y.*, **158**, 1001–1011.                                                                                                    [213]

HEALY, J. H. (1963). Crustal structure along the coast of California from seismic-refraction measurements. *J. geophys. Res.*, **68**, 5777–5787.                                                                                             [44]

HEEZEN, B. C.                                                                                                                 [84]

HEEZEN, B. C. (1962). The deep-sea floor. In *Continental drift*, pp. 235–288, edited by Runcorn, S. K., Academic Press, New York and London.                                                                                               [*71*]

HEIRTZLER, J. R., LE PICHON, X. and BARON, J. G. (1966). Magnetic anomalies over the Reykjanes ridge. *Deep-sea Res.*, **13**, 427–443.                                                                                                     [*81*]

HEIRTZLER, J. R., DICKSON, G. O., HERRON, E. M., PITMAN, W. C., III and LE PICHON, X. (1968). Marine magnetic anomalies, geomagnetic field reversals, and motions of the ocean floor and continents. *J. geophys. Res.*, **73**, 2119–2136.                                                                             [214, 215, *217*, 217, *218, 219*, 220]

HEISKANEN, W. A. and VENING MEINESZ, F. A. (1958). *The Earth and its gravity field*, McGraw-Hill Book Company, New York, Toronto and London, 470 pp.                                                                                     [*61*, 62, 241]

HERRIN, E. and TAGGART, J. (1962). Regional variations in $P_n$ velocity and their effect on the location of epicenters. *Bull. seism. Soc. Am.*, **52**, 1037–1046.                                                                        [112, 114]

HERZENBERG, A. (1958). Geomagnetic dynamos. *Phil. Trans. R. Soc.*, **250A**, 543–583.                                        [171]

HESS, H. H.                                                                                    [77, 130, 134, 208]
HESS, H. H. (1962). History of ocean basins. In *Petrologic studies: a volume in honor of A. F. Buddington*, pp. 599–620, Geological Society of America.                                                              [82, 278]
HESS, H. H. (1964). Seismic anisotropy of the uppermost mantle under oceans. *Nature, Lond.*, **203**, 629–631. [114]
HESS, H. H. (1965). Mid-oceanic ridges and tectonics of the sea-floor. In *Submarine geology and geophysics*, pp. 317–333, edited by Whittard, W. F. and Bradshaw, R., Colston Papers No. 17, Butterworths, London.      [83]
HILGENBERG, O. C. (1933). *Vom wachsendem Erdball*. Giessmann and Bartsch, Berlin.                      [270]
HILGENBERG, O. C. (1962). Rock magnetism and the Earth's palaeopoles. *Geofis. pura appl.*, **53**, 52–54.   [271]
HILL, D. P. and PAKISER, L. C. (1966). Crustal structure between the Nevada test site and Boise, Idaho, from seismic refraction measurements. In *The Earth beneath the continents*, pp. 391–419, edited by Steinhart, J. S. and Smith, T. J., *Geophys. Monogr.* No. 10, American Geophysical Union, Washington, D.C.          [43, 44]
HILL, M. N.                                                                                              [72]
HILL, M. N. (1963). Single-ship seismic refraction shooting. In *The sea*, volume 3, pp. 39–46, edited by Hill, M. N., Interscience Publishers, New York and London.                                                     [72]
HOFFMAN, J. P., BERG, J. W., JR. and COOK, K. L. (1961). Discontinuities in the Earth's upper mantle as indicated by reflected seismic energy. *Bull. seism. Soc. Am.*, **51**, 17–27.                                   [107]
HOLMES, A. (1928). Radioactivity and continental drift (Report of Glasgow Geological Society meeting 12th January, 1928). *Geol. Mag.*, **65**, 236–238.                                                            [273]
HOLMES, A. (1931). Radioactivity and Earth movements. *Trans. geol. Soc. Glasg.*, **18**, 559–606.      [273]
HOLMES, A. (1933). The thermal history of the Earth. *J. Wash. Acad. Sci.*, **23**, 169–195.        [273, *274*]
HOLMES, A. (1965). *Principles of physical geology*, Second edition, Nelson, London and Edinburgh, 1288 pp.
                                                                                              [13, *205, 207*]
HOLOPAINEN, P. E. (1947). On the gravity field and the isostatic structure of the Earth's crust in the East Alps. *Publs isostatic Inst. int. Ass. Geod.*, No. 16.                                                  [*55*]
HOPKINS, W.                                                                                            [273]
HORAI, K. (1969). Effect of past climatic changes on the thermal field of the Earth. *Earth & planet. Sci. Lett.* (*Neth.*), **6**, 39–42.                                                                              [179]
HOSPERS, J. (1951). Remanent magnetism of rocks and the history of the geomagnetic field. *Nature, Lond.*, **168**, 1111–1112.                                                                                         [166]
HOUTZ, R., EWING, J. and LE PICHON, X. (1968). Velocity of deep-sea sediments from sonobuoy data. *J. geophys. Res.*, **73**, 2615–2641.                                                                                   [76]
HUBBERT, M. K. and RUBEY, W. W. (1959). Role of fluid pressure in mechanics of overthrust faulting I. Mechanics of fluid-filled porous solids and its application to overthrust faulting. *Bull. geol. Soc. Am.*, **70**, 115–166.
                                                                                                 [233, 262]
HUGHES, D. S. and MAURETTE, C. (1956). Variation of elastic wave velocities in granites with pressure and temperature. *Geophysics*, **21**, 277–284.                                                                   [64]
HUGHES, D. S. and MAURETTE, C. (1957). Variation of elastic wave velocities in basic igneous rocks with pressure and temperature. *Geophysics*, **22**, 23–31.                                                          [64]

IRVING, E. (1964). *Paleomagnetism and its application to geological and geophysical problems*, John Wiley & Sons, New York, London and Sydney, 399 pp.                                                          [165, *166, 272*]
IRVING, E. and ROBERTSON, W. A. (1969). Test for polar wandering and some possible implications. *J. geophys. Res.*, **74**, 1026–1036.                                                                                   [250]
ISACKS, B. and MOLNAR, P. (1969). Mantle earthquake mechanisms and the sinking of the lithosphere. *Nature, Lond.*, **223**, 1121–1124.                                                                             [268, *268*]
ISACKS, B., OLIVER, J. and SYKES, L. R. (1968). Seismology and the new global tectonics. *J. geophys. Res.*, **73**, 5855–5899.                                                                                    [*212*, 263, 266]
ISACKS, B., SYKES, L. R. and OLIVER, J. (1969). Focal mechanisms of deep and shallow earthquakes in the Tonga-Kermadec region and the tectonics of island arcs. *Bull. geol. Soc. Am.*, **80**, 1443–1469.             [267]

JAEGER, J. C. (1960). Shear failure of anisotropic rocks. *Geol. Mag.*, **97**, 65–72.                  [233]
JAEGER, J. C. (1962). *Elasticity, fracture and flow*, Second edition, Methuen, London; John Wiley, New York, 208 pp.                                                                                             [228, *230*]
JAMES, D. E. and STEINHART, J. S. (1966). Structure beneath continents: a critical review of explosion studies 1960–1965. In *The Earth beneath the continents*, pp. 293–333, edited by Steinhart, J. S. and Smith, T. J., *Geophys. Monogr.* No. 10, American Geophysical Union, Washington, D.C.               [28, *38*, 39, 114]
JASTROW, R. and CAMERON, A. G. W. (editors) (1963). *Origin of the solar system*, Academic Press, New York and London, 176 pp.                                                                                             [16]
JEFFERY, R. N., BARNETT, J. D., VANFLEET, H. B. and HALL, H. T. (1966). Pressure calibration to 100 kbar based on the compression of NaCl. *J. Appl. Phys.*, **37**, 3172–3180.                                      [**143**]

JEFFREYS, SIR HAROLD                                                                                          [26]
JEFFREYS, H. (1930). The instability of a compressible fluid heated below. *Proc. Camb. phil. Soc. math. phys. Sci.*,
    **26**, 170–172.                                                                                          [275]
JEFFREYS, H. (1939a). The times of *P, S* and *SKS*, and the velocities of *P* and *S. Mon. Not. R. astr. Soc. geophys.*
    *Suppl.*, **4**, 498–533.                                                                              [9, 152]
JEFFREYS, H. (1939b). The times of *PcP* and *ScS. Mon. Not. R. astr. Soc. geophys. Suppl.*, **4**, 537–547.  [152]
JEFFREYS, H. (1959). *The Earth, its origin history and physical constitution*, Fourth edition, Cambridge University
    Press, London and New York, 420 pp.                                        [2, 3, 41, 235, 245, 269]
JOHNSON, G. A. L.                                                                                            [*177*]
JORDAN, P. (1962). Geophysical consequences of Dirac's hypothesis. *Rev. mod. Phys.*, **34**, 596–600.       [271]
JULIAN, B. R. and ANDERSON, D. L. (1968). Travel times, apparent velocities and amplitudes of body waves. *Bull.*
    *seism. Soc. Am.*, **58**, 339–366.                                                                 [*109, 110*]

KANAMORI, H., FUJII, N. and MIZUTANI, H. (1968). Thermal diffusivity measurement of rock-forming minerals from
    300° to 1100°K. *J. geophys. Res.*, **73**, 595–605.                                                      [192]
KANT, I.                                                                                                       [16]
KAULA, W. M. (1963). Elastic models of the mantle corresponding to variations in the external gravity field. *J.*
    *geophys. Res.*, **68**, 4967–4978.                                                                 [245, 246]
KELVIN, LORD                                                                                        [12, 185, 270]
KING-HELE, D. (1967). The shape of the Earth. *Scient. Am.*, **217**, No. 4, 67–76.                        [*3, 3*]
KING-HELE, D. G. (1969). The shape of the Earth. *Royal Aircraft Establishment Technical Memorandum Space*
    130, 1–10.                                                                                                [*3*]
KLEINMANN, D. E. and LOW, F. J. (1967). Discovery of an infrared nebula in Orion. *Astrophys. J.*, **149**, L1–L4. [16]
KNOPOFF, L. (1964a). The convection current hypothesis. *Rev. Geophys.*, **2**, 89–122.                  [275, 276]
KNOPOFF, L. (1964b). *Q. Rev. Geophys.*, **2**, 625–660.                                                      [247]
KNOPOFF, L. (1967). Density-velocity relations for rocks. *Geophys. J. R. astr. Soc.*, **13**, 1–8.            [65]
KNOPOFF, L. and MACDONALD, G. J. F. (1960). An equation of state for the core of the Earth. *Geophys. J. R. astr.*
    *Soc.*, **3**, 68–77.                                                                                     [161]
KOSMINSKAYA, I. P., BELYAEVSKY, N. A. and VOLVOVSKY, I. S. (1969). Explosion seismology in the USSR. In *The*
    *Earth's crust and upper mantle*, pp. 195–208, edited by Hart, P. J., *Geophys. Monogr.* No. 13, American
    Geophysical Union, Washington, D.C.                                                                        [39]
KOVACH, R. L. (1967). Attenuation of seismic body waves in the mantle. *Geophys. J. R. astr. Soc.*, **14**, 165–170.
                                                                                                             [250]
KOVACH, R. L. and ANDERSON, D. L. (1964). Attenuation of shear waves in the upper and lower mantle. *Bull. seism.*
    *Soc. Am.*, **54**, 1855–1864.                                                                            [248]
KOZAI, Y. (1964). New determination of zonal harmonic coefficient of the Earth's gravitational potential. *Smith-*
    *sonian Astrophys. Obs. Spec. Rept.* 165.                                                                   [2]
KUHN, W. and RITTMANN, A. (1941). Über den Zustand des Erdinnern und seine Entstehung aus einem homogenen
    Urzustand. *Geol. Rdsch.*, **32**, 215–256.                                                               [159]
KUNO, H. (1967). Volcanological and petrological evidences regarding the nature of the upper mantle. In *The*
    *Earth's mantle*, pp. 89–110, edited by Gaskell, T. F., Academic Press, London and New York.
                                                                                          [132, *133*, **134**]
KUSHIRO, I. and KUNO, H. (1963). Origin of primary basalt magmas and classification of basaltic rocks. *J. Petrology*,
    **4**, 75–89.                                                                                             [132]

LAHIRI, B. N. and PRICE, A. T. (1939). Electromagnetic induction in non-uniform conductors, and the determination
    of the conductivity of the Earth from terrestrial magnetic variations. *Phil. Trans. R. Soc.*, **237A**, 509–540.
                                                                                                             [120]
LANDISMAN, M., SATÔ, Y. and NAFE, J. (1965). Free vibrations of the Earth and the properties of its deep interior
    regions Part 1: Density. *Geophys. J. R. astr. Soc.*, **9**, 439–502.                                      [146]
LANGSETH, M. G. (1965). Techniques of measuring heat flow through the ocean floor. In *Terrestrial heat flow*,
    pp. 58–77, edited by Lee, W. H. K., *Geophys. Monogr.* No. 8, American Geophysical Union, Washington,
    D.C.                                                                                                     [*179*]
LANGSETH, M. G., GRIM, P. J. and EWING, M. (1965). Heat-flow measurements in the East Pacific Ocean. *J. geophys.*
    *Res.*, **70**, 367–380.                                                                                 [*198*]
LAPADU-HARGUES, P. (1953). Sur la composition chimique moyenne des amphibolites. *Bull. Soc. géol. Fr.*, sér. 6,
    **3**, 153–173.                                                                                          [**63**]
LAPLACE, P. S.                                                                                             [18, 20]
LARMOR, J. (1920). How could a rotating body such as the Sun become a magnet? *Rep. Br. Ass. Advmt Sci.* (for
    1919), 159–160.                                                                                          [170]

LATHAM, G. V. and others (1970). Passive seismic experiment. *Science, N. Y.*, **167**, 455–457.  [25, *25*]

LEE, W. H. K. (1963). Heat flow data analysis. *Rev. Geophys.*, **1**, 449–479.  [178]

LEE, W. H. K. and MACDONALD, G. J. F. (1963). The global variation of terrestrial heat flow. *J. geophys. Res.*, **68**, 6481–6492.  [182]

LEE, W. H. K. and UYEDA, S. (1965). Review of heat flow data. In *Terrestrial heat flow*, pp. 87–190, edited by Lee, W. H. K., *Geophys. Monogr.* No. 8, American Geophysical Union, Washington, D.C.
[178, *180, 181*, **182**, *183, 184, 200*]

LEHMANN, I. (1936). *P̂. Bur. Centr. seism. Internat. A.*, **14**, 3–31.  [10, 152]

LEHMANN, I. (1962). The travel times of the longitudinal waves of the Logan and Blanca atomic explosions and their velocities in the upper mantle. *Bull. seism. Soc. Am.*, **52**, 519–526.  [107]

LEHMANN, I. (1964). On the travel times of *P* as determined from nuclear explosions. *Bull. seism. Soc. Am.*, **54**, 123–139.  [106]

LEHMANN, I. (1967). Low-velocity layers. In *The Earth's mantle*, pp. 41–61, edited by Gaskell, T. F., Academic Press, London and New York.  [107]

LE PICHON, X. (1968). Sea-floor spreading and continental drift. *J. geophys. Res.*, **73**, 3661–3697.
[217, *221*, 223, *224*, **225**, 273]

LE PICHON, X., EWING, J. and HOUTZ, R. E. (1968). Deep-sea sediment velocity determination made while reflection profiling. *J. geophys. Res.*, **73**, 2597–2614.  [76]

LEVIN, B. JU. (1969). Origin of meteorites and planetary cosmogony. In *Meteorite research*, pp. 16–30, edited by Millman, P. M., D. Reidel Publishing Company, Dordrecht, Holland.  [17]

LIEBSCHER, H. J. (1964). Deutungsversuche für die Struktur der tieferen Erdkruste nach reflexionsseismischen und gravimetrischen Messungen im deutschen Alpenvorland. *Z. Geophys.*, **30**, 51–96.  [37, *38*, 40]

LOVERING, J. F. (1958). The nature of the Mohorovičić discontinuity. *Trans. Am. geophys. Un.*, **39**, 947–955. [130]

LOWES, F. J. and RUNCORN, S. K. (1951). The analysis of the geomagnetic secular variation. *Phil. Trans. R. Soc.*, **243A**, 525–546.  [174]

LOWES, F. J. and WILKINSON, I. (1963). Geomagnetic dynamo: a laboratory model. *Nature, Lond.*, **198**, 1158–1160.  [171]

LUBIMOVA, E. A. (1967). Theory of thermal state of the Earth's mantle. In *The Earth's mantle*, pp. 231–323, edited by Gaskell, T. F., Academic Press, London and New York.  [181, 193]

LUBIMOVA, E. A., VON HERZEN, R. P. and UDINTSEV, G. B. (1965). On heat transfer through the ocean floor. In *Terrestrial heat flow*, pp. 78–86, edited by Lee, W. H. K., *Geophys. Monogr.* No. 8, American Geophysical Union, Washington, D.C.  [178]

LUNAR SAMPLE ANALYSIS PLANNING TEAM (1970). Summary of Apollo 11 lunar science conference. *Science, N. Y.*, **167**, 449–451.  [25]

LUNAR SAMPLE PRELIMINARY EXAMINATION TEAM (1969). Preliminary examination of lunar samples from Apollo 11. *Science, N. Y.*, **165**, 1211–1227.  [24]

LUNAR SAMPLE PRELIMINARY EXAMINATION TEAM (1970). Preliminary examination of lunar samples from Apollo 12. *Science, N. Y.*, **167**, 1325–1339.  [25]

LYTTLETON, R. A. (1960). Dynamical calculations relating to the origin of the solar system. *Mon. Not. R. astr. Soc.*, **121**, 551–569.  [26]


MAO, H., TAKAHASHI, T., BASSETT, W. A., WEAVER, J. S. and AKIMOTO, S. (1969). Effect of pressure and temperature on the molar volumes of wüstite and of three (Fe,Mg)$_2$SiO$_4$ spinel solid solutions. *J. geophys. Res.*, **74**, 1061–1069.  [150]

MACDONALD, G. J. F. (1959). Calculations on the thermal history of the Earth. *J. geophys. Res.*, **64**, 1967–2000.  [*190*]

MACDONALD, G. J. F. (1963). The deep structure of continents. *Rev. Geophys.*, **1**, 587–665.  [195, 246]

MACDONALD, G. J. F. (1964). Tidal friction. *Rev. Geophys.*, **2**, 467–541.  [189]

MACDONALD, G. J. F. (1965). Geophysical deductions from observations of heat flow. In *Terrestrial heat flow*, pp. 191–210, edited by Lee, W. H. K., *Geophys. Monogr.* No. 8, American Geophysical Union, Washington, D.C.  [186, **187**, 195, *196*]

MACDONALD, G. J. F. (1966). Origin of the Moon: dynamical considerations. In *The Earth–Moon system*, pp. 165–209, edited by Marsden, B. G. and Cameron, A. G. W., Plenum Press, New York.  [22, 23, 26]

MACDONALD, G. J. F. and NESS, N. F. (1960). Stability of phase transitions within the Earth. *J. geophys. Res.*, **65**, 2173–2190.  [137]

MANSINHA, L. and SMYLIE, D. E. (1968). Earthquakes and the Earth's wobble. *Science, N. Y.*, **161**, 1127–1129. [157]

MASON, R. G. and RAFF, A. D. (1961). Magnetic survey off the west coast of North America, 32° N. latitude to 42° N. latitude. *Bull. geol. Soc. Am.*, **72**, 1259–1266.  [80]

MAXWELL, A. E. and others (1970). Deep sea drilling in the South Atlantic. *Science, N. Y.*, **168**, 1047–1059
[220, *220*, **220**,]

MCCONNELL, R. K., JR. (1965). Isostatic adjustment in a layered Earth. *J. geophys. Res.*, **70**, 5171–5188.  [241]

MCCONNELL, R. K., JR. GUPTA, R., N. and WILSON, J. T. (1966). Compilation of deep crustal seismic refraction profiles. *Rev. Geophys.*, **4**, 41–100.                                                                                                    [39]

MCDONALD, K. L. (1957). Penetration of the geomagnetic secular field through a mantle with variable conductivity. *J. geophys. Res.*, **62**, 117–141.                                                                                       [121, *121*, 128]

MCDOUGALL, I. and CHAMALAUN, F. H. (1966). Geomagnetic polarity scale of time. *Nature, Lond.*, **212**, 1415–1418.                                                                                                                          [213]

MCKENZIE, D. P. (1966). The viscosity of the lower mantle. *J. geophys. Res.*, **71**, 3995–4010.                    [245, 246]

MCKENZIE, D. P. (1967). The viscosity of the mantle. *Geophys. J. R. astr. Soc.*, **14**, 297–305.     [240, *240*, 246, 250]

MCKENZIE, D. P. (1968). The geophysical importance of high-temperature creep. In *The history of the Earth's crust*, pp. 28–44, edited by Phinney, R. A., Princeton University Press, Princeton, New Jersey.                              [239, *240*]

MCKENZIE, D. P. and PARKER, R. L. (1967). The north Pacific: an example of tectonics on a sphere. *Nature, Lond.*, **216**, 1276–1280.                                                                                                        [220]

MCNITT, J. R. (1965). Review of geothermal resources. In *Terrestrial heat flow*, pp. 240–266, edited by Lee, W. H. K., *Geophys. Monogr.* No. 8, American Geophysical Union, Washington, D.C.                                                [197]

MCQUEEN, R. G., FRITZ, J. N. and MARSH, S. P. (1964). On the composition of the Earth's interior. *J. geophys. Res.*, **69**, 2947–2965.                                                                                                      [159]

MELCHIOR, P. J. (1957). Latitude variation. *Physics Chem. Earth*, **2**, 212–243.                                   [*152*]

MELCHIOR, P. (1966). *The Earth tides*, Pergamon Press, Oxford, London, Edinburgh, New York, Paris and Frankfurt, 458 pp.                                                                                                       [155, *156*, 157]

MENARD, H. W. (1964). *Marine geology of the Pacific*, McGraw-Hill Book Company, New York, San Francisco, Toronto and London, 271 pp.                                                                                       [71, 74, 75, 77, 78, *79*, 148]

MISRA, A. K. and MURRELL, S. A. F. (1965). An experimental study of the effect of temperature and stress on the creep of rocks. *Geophys. J. R. astr. Soc.*, **9**, 509–535.                                                                 [234, 239]

MOHOROVIČIĆ, A. (1909). Das Beben vom 8.x.1909. *Jb. met. Obs. Zagreb (Agram.)*, **9**, 1–63.                        [8, 29]

MOORBATH, S. (1965). Isotopic dating of metamorphic rocks. In *Controls of metamorphism*, pp. 235–267, edited by Pitcher, W. S. and Flinn, G. W., Oliver and Boyd, Edinburgh and London.                                                     [*136*]

MORGAN, W. J. (1968). Rises, trenches, great faults, and crustal blocks. *J. geophys. Res.*, **73**, 1959–1982.      [220]

MUELLER, S. and LANDISMAN, M. (1966). Seismic studies of the Earth's crust in continents I: Evidence for a low-velocity zone in the upper part of the lithosphere. *Geophys. J. R. astr. Soc.*, **10**, 525–538.                          [40, *41*]

MUNK, W. (1968). Once again—tidal friction. *Q. Jl R. astr. Soc.*, **9**, 352–375.                                   [22]

MUNK, W. H. and MACDONALD, G. J. F. (1960). *The rotation of the Earth.* Cambridge University Press, London and New York, 323 pp.                                                                                            [*19, 21, 245, 250, 251*]

NAFE, J. E. and DRAKE, C. L. (1963). Physical properties of marine sediments. In *The sea*, volume 3, pp. 794–815, edited by Hill, M. N., Interscience Publishers, New York and London.                                                      [64, *66*]

NAGATA, T. (1953). Self-reversal of thermo-remanent magnetization of igneous rocks. *Nature, Lond.*, **172**, 850–852.                                                                                                                       [167]

NAGATA, T. (1961). *Rock magnetism*, Revised edition, Maruzen Company, Tokyo, 350 pp.                                 [165]

NAIRN, A. E. M. (editor) (1961). *Problems in palaeoclimatology*. Interscience Publishers, London, New York and Sydney, 705 pp.                                                                                                              [204]

NÉEL, L. (1951). L'inversion de l'aimantation permanente des roches. *Annls Géophys.*, **7**, 90–102.                [167]

NEWTON, SIR ISAAC                                                                                                     [1, 2]

NIAZI, M. and ANDERSON, D. L. (1965). Upper mantle structure of western North America from apparent velocities of *P* waves. *J. geophys. Res.*, **70**, 4633–4640.                                                                         [107, *108*]

NIBLETT, E. R. and SAYN-WITTGENSTEIN, C. (1960). Variation of electrical conductivity with depth by the magneto-telluric method. *Geophysics*, **25**, 998–1008.                                                                            [121]

NOCKOLDS, S. R. (1954). Average chemical compositions of some igneous rocks. *Bull. geol. Soc. Am.*, **65**, 1007–1032.                                                                                                                      **[63, 67]**

O'KEEFE, J. A. (1965). Discussion. In *A symposium on continental drift, Phil. Trans. R. Soc.*, **258A**, 272–275.   [4]

OLDHAM, R. D. (1906). The constitution of the interior of the Earth, as revealed by earthquakes. *Q. Jl geol. Soc. Lond.*, **62**, 456–475.                                                                                                  [7, 159]

OLIVER, J. and ISACKS, B. (1967). Deep earthquake zones, anomalous structures in the upper mantle, and the lithosphere. *J. geophys. Res.*, **72**, 4259–4275.                                                                            [106, 267, *267*]

OPDYKE, N. D., GLASS, B., HAYS, J. D. and FOSTER, J. (1966). Palaeomagnetic study of Antarctic deep-sea cores. *Science, N.Y.*, **154**, 349–357.                                                                                          [213, *214*]

OROWAN, E. (1960). Mechanism of seismic faulting. In *Rock deformation (a symposium)*, pp. 323–345, edited by Griggs, D. and Handin, J., *Mem. geol. Soc. Am.*, **79**,                                                                    [262]

OROWAN, E. (1965). Convection in a non-Newtonian mantle, continental drift, and mountain building. In *A symposium on continental drift, Phil. Trans. R. Soc.*, **258A**, 284–313.                                                        [*230*, 239 283]

OROWAN, E. (1969). Density of the Moon and nucleation of planets *Nature, Lond.*, **222**, 867.    [191]

OSEMEIKHIAN, J. E. A. and EVERETT, J. E. (1968). Anomalous magnetic variations in southwestern Scotland. *Geophys. J. R. astr. Soc.*, **15**, 361–366.    [122]

PAKISER, L. C. (1963). Structure of the crust and upper mantle in the western United States. *J. geophys. Res.*, **68**, 5747–5756.    [38, 39, *39*, 51, 59, 112]

PARKINSON, W. D. (1962). The influence of continents and oceans on geomagnetic variations. *Geophys. J. R. astr. Soc.*, **6**, 441–449.    [122]

PEACOCK, J. H.    [75]

PENTTILÄ, E. (1969). A report summarizing on the velocity of earthquake waves and the structure of the Earth's crust in the Baltic shield. *Geophysica*, **10**, 7–23.    [114]

PETERSON, M. N. A. and others (1970). *Initial reports of the Deep Sea Drilling Project*, volume II. Washington, D.C. (U.S. Government Printing Office), 491 pp.    [78]

PHINNEY, R. A. and ALEXANDER, S. S. (1967). Diffraction of *P*-waves and the structure of the core-mantle boundary. *Geophys. J. R. astr. Soc.*, **13**, 371.    [152]

PITMAN, W. C., III and HEIRTZLER, J. R. (1966). Magnetic anomalies over the Pacific–Antarctic ridge. *Science, N.Y.*, **154**, 1164–1171.    [213, 214, *216*]

PLUMSTEAD, E. P. (1965). Evidence of vast lateral movements of the Earth's crust provided by fossil floras. In *The upper mantle symposium, New Delhi, 1964*, pp. 65–74, edited by Smith, C. H. and Sorgenfrei, T., International Union of Geological Sciences, Det Berlingske Bogtrykkeri, Copenhagen.    [204]

POLDERVAART, A. (1955). Chemistry of the Earth's crust. *Spec. Pap. geol. Soc. Am.*, **62**, 119–144.    [28, *67*]

PRATT, J. H. (1855). On the attraction of the Himalaya mountains, and of the elevated regions beyond them, upon the plumb-line in India. *Phil. Trans. R. Soc.*, **145**, 53–100.    [46]

PRESS, F. (1964). Seismic wave attenuation in the crust. *J. geophys. Res.*, **69**, 4417–4418.    [250]

PRESS, F. (1968). Density distribution in Earth. *Science, N.Y.*, **160**, 1218–1221.    [143]

PRESS, F. and EWING, M. (1955). Earthquake surface waves and crustal structure. *Spec. Pap. geol. Soc. Am.*, **62**, 51–60.    [31]

PREY, A. (1922). Darstellung der Höhen- und Tiefenverhältnisse der Erde durch eine Entwicklung nach Kugelfunktionen bis zur 16. *Abh. K. Ges. Wiss. Göttingen*, Math.-physik Kl., **11**, No. 1.    [278]

PRICE, A. T. (1967). Magnetic variations and telluric currents. In *The Earth's mantle*, pp. 125–170, edited by Gaskell, T. F., Academic Press, London and New York.    [121]

RAFF, A. D. and MASON, R. G. (1961). Magnetic survey off the west coast of North America, 40°N. latitude to 52°N. latitude. *Bull. geol. Soc. Am.*, **72**, 1267–1270.    [80]

RAIT, R. W.    [114]

RAITT, R. W., SHOR, G. G., JR., FRANCIS, T. J. G. and MORRIS, G. B. (1969). Anisotropy of the Pacific upper mantle. *J. geophys. Res.*, **74**, 3095–3109.    [114]

RALEIGH, C. B. and PATERSON, M. S. (1965). Experimental deformation of serpentinite and its tectonic implications. *J. geophys. Res.*, **70**, 3965–3985.    [262]

RAMBERG, H. and STEPHANSSON, O. (1964). Compression of floating elastic and viscous plates affected by gravity, a basis for discussing crustal buckling. *Tectonophysics*, **1**, 101–120.    [238]

RAMSEY, A. S. (1940). *An introduction to the theory of Newtonian attraction*, Cambridge University Press, London and New York, 184 pp.    [4]

RAMSEY, W. H.    [270]

RAMSEY, W. H. (1949). On the nature of the Earth's core. *Mon. Not. R. astr. Soc. geophys. Suppl.*, **5**, 409–426.    [159]

RAPP, R. H. (1968). Gravitational potential of the Earth determined from a combination of satellite, observed, and model anomalies. *J. geophys. Res.*, **73**, 6555–6562.    [4, *5*]

RAYLEIGH, J. W. S. (LORD) (1916). On convection currents in a horizontal layer of fluid, when the higher temperature is on the under side. *Phil. Mag.*, ser. 6, **32**, 529–546.    [275]

RICHTER, C. F. (1935). An instrumental earthquake magnitude scale. *Bull. seism. Soc. Am.*, **25**, 1–32.    [254]

RICHTER, C. F. (1958). *Elementary seismology*, W. H. Freeman and Company, San Francisco, 768 pp.    [255]

RIEDEL, W. R., LADD, H. S., TRACEY, J. I., JR. and BRAMLETTE, M. N. (1961). Preliminary drilling phase of Mohole project II. Summary of coring operations (Guadalupe site). *Bull. Am. Soc. Pet. Geol.*, **45**, 1793–1798.    [77]

RIKITAKE, T. (1964). Outline of the anomaly of geomagnetic variations in Japan. *J. Geomagn. Geoelect., Kyoto*, **15**, 181–184.    [122]

RIKITAKE, T. (1966). *Electromagnetism and the Earth's interior*, Elsevier, Amsterdam, London and New York, 308 pp.    [122]

RINGWOOD, A. E. and GREEN, D. H. (1964). Experimental investigations bearing on the nature of the Mohorovičić discontinuity. *Nature, Lond.*, **201**, 566–567.    [130]

RINGWOOD, A. E. and GREEN, D. H. (1966). Petrological nature of the stable continental crust. In *The Earth beneath the continents*, pp. 611–619, edited by Steinhart, J. S. and Smith, T. J., *Geophys. Monogr.* No. 10, American Geophysical Union, Washington, D.C. [68, *69*, 70, **70**]

ROBSON, G. R., BARR, K. G. and LUNA, L. C. (1968). Extension failure: an earthquake mechanism. *Nature, Lond.*, **218**, 28–32. [262]

RUNCORN, S. K. [165, 190]

RUNCORN, S. K. (1956). Palaeomagnetic comparisons between Europe and North America. *Proc. geol. Assoc. Canada*, **8**, 77–85. [206]

RUNCORN, S. K. (1962). Palaeomagnetic evidence for continental drift and its geophysical cause. In *Continental drift*, pp. 1–40, edited by Runcorn, S. K., Academic Press, New York and London. [207, 285, *286*]

RUNCORN, S. K. (1964). Changes in the Earth's moment of inertia. *Nature, Lond.*, **204**, 823-825. [22, 273]

RUNCORN, S. K. (1965). Palaeomagnetic comparisons between Europe and North America. In *A symposium on continental drift, Phil. Trans. R. Soc.*, **258A**, 1–11. [204, 207]

RUNCORN, S. K., BENSON, A. C., MOORE, A. F. and GRIFFITHS, D. H. (1951). Measurements of the variation with depth of the main geomagnetic field. *Phil. Trans. R. Soc.*, **244A**, 113–151. [170]

SABU, D. D. and KURODA, P. K. (1967). Plutonium-244 in the early solar system and concordant plutonium/xenon and iodine/xenon decay intervals of achondrites. *Nature, Lond.*, **216**, 442–446. [17]

SASS, J. H. (1964). Heat-flow values from eastern Australia. *J. geophys. Res.*, **69**, 3889–3893. [199]

SATO, R. and EPINOSA, A. F. (1967). Dissipation in the Earth's mantle and rigidity and viscosity in the Earth's core determined from waves multiply reflected from the mantle-core boundary. *Bull. seism. Soc. Am.*, **57**, 829–856. [248]

SCHMUCKER, U. (1964). Anomalies of geomagnetic variations in the southwestern United States. *J. Geomagn. Geoelect., Kyoto*, **15**, 193–221. [122]

SCRUTTON, C. T. (1964). Periodicity in Devonian coral growth. *Palaeontology*, **7**, 552–558. [21]

SCRUTTON, C. T. (1967). Absolute time data from palaeontology. In *International dictionary of geophysics*, volume 1, p. 1, edited by Runcorn, S. K. and others, Pergamon Press, Oxford, London, Edinburgh, New York, Toronto, Sydney, Paris and Braunschweig. [*22*]

SHOR, G. G. [114]

SHOR, G. G., JR. (1963). Refraction and reflection techniques and procedure. In *The sea*, volume 3, pp. 20–38, edited by Hill, M. N., Interscience Publishers, New York and London. [72]

SIMMONS, G. [64]

SMITH, A. G. and HALLAM, A. (1970). The fit of the southern continents. *Nature, Lond.*, **225**, 139–144. [202]

SMITH, P. J. (1967). The intensity of the Tertiary geomagnetic field. *Geophys. J. R. astr. Soc.*, **12**, 239–258. [165]

SMITH, S. W. (1966). Free oscillations excited by the Alaskan earthquake. *J. geophys. Res.*, **71**, 1183–1193. [*104*]

SMITH, T. J., STEINHART, J. S. and ALDRICH, L. T. (1966). Crustal structure under Lake Superior. In *The Earth beneath the continents*, pp. 181–197, edited by Steinhart, J. S. and Smith, T. J., *Geophys. Monogr.* No. 10, American Geophysical Union, Washington, D.C. [*44*]

SPITZER, L., JR. (1939). The dissipation of planetary filaments. *Astrophys. J.*, **90**, 675–688. [16]

SPITZER, L., JR. (1963). Star formation. In *Origin of the solar system*, pp. 39–53, edited by Jastrow, R. and Cameron, A. G. W., Academic Press, New York and London. [16]

STAUDER, W. (1962). The focal mechanism of earthquakes. *Adv. Geophys.*, **9**, 1–76. [258]

STAUDER, W. (1968). Tensional character of earthquake foci beneath the Aleutian trench with relation to sea-floor spreading. *J. geophys. Res.*, **73**, 7693–7701. [267]

STEINHART, J. S. and MEYER, R. P. (1961). *Explosion studies of continental structure*, Carnegie Institution of Washington Publication 622, 409 pp. [*33*, 38, 39, 51]

STEINHART, J. S. and SMITH, T. J. (editors) (1966). *The Earth beneath the continents*, a volume of geophysical studies in honor of Merle A. Tuve, *Geophys. Monogr.* No. 10, American Geophysical Union, Washington, D.C., 663 pp. [38]

STRAHOW, N. M. [270–271]

STRONG, H. M. (1959). The experimental fusion curve of iron to 96,000 atmospheres. *J. geophys. Res.*, **64**, 653–659. [128, 159]

SYKES, L. R. (1966). The seismicity and deep structure of island arcs. *J. geophys. Res.*, **71**, 2981–3006. [264, *265*]

SYKES, L. R. (1967). Mechanism of earthquakes and nature of faulting on the mid-oceanic ridges. *J. geophys. Res.*, **72**, 2131–2153. [260, *260*, 263, *264*]

TAKEUCHI, H. (1950). On the Earth tide of the compressible Earth of variable density and elasticity. *Trans. Am. geophys. Un.*, **31**, 651–689. [158]

TAKEUCHI, H. and HASEGAWA, Y. (1965). Viscosity distribution within the Earth. *Geophys. J. R. astr. Soc.*, **9**, 503–508. [241]

TALWANI, M. (1964). A review of marine geophysics. *Marine Geol.*, **2**, 29–80.                    [72, *73*

TALWANI, M., LE PICHON, X. and EWING, M. (1965). Crustal structure of the mid-ocean ridges 2. Computed model from gravity and seismic refraction data. *J. geophys. Res.*, **70**, 341–352.    [*86, 87, 88, 88*]

TALWANI, M., SUTTON, G. H. and WORZEL, J. L. (1959). A crustal section across the Puerto Rico trench. *J. geophys. Res.*, **64**, 1545–1555.    [90, *91*]

TATEL, H. E. and TUVE, M. A. (1955). Seismic exploration of a continental crust. *Spec. Pap. geol. Soc. Am.*, **62**, 35–50.    [38, 41, 43]

TAYLOR, F. B. (1910). Bearing of the Tertiary mountain belt on the origin of the Earth's plan. *Bull. geol. Soc. Am.*, **21**, 179–226.    [201]

TER HAAR, D. and CAMERON, A. G. W. (1963). Historical review of theories of the origin of the solar system. In *Origin of the solar system*, pp. 1–37, edited by Jastrow, R. and Cameron, A. G. W., Academic Press, New York and London.    [16]

TERMIER, H.    [270–271]

THIRLAWAY, H. I. S.    [97]

TOKSÖZ, M. N. and ANDERSON, D. L. (1966). Phase velocities of long-period surface waves and structure of the upper mantle 1. Great-circle Love and Rayleigh wave data. *J. geophys. Res.*, **71**, 1649–1658.    [117]

TOKSÖZ, M. N., CHINNERY, M. A. and ANDERSON, D. L. (1967). Inhomogeneities in the Earth's mantle. *Geophys. J. R. astr. Soc.*, **13**, 31–59.    [*105*, 111, *111, 112, 113, 116*]

TOZER, D. C. (1959). The electrical properties of the Earth's interior. *Physics Chem. Earth*, **3**, 414–436. [124, 128]

TOZER, D. C. (1965*a*). Heat transfer and convection currents. In *A symposium on continental drift, Phil. Trans. R. Soc.*, **258A**, 252–271.    [278]

TOZER, D. C. (1965*b*). Thermal history of the Earth I. The formation of the core. *Geophys. J. R. astr. Soc.*, **9**, 95–112.    [189, 190]

TRUSCOTT, J. R. (1964). The Eskdalemuir seismological station. *Geophys. J. R. astr. Soc.*, **9**, 59–68.    [*98*]

TRYGGVASON, E. (1964). Arrival times of *P* waves and upper mantle structure. *Bull. seism. Soc. Am.*, **54**, 727–736.    [114, 115]

TUCKER, W., HERRIN, E. and FREEDMAN, H. W. (1968). Some statistical aspects of the estimation of seismic travel times. *Bull. seism. Soc. Am.*, **58**, 1243–1260.    [*113*]

UFFEN, R. J. (1952). A method of estimating the melting-point gradient in the Earth's mantle. *Trans. Am. geophys. Un.*, **33**, 893–896.    [127]

UREY, H. C. (1952). *The planets, their origin and development*, Yale University Press, 245 pp.    [17, 190]

VACQUIER, V. (1965). Transcurrent faulting in the ocean floor. In *A symposium on continental drift, Phil. Trans. R. Soc.*, **258A**, 77–81.    [210]

VANZIJL J. S. V., GRAHAM, K. W. T. and HALES, A. L. (1962). The palaeomagnetism of the Stormberg Lavas, II. The behaviour of the magnetic field during a reversal. *Geophys. J. R. astr. Soc.*, **7**, 169-182.    [167]

VENING MEINESZ, F. A.    [47, 62, 90]

VENING MEINESZ, F. A. (1948). *Gravity expeditions at sea, 1923–1938*, volume IV, *Publ. Netherlands Geod. Comm., Delft*, 233 pp.    [71, 86]

VERHOOGEN, J. (1960). Temperatures within the earth. *Am. Scient.*, **48**, 134–159.    [188]

VERHOOGEN, J. (1965). Phase changes and convection in the Earth's mantle. In *A symposium on continental drift, Phil. Trans. R. Soc.*, **258A**, 276–283.    [276]

VESTINE, E. H. (1962). Influence of the Earth's core upon the rate of the Earth's rotation. In *Benedum Earth magnetism symposium*, pp. 57–67, edited by Nagata, T., University of Pittsburgh Press.    [*175*]

VESTINE, E. H., LAPORTE, L., COOPER, C., LANGE, I. and HENDRIX, W. C. (1947). *Description of the Earth's main magnetic field and its secular change, 1905–1945*, Carnegie Institution of Washington Publication 578, 532 pp.    [*162, 163*]

VINE, F. J. (1966). Spreading of the ocean floor: new evidence. *Science, N.Y.*, **154**, 1405–1415.    [211, 213, 214, *215*]

VINE, F. J. and MATTHEWS, D. H. (1963). Magnetic anomalies over oceanic ridges. *Nature, Lond.*, **199**, 947–949. [83]

VON HERZEN, R. P. and MAXWELL, A. E. (1964). Measurements of heat flow at the preliminary Mohole site off Mexico. *J. geophys. Res.*, **69**, 741–748.    [178]

VON WEIZSÄCKER, C. F. (1944). Über die Entstehung des Planetensystems. *Z. Astrophys.*, **22**, 319–355.    [17]

WADE, N. (1969). Three origins of the Moon. *Nature, Lond.*, **223**, 243–246.    [26]

WANG, C. (1966). Earth's zonal deformations. *J. geophys. Res.*, **71**, 1713–1720.    [245]

WARD, M. A. (1963). On detecting changes in the Earth's radius. *Geophys. J. R. astr. Soc.*, **8**, 217–225.    [272]

WEGENER, A.    [201, 202]

WEGENER, A. (1912). Die Entstehung der Kontinente. *Peterm. Mitt.*, 1912, 185–195, 253–256, 305–309.    [201]

WELLS, J. W. (1963). Coral growth and geochronometry. *Nature, Lond.*, **197**, 948–950.                    [21]

WESTOLL, T. S. (1965). Geological evidence bearing upon continental drift. In *A symposium on continental drift*, *Phil. Trans. R. Soc.*, **258A**, 12–26.                    [202, 204]

WHEILDON, J.                    [*177*]

WHITHAM, K. (1963). An anomaly in geomagnetic variations at Mould Bay in the Arctic archipelago of Canada. *Geophys. J. R. astr. Soc.*, **8**, 26–43.                    [122]

WIECHERT, E.                    [159]

WILLMORE, P. L. and BANCROFT, A. M. (1960). The time term approach to refraction seismology. *Geophys. J. R. astr. Soc.*, **3**, 419–432.                    [31]

WILSON, J. T. (1965). A new class of faults and their bearing on continental drift. *Nature, Lond.*, **207**, 343–347.                    [*209*, 210, *210*, *212*]

WISEMAN, J. D. H. (1966). St Paul rocks and the problem of the upper mantle. *Geophys. J. R. astr. Soc.*, **11**, 519–525.                    [134]

WOOD, H. O. and RICHTER, C. F. (1931). A study of blasting recorded in southern California. *Bull. seism. Soc. Am.*, **21**, 28–46.                    [40]

WOOD, H. O. and RICHTER, C. F. (1933). A second study of blasting recorded in southern California. *Bull. seism. Soc. Am.*, **23**, 95–110.                    [40]

WOOLLARD, G. P. (1966). Regional isostatic relations in the United States. In *The Earth beneath the continents*, pp. 557–594, edited by Steinhart, J. S. and Smith, T. J., *Geophys. Monogr.* No. 10, American Geophysical Union, Washington, D.C.                    [54]

WORZEL, J. L. (1965*a*). *Pendulum gravity measurements at sea 1936–1959*. John Wiley & Sons, New York, London and Sydney, 422 pp.                    [71]

WORZEL, J. L. (1965*b*). Deep structure of coastal margins and mid-oceanic ridges. In *Submarine geology and geophysics*, pp. 335–361, edited by Whittard, W. F. and Bradshaw, R., Colston Papers No. 17, Butterworths, London.                    [89, *89*]


YODER, H. S., JR. and TILLEY, C. E. (1962). Origin of basalt magmas: an experimental study of natural and synthetic rock systems. *J. Petrology*, **3**, 342-532.                    [132, *133*]

# Subject Index

Page numbers given in italics refer to an entire section on the topic indexed.